D0946980

Yale Agrarian Studies Series
James C. Scott, series editor

STEPHEN B. BRUSH

Farmers' Bounty

LOCATING CROP DIVERSITY IN THE CONTEMPORARY WORLD

Yale University Press
New Haven &
London

Published with assistance from the foundation established in memory of
Philip Hamilton McMillan of the Class of 1894, Yale College.

Set in Sabon type by Keystone Typesetting, Inc.
Printed in the United States of America by Sheridan Books, Ann Arbor, Michigan.

Library of Congress Cataloging-in-Publication Data
Brush, Stephen B., 1943–
Farmers' bounty : locating crop diversity in the contemporary world /
Stephen B. Brush.
p. cm. — (Yale agrarian studies)
Includes bibliographical references (p.).
ISBN 0-300-10049-3 (hardcover : alk. paper)
1. Crops — Germplasm resources. 2. Germplasm resources, Plant. I. Title.
II. Series.
SB123.3.B78 2004
631.5'23 — dc22
2003020235

A catalogue record for this book is available from the British Library.
The paper in this book meets the guidelines for permanence and durability of the Committee on Production Guidelines for Book Longevity of the Council on Library Resources.
10 9 8 7 6 5 4 3 2 1

To my children, Jason and Amanda,
who have patiently watched the long gestation of this book;
and
to my students, from whom I have learned so much

Contents

List of Tables and Figures

Tables

Figures

Preface

Crop diversity holds intrinsic pleasures as well as scientific ones. Variations in the simplest staples are essential to all great cuisine, whether low or high: breads made with different flours; fragrant and sticky rice; tortillas made from blue, yellow, and white maize; and purple potatoes. For a thousand generations the contrasting tastes, textures, colors, and shapes of the world's staples have enlivened humble meals under low ceilings and near cook fires. Crops tell stories about their places and peoples. Diversity informs people of their uniqueness in a wide and alien world. It reminds us of our connection to kinsmen and neighbors, and it roots us to the landscape.

For an anthropologist or traveler to exotic places, crop diversity is a passport to different lives. Naming and categorizing types of crops are popular pastimes that open doors to many aspects of rural life: knowledge of the environment, connections with other people and places, the difficulties of wresting subsistence from the Earth, ingenuity in adapting different technologies to a specific environment. Rare is the farmer who will not stop to talk about the different qualities of seed or be eager to learn about unfamiliar seed. Curiosity is universal about the origin and nature of the various foods that greet us at mealtime.

Since antiquity, natural historians have noticed that crops and their many varieties change in kind and number across different regions; there are cultures that prefer sticky rice over hard; bitter cassava over sweet; white maize over yellow. This diversity marks both social and environmental boundaries. Indeed, because biophysical and social landscapes are naturally varied, diversity in crops is not surprising. What is unexpected is the uneven distribution of agricultural diversity and its occurrence in patterns that are not entirely explained by different combinations of soil and climate or by social groups. Consequently, variation in domesticated plants and animals has remained a puzzle that has attracted such naturalists as Charles Darwin. Crop diversity is today a scientific object, an economic resource, and a source of political conflict. Diversity directed botanists to the cradles of crop domestication, and it enabled the diffusion of crops beyond their original habitats. Crop diversity remains the single most important biological resource in the worldwide effort to meet the food demands of the rapidly expanding human population and for farmers who aspire to a better life.

Numerous issues circulate around the nucleus of crop diversity and draw scholars from many perspectives. What were a crop's wild ancestors, and how were they transformed? If a biological bridge remains, how important is it to the performance of a crop? How did the diffusion of crops affect them? Is diversity an asset or a liability to poor farmers or to farmers in marginal agricultural environments? How important is crop diversity in the worldwide effort to increase food production and to do so with less environmental impact? Will diversity be important in coping with drastic climate change? How can we best maintain crop diversity so that it is available to future generations?

Responding to these questions requires the talents and perspectives of many specialists, and the available answers are neither easy nor definitive. Indeed, we may never have satisfactory responses. Distinct areas of biology, ecology, and social science that are relevant to addressing these questions include agricultural botany and genetics, biogeography, population biology, archaeology, economics, cultural geography, and human ecology. Generalization and loss of professional identity are risks worth taking, but ultimately the answers must be a fabric woven from specialized contributions.

My involvement with the issue of crop diversity began in the fields and kitchens of an Andean village where I was doing field research as a graduate student in anthropology. My initial quest to understand the relationship between Andean culture and the environment of its mountainous home eventually led me to study one of the great achievements of Andean people, the domestication and diversification of the potato. This living accomplishment is in every way as impressive as the better-known monuments of the Andes, such as Machu Picchu, yet the tuber is a humble crop long associated with poverty and submission in both Andean and European cultures that celebrated cereals as more noble foods.

As a young anthropologist in the village of Uchucmarca on the rim of the gorge carved by the Marañon River in northern Peru, I was shown a trove of local beauties in their potato larders. The various potato names were heavy with local knowledge about Andean life: for example, *la sapa,* or "the toad," harbinger of rain which resists frost, is slow growing and cooking, with a prized floury consistency and flavor. During my initial ethnographic fieldwork, I stood or knelt in numerous potato storerooms as I was tutored on names and qualities of the many distinct types. I plowed, planted, and harvested potato fields and knew of the effort and the local technology invested in the crop. Many meals consisted entirely of potatoes "washed down" with ground-chili sauce. I visited the nascent International Potato Center and learned about the impending Green Revolution in Andean agriculture, as modern, high yielding varieties, such as the *renacimiento* (renaissance), made their way into potato fields. In the half-light of Andean storerooms lay wealth greater than the gold pillaged by the conquistadors as well as a bridge between these mountain folk and the world.

People like me from industrial and wealthy nations are beholden to Andean farmers for sharing this precious resource. The potatoes in Milciades Rojas' back room sparked questions that have propelled me for most of my career, leading me into the communities and farms of very different places and people, and triggering questions that were unimaginable to a novice anthropologist. The first questions are still among the most difficult to answer:

- Why is so much diversity concentrated in particular crops and places?
- What happens to diversity as farming changes and becomes more modern?
- Can industrialized countries somehow repay their debt to the farmers of less developed countries that have provided genetic resources for our crops?

These questions can be answered only by broad, interdisciplinary research; answers to date are roughly conceived and not definitive, yet these answers are evidence of scientific progress.

My objective is to pull together a quarter century of research in different places and on different themes relating to the puzzles and pleasures of crop diversity. Four themes have organized my research: to explain the patterns of crop diversity, to ponder the fate of that diversity, to imagine new means of conserving it, and to address equity issues. A large literature exists on these themes, published mostly as individual articles. Although these themes are interconnected and may, as a metatheme, inform any one article, the canvas provided by scientific journals hampers efforts to explore connections or to expose the robust whole. This book is not a recapitulation of a scientific enterprise that has been concluded but rather a work in progress. I try to show how my thought and approach have developed to reflect both personal discovery and the evolution of an interdisciplinary field of inquiry.

Farmers' Bounty has three general sections. Chapters 1–4 seek to define the dimensions of crop diversity and questions surrounding it. Chapter 1 explains how human ecology came to the study of crop diversity. Chapter 2 describes why crop diversity is a puzzle and some of the ways it has been used to address larger issues about human and agricultural evolution. Chapter 3 examines various ways of defining and measuring crop diversity. Chapter 4 introduces the three crops and farming regions that I have worked in: potatoes in the Peruvian Andes, maize in Mexico, and wheat in Turkey. Chapters 5–8 explore four scientific issues that have occupied my research efforts. Chapter 5 describes the ethnobiology of Andean potatoes as an example of how anthropological research can contribute to an overall understanding of the ecology and evolution of a crop in its center of origin. Chapter 6 examines the nature of farmer selection, using material from research on wheat diversity in Turkey. Chapter 7 discusses the issue of genetic erosion and attempts to deconstruct the history of this scientific and agricultural policy problem. Chapter 8 builds on the problems explored in the preceding chapter and offers a theoretical framework to improve our understanding of genetic erosion. The book's last section explores three broad policy issues involving crop diversity. Chapter 9 discusses the conservation of crop resources, with a focus on in situ, or on-farm, conservation. Chapter 10 addresses the issues of equity in a world divided by economic wealth and poverty, uneven distribution and trade of genetic resources, and ownership of these resources as intellectual property. Finally, Chapter 11 meditates on the future of crop diversity in a world with eleven billion people, reading case histories of crop diversity in developed societies as optimistic signs for the future of diversity elsewhere.

Acknowledgments

A product so long in process has drawn on the insight and help of numerous friends, students, and colleagues. Inevitably, I will omit names of persons that I should acknowledge, and for this I am regretful.

Among the many Andean scholars who have helped me is Ben Orlove, a frequent and helpful listener and interlocutor on many thoughts on life in the Andes and elsewhere. Billie Jean Isbell was generous with her understanding of the Andean people and anthropology. Rob Werge and I enjoyed playing pioneer "potato anthropologists" at the International Potato Center (CIP). Heath Carney helped me develop the field methods of my initial foray into agricultural ecology. Deborah Rabinowitz and I shared lessons from ecological fieldwork in Peru, and her premature death was deeply felt. Zosimo Huamán, Carlos Ochoa, Maria Mayer, and Doug Horton of CIP provided critical support and information about the complexity of the Andean potato crop. My continued studies on potato diversity were done in collaboration with Enrique Mayer and the late Cesar Fonseca, both marvelous friends to have in the field and beyond. Cesar died on his final trip out of Cusco, and we Andeanists lost a brilliant light when he fell. Karl Zimmerer, then a graduate student, stepped into the breach and helped us complete a successful project. Carlos Quiros at the University of California, Davis, and Ramiro Ortega of the Universidad Nacional San Antonio Abad in Cusco continue to inform my work. Graduate students from Peru were indefatigable field-workers in the Tulumayo and Paucartambo valleys.

My work with maize and maize farmers in Mexico began with the help of Mauricio Bellon, who has continued his insightful engagement with issues that brought us together in 1988. Other Mexican and North American students and colleagues have helped me immeasurably. These include participants in the Milpa Project: Fernando Castillo, Rafael Ortega-Pazcka, Robert Bye, Cal Qualset, Major Goodman, Jesus Sanchez, Ed Taylor, Laura Merrick, and Alfonso Delgado. Graduate students on that project were unusually creative in pushing us in new directions: Eric VanDusen, George Dyer, and especially Hugo Perales. Hugo has continued on several other projects befitting men made of maize, and I have been fortunate to go along. Bruce Benz has shored up my excitement in the possibility of showing a link between the human and crop diversity of Mesoamerica.

Cal Qualset and Jack Harlan fueled my interest to look at wheat in Turkey, within its center of diversity in the Old World. Jack and Cal were patient and infinitely helpful in schooling a novice in some of the basics of the biology and agronomy of crop diversity and crop evolution. Jack's unflagging energy was powered by sharp memories of the people, places, and crops he had encountered on the same routes forty-five years before our trip. The fieldwork on wheat was made possible with the unstinting support of colleagues at Ege University in Bornova, Turkey, in particular Emin Işikli, Ayşen Olgun, and Íbrahim Demir. Emin and Ayşen were brilliant and creative in mounting a large survey with a talented group of students and staff colleagues. In addition, we are obliged to the administration and scientists at the Plant Genetic Resources Research Center in Menemen, particularly Ayfer Tan and Ertuğ Firat, and to Fahri Altay of the Ministry of Agriculture station in Eskişehir. Again our work benefited from the input of talented graduate students, here Erika Meng and Ana Cristina Zanatta.

My efforts to understand the complexities of conservation were greatly aided by Calvin Sperling, another brilliant colleague whose early death impoverished us all. A year's leave at the International Plant Genetic Resources Institute (IPGRI) was instrumental in applying lessons from my fieldwork to conserving genetic resources. There, Toby Hodgkin was a provocative mentor, and I gained important insights from Jan Engles, Masa Iwanaga, and Pablo Eyzaguirre. I have learned from IPGRI's partners in the project Strengthening the Scientific Basis of *In Situ* Conservation, including Tony Brown, Rene Salazar, Salvatore Ceccarelli, and many crop scientists working on the project. Cary Fowler provided useful insights about the politics of science and genetic resources before, during, and after my time at IPGRI.

Several Italian colleagues showed me that crop diversity is also found in modern European agriculture and is a key to understanding the soul of a place. Roberto Papa, Cristina Papa, Fabio Veronesi, Paolo Parinni, and Donato Silveri have taken me into the agriculturally diverse regions of central and northern Italy.

My research has been generously funded from various sources. Most importantly, the National Science Foundation has supported various projects in three countries over twenty-five years. In addition, my employers, the College of William and Mary, the University of California, Davis, and the International Plant Genetics Resources Institute have helped support my investigations. The McKnight Foundation, UC Mexus, USAID, and the Ford Foundation have also contributed to this research.

This volume has benefited from the direct input of several people. Tony Brown, Subodh Jain, Toby Kiers, and Eric VanDusen read draft chapters and provided helpful comments. Dawit Tadesse, Erika Meng, and Mauricio Bellon, former graduate students, have allowed me to use some of their creative products: graphs, concepts, and analysis. Laura Lewis assisted me in preparing graphics for the text, and Fernanda Zermoglio helped out with the maps. Donald Ugent and Christopher Donnon graciously allowed me to use their photographs of Andean ceramics that depict the potato, and Cal Qualset provided me with photographs of wheat in Turkey. Oriana Porfiri was especially helpful and generous with her knowledge of *farro*.

Throughout my various forays into the fields and storerooms of farmers in Peru, Mexico, and Turkey, and along the long learning curve on the nature of crop diversity, I have been supported, counseled, and loved by my wife, Peggy. Her editorial skills and suggestions have helped me throughout the writing of *Farmers' Bounty*. I owe her more that I can ever express.

Of course, my research never would have been possible if it weren't for the hospitality, understanding, patience, and insights of numerous farmers in villages along my research transects. Most will never read the results of my time with them, but I hope that my knowledge reflects their wisdom and somehow reciprocates for their kindness. The urban and industrial sectors of the world owe an immense debt to the stewards of crop resources, and two of my long-term objectives in undertaking this research were to help illuminate this debt and to work in the collective enterprise to repay some of this debt. I hope that the product of my efforts contributes to recognizing the benefit that we have received from the stewards of the Neolithic legacy.

1

Encountering Crop Diversity

The diversity of crops is testimony to individual and collective creativity. The myriad forms of simple grains, fruits, and tubers are a singular human accomplishment and measure of social identity. As early as the fourth century B.C.E., Theophrastus (1916) described the distinguishing characters and geographic origin of the major wheat types of the Mediterranean area — hard and soft, naked and hulled, fall- or spring-sown, short season and long, Libyan, Pontic, Thracian, Assyrian, Egyptian, Sicilian. Since the latter half of the nineteenth century, the scope of crop diversity has become even clearer through systematic collection, recovery, conservation, and genetic analysis. Darwin (1896, I:332) observed: "Although few of the varieties of wheat present any conspicuous difference, their number is great. Dalbert cultivated during thirty years from 150 to 160 kinds, and excepting in the quality of the grain, they all kept true; Colonel Le Couteur possessed upwards of 150 and Philippar 332 varieties." After 1900, agronomists and plant geneticists working in the new seed industry began to use diversity to breed new crop varieties, and by midcentury conservationists had organized systematic programs to collect and preserve crop diversity.

These scientific efforts quickly revealed the astounding accomplishment of generations of farmers in amassing diversity. In 1905, W. H. Ragan published the *Nomenclature of the Apple*, a catalog of 14,000 varieties referred to in

American publications. In 1987, 125,000 distinct wheat samples and 90,000 distinct rice samples were reported among the world's gene banks (Plucknett et al. 1987, 111). The Andes are home to thousands of potato varieties and several potato species. In Mexico we find thousands of maize types, and in Turkey tens of thousands of wheat varieties. These places, known for their biological wealth in particular crops, are labeled as "cradle areas," "Vavilov centers," or "centers of origin" (Harlan 1992).

Understanding the nature of this diversity and its fate in the modern world is an international scientific enterprise that draws scientists from many different disciplines—archaeology, geography, botany, genetics, anthropology, economics. Since the mid-nineteenth century, many investigators have dealt with this topic and have defined an array of scientific, industrial, and political issues that reach far beyond the original investigations of botanists and natural historians. Geneticists and social scientists study diversity in agriculture for different reasons—for instance, to understand gene flow or the effect of the industrial seed industry. Farmers and crop breeders depend on diversity as a resource for exploiting heterogeneous environments, managing risk, or finding commercially valuable traits. Indigenous people, private companies, and nation-states all claim different types of rights over crop genetic resources, while formal and informal flows of crop resources as public goods continue. The many interests and different perspectives on crop diversity lead to opposing interpretations and conclusions on fundamental issues such as why diversity exists, whether it will persist, its value, and how best to conserve it.

Crop Evolution and Diversity

Like natural evolution, crop evolution is described by the universal processes that generate genetic diversity and regulate it through natural selection. Crop evolution differs because natural selection does not act alone but rather in concert with human ("artificial") selection (Donald and Hamlin 1984). Identifying the right seed has been an imperative of human survival since people first developed the arts of producing rather than gathering food. A Quechua peasant in southern Peru separates "seed" potatoes from the harvest; a suburban gardener in California orders potato seed from catalog; a commercial potato grower in Canada consults with an industrial seed company. All of them follow similar routines of identifying differences between potato varieties, weighing these against the individual's experience and resources, and making choices for next season. Despite their differences, peasants, gardeners, and commercial growers share interests as farmers and seed selectors—to find the best variety for their land, to meet the vagaries of weather, disease, and pests, and to produce food that meets standards of the farmer and others in the same food system.

However, similarities in routine and goals of seed selection belie enormous differences in the amount of biological diversity in different farming systems and crop populations. The commercial producer who obtains seed from an industrial seed company is choosing among a few varieties whose lineages are well known but with similar, often complex pedigrees. The peasant producer in the Andean homeland of potato manages potatoes of different species, subspecies, numbers of chromosomes, genotypes, and plant types. A handful of farms in a single Andean valley harbors as much genetic diversity in potatoes as that found in large areas of North America (Quiros et al. 1990).

We have studied biological diversity in agriculture for 150 years, during which time two puzzles have emerged: (1) why is there so much variation, and (2) why is variation within crop species distributed unevenly? Other questions are embedded in these puzzles—how did such crop diversity arise, does variation serve some purpose, will it survive changing conditions? Answering these questions requires that we delve into the human and biological ecology of different farming systems and into the population biology of crops.

Crop evolution since the Neolithic has produced similar outcomes in different crops (Evans 1993). Enhancement of the plant parts that are used by humans, such as tubers or seed spikes, has led to increased yields. Diffusion beyond the original geographic range of crop ancestors is another outcome of crop evolution (Sauer 1952). Diversification in plant phenotype of crops is a third, general outcome. The wild progenitor of maize, *teosinte* (*Zea mexicana*), appears uniform while its domesticated progeny is remarkably variable—white, yellow, blue, or red; peg-like pop corn or broad-grained hominy; and ears with delicate or plump cobs. Crop evolution involves contradictory forces to both decrease and increase diversity. For many crops, it is rather easy to argue that diversity should be far less than it is. The domestication of wild plants created a bottleneck whereby only a small amount of the total diversity of a wild species was passed into a domesticated form (e.g., Gepts 1998). Constriction occurred because domestication happened in only a few places and over a few generations (Hillman and Davies 1990) or because gene flow between different species was greatly restricted by domestication. Many crops, including wheat, barley, and rice, are largely self-pollinating, and vegetative propagation in others, e.g., potato, limits gene flow. Certain characteristics were strongly favored in many crop species that isolated them biologically from their wild ancestors, for instance changes in the plant's life cycle that synchronized flowering and maturation (Evans 1993).

Moreover, diversity must also have been constantly restricted by the normal processes of natural and conscious selection. Human management smoothes out differences in soil fertility and water availability, and makes agricultural environments less heterogeneous than natural environments. The ability of farmers to identify and multiply an outstanding cultivar is matched by the ability of other farmers to learn about and acquire it. Common farming practices, such as rotating the same crop through different fields and the exchange of seed, favor crops that are broadly adapted (Louette 1999, Brush et al. 1995). Varieties showing overall superiority over a range of different environments should naturally rise to dominance in any farming system. In sum, the normal course of agricultural evolution involves processes that reduce diversity in all farming systems, whether "traditional" or "modern."

Forces that increased diversity include dispersal outside cradles of domestication, adaptation to new environments, cultural change, and population growth (Rindos 1989). Diversity also arose because of conscious selection for special traits that were present at domestication or arose during evolution through mutation and recombination. These traits include adaptation to different soils, resistance to pests and pathogens, and different tastes and cooking qualities. Domestication in many, perhaps most, crops is an ongoing phenomenon in which gene flow from wild to domesticated plants continues to provide new material (e.g., Johns and Keen 1986; Elias, Rival, and McKey 2000). An example of traits that contributed to crop diversification after domestication includes the addition of proteins in different wheat species that are carried on added genomes and confer baking and other qualities to the crop (Zohary and Hoph 2000). Likewise, the selection for differing levels of alkaloids in potatoes (Johns 1990) and cassava (Wilson and Dufour 2002) has diversified the crops. Diversity undoubtedly reflects a human fascination with and desire for variation in the goods and tastes of everyday life, along with the intelligence of how to tease variation out of a seemingly uniform group of plants (Boster 1986). Diversity signifies social identity such as membership in a particular kin group (Freeman 1955), and it often is associated with the sacred status of crops (Shigeta 1990, Iskandar and Ellen 1999).

Historically a scientific curiosity, the issue of crop diversity has recently emerged as an important policy area with implications for agricultural development, environmental protection, resource conservation, and the rights of cultural minorities and poor farmers. Crop diversity is now perceived as a fundamental resource for crop improvement in modern agriculture (Plucknett et al. 1987), and this resource has become more valuable because of record population numbers, the exodus from agriculture, and the threat of climate change. Improved varieties, fertilizers and pesticides, and markets are widely available and now help meet the demands of farming that diversity used to fulfill. Farming in North America and Europe experienced a drastic reduction in the number of crop varieties, and this reduction has also occurred in parts of the tropical world where diversity has historically been greatest (Fowler and Mooney 1990).

Our enhanced ability to use diversity and the possibility of decreasing availability of crop genetic resources have redirected our attention to conservation. To be successful, however, we must understand not only the biology and genetics of crops but also the cultures and economies of farmers who are stewards of crop diversity. Crop genetic resources exist in two complementary and intertwined forms—crop genes and human knowledge about the species, including the knowledge that has been transmitted over generations of farmers. Indigenous knowledge, as much as crop genes, is part of the evolutionary system of a crop species, determining traits that will or will not be passed on. By understanding the nature of both biological and cultural diversity in agriculture, we should be better equipped to conserve and use diversity to help feed the billions of additional people who will live on Earth, perhaps in a period of drastic climate change.

Human Ecology and Crop Diversity

Understanding the relationship between humans and their biophysical environment has been a staple of natural history since antiquity (Glacken 1967), and it now is a key part of all modern disciplines of social science. Anthropology and geography are direct descendants of the natural history that coalesced during the colonial expansion of Europe, a fact reflected by the ethnographic exhibitions in many natural history museums. The material basis of cultural systems was emphasized by an important segment of the mid-twentieth-century anthropologists in America. Between 1930 and 1980, anthropology and geography shared in a project of describing and analyzing human/environment interactions, represented by the cultural ecology movement and the work of the anthropologist Julian Steward and the geographer Carl Sauer. Cultural ecology gave strong emphasis to two themes: (1) cultural adaptation to different environments, and (2) cultural evolution and social change as embodied in the growth of complex societies. As a subfield of anthropology and geography, cultural ecology arose prior to the full development of systems-based ecology represented by the work of Odum (1953). Steward, the nominal "father of cultural ecology" in anthropology, drew on other social scientists, such as Hawley (1950), but his major sources came from within anthropology itself and from his own fieldwork in the American West. Cultural ecology's apogee in anthropology occurred around 1970 and was marked by the establishment of the journal *Human Ecology* in 1972. Since 1990, cultural ecology has diminished within anthropology as symbolic studies have become more prominent. Ironically, the waning of cultural ecology as a specialized field in anthropology coincides with the maturation of ecology across biological and social sciences, as evidenced, for instance, by the rise of ecological and natural resource economics.

The study of cultural adaptation to the environment was central to the rise of modern anthropology at the turn of the twentieth century, and continues as a background to such areas within anthropology as ethnobiology and economic anthropology. Like its counterparts in biology and ecology, the study of cultural adaptation in anthropology is largely a descriptive exercise to link specific environmental aspects and cultural features. Environmental aspects that act as strong limiting factors to human health and welfare have provided particularly fertile mediums for studying cultural adaptation (e.g., Baker and Little 1976). Thus, a significant amount of research in cultural ecology has been extended to regions of protein deficiency, aridity, poor soils, and high altitude (Moran 1982). Likewise, environments that offer unique opportunities, such as great heterogeneity of useful habitats or isolation, have also attracted studies of cultural adaptation (e.g., Padoch et al. 1999).

From the inception of cultural ecology (Steward 1955), agriculture was perceived as a definitive element of culture, in Steward's terms a "culture core." Food production is both the human watershed that allowed the organization of complex societies and a key element in the vast majority of extant cultures. The study of agriculture drew prehistorians and ethnographers into the common endeavor of understanding the origin and evolution of agriculture and its relationship to other social elements such as demography, migration, and stratification. Geertz's *Agricultural Involution* (1963), Netting's *Hill Farmers of Nigeria* (1968), and Rappaport's *Pigs for the Ancestors* (1967) epitomize ethnographies with specific focus on agriculture. The intensification of agriculture, represented by the contrast between shifting and permanent cultivation, is a dominant theme in "agrarian ecology" (Netting 1974). Since Steward's (1955) contrasts between subsistence systems, intensification had been a preeminent concern of the cultural ecology of agriculture, and this concern was theoretically framed by two economists—Boserup (1965) and Chayanov (1966). As a unifying theme in cultural ecology, the intensification of agriculture attracted scholars with very diverse interests (Brush and Turner 1987, Netting 1993).

At the zenith of anthropology's interest in agrarian ecology, Vayda and Rappaport (1968) issued a broadside critique of cultural ecology, depicting it as nonecological and questioning its methods and conclusions. One of Vayda and Rappaport's (1968) recommendations is that anthropologists adopt an ecologist's framework by studying the way human populations interact with other populations to form food webs, biotic communities, and ecosystems. Vayda and Rappaport's recommendations resonated with intellectual movements outside anthropology—the strengthening of systems analysis and formal ecology. The initial response to the challenge of formalizing human ecology was to study subsistence systems as energy flows (e.g., Winterhalder and Thomas 1978), but the complexity of describing energy flows and the difficulty of linking them to behavior and farm management soon led human ecologists to turn to other problems in ecology. A burgeoning area was the study of biological diversity and conservation biology (Soulé 1986), a field that welcomed human ecologists (Redford and Padoch 1992).

The extravagance of crop diversity would seem to be a natural and compelling topic for human ecologists, but this has not usually been the case. The uneven distribution of diversity has been recognized since the nineteenth century, and diversity is concentrated in areas and among people that are familiar to anthropologists. Diversity in certain agricultural systems is linked to crop domestication, which is a central topic in archaeology, one of anthropology's primary subfields. Crop diversity would seem to invite the skills and interests of anthropologists such as ethnolinguistic investigation, cultural adaptation, and cross-cultural comparison. Geographers would likewise appear to be naturally drawn to studying crop diversity after the pioneering work of Carl Sauer (1952), who saw crop domestication and dispersal as crucial topics. Moreover, geography's tradition of working in the interstices of social and biophysical science would likewise seem to draw geographers to crop diversity (Zimmerer and Young 1998).

Despite the seemingly natural attractiveness of crop diversity for anthropological and geographical investigation, occurrences of research are rare and dispersed. Instances of perceiving the potential relevance of crop diversity are present, but these have not coalesced to make it a topic for either discipline until recently. Only since the 1980s has human ecology of crop diversity achieved a critical number of case studies or researchers who would define theory and method to make this a significant area linking social science and agricultural ecology.

An example of the suggestive but unrealized presence of crop diversity in studies of traditional farming is the case of rice in Asia. The cradle area of rice domestication follows the southern flanks of the Himalaya from northern India to the Gulf of Tonkin and into southwestern China (Chang 1985). The two major ecogeographic races of cultivated rice recognized by modern geneticists, *indica* and *japonica* (*sinica*), were differentiated in China two thousand years ago, and the 1742 agricultural compendium, the *Shou Shih Thung Khao,* recorded over three thousand named varieties (Bray 1984). Modern researchers from the West observed the same astonishing numbers. In his monograph *Rice,* Grist (1986) observes that great diversification is characteristic of the plant, referring to sixty-seven thousand Asian cultivars in 1982. Even for improved varieties, Grist (1986) cites the tendency to accumulate large numbers in single countries such as India with over six hundred improved varieties in 1962. Bray (1986) comments that the diversity of rice is several fold larger than that of wheat, another Asian cereal.

Research in Asian rice systems has a long and accomplished record (e.g., Marten 1986), albeit with a curious lacuna of studies on the social and ecological basis of rice diversification. Indeed, some of the most enduring interests in the human ecology of agriculture were defined by research in Southeast Asia, for example the nature of intensification in shifting cultivation in response to population growth (Hanks 1972, Spencer 1966). Asian rice is associated with an unusually intricate social apparatus and cultural identification, a treatment that is rarely matched in other regions and crop complexes and is uncharacteristic of other Asian cultigens. Perhaps no region has been as thoroughly transformed by its major crop (*Oryza sativa*) as Asia's rice lands. The emblematic and pivotal role of rice in Asian society has been well documented and extensively studied (Geertz 1963, Bray 1986, Ohnuki Tierney 1993). Spencer (1954) describes Asia as a "rice landscape" sculpted by millennia of molding land to serve as paddies, and other regional analyses refer to Asia's "rice societies" (Nørlund, Cedrroth, and Gerdin 1986) and "rice economies" (Bray 1986). Adams (1948) called anthropologists' attention to the complexity of rice systems, terminology, and varieties. Two primary issues have subsequently framed human ecology research on rice systems. First, the nature and processes of water control has driven investigation on whether Asian nations are to be understood as "hydraulic" societies (e.g., Bray 1986, Lansing 1991). Second, the intensification of rice production has framed numerous studies. Geertz (1963) described the two traditional rice ecosystems of Indonesia, swidden and *sawah* (irrigated paddy), that are common throughout much of Asia. He noted that both seem to exist in equilibrium but that the latter is highly specialized, dependent on complex technology, and able to support large populations and absorb labor. Following *Agricultural Involution* (Geertz 1963), intensification studies have examined land use changes within both swidden systems (e.g., Kundstadter, Chapman, and Sabharsi 1978) and paddy (e.g., Moerman 1968).

However, Geertz (1963) also noted lack of studies of the microecology of aquarium-like rice paddies of Asia, despite their role in Asian society and their evolution as one of the world's most intensive agricultures. The lack of agro-ecological research on wet-rice in Asia reflects a similar absence of rice diversity in the literature, and the crop's diversity did not become an important subject for research until the 1980s. The staple's diversity is perhaps less visible than its terraced landscape, but diversity is likewise a testimony to the skill and diligence of generations of farmers in molding the crop as they have molded the landscape.

Both the stunning accomplishment of traditional farmers in diversifying rice and the relationship of this diversity to the social and physical landscape are compelling topics for research. But a third rationale is the threat to rice diversity and traditional management posed by the diffusion of modern, high yielding varieties (HYV) of the Green Revolution. By 1970, crop scientists mounted a highly visible campaign to meet the threat that HYV rice would wipe out the vast stores of local varieties (Frankel and Bennett 1970). A consequence of this campaign was to establish gene banks for major stables, such as the one at the International Rice Research Institute established in 1960 (Plucknett et al. 1987). Social scientists (e.g., Scott 1976, 1985; Farmer 1977) began to argue that besides the inequities related to the new rice types, they also replaced locally adapted varieties that represented generations of farmer investment, knowledge, and ability. This agrarian version of deskilling technological change might have also prompted ecologists to turn to the urgent task of documenting the traditional ecology of rice agriculture through a lens of rice diversity.

Among the few agriculturally focused monographs set in rice's cradle area is Izikowitz's (1951) study of the Lamet, swidden farmers of highland Laos. He records that the Lamet have many named varieties of rice, distinguished by color, maturity period, and whether they are glutinous or hard. Gourou (1955, original 1936) reported "at least 300 varieties of rice grown in the (Tonkin) Delta, two hundred of ten month rice and one hundred of fifth month rice. These varieties are not grown without discernment by the peasant: each one of them has qualities which are appreciated in particular circumstances."

Fieldworkers in Malaysia, Indonesia, and the Philippines, on the southern periphery of rice's cradle area, likewise noted high levels of rice diversity. Grist (1936), an agricultural economist, reported that thirteen hundred Malayan varieties had been collected by 1915. He describes the classification of six major categories for Malayan rice based on grain type, maturity period, hardness, and color, and he notes that some types are grown exclusively for ceremonial purposes. Freeman's (1955) monograph on Iban agriculture focused on rice production. The Iban kept numerous strains of rice, although Freeman did not make a systematic inventory. Freeman (1955, 28) found that "every Iban *bilek* family possesses a strain of sacred rice called *padi pun,*" and he noted that individual rice strains are roughly grouped according to maturation period. One family had eighteen different strains. Conklin's (1957) famous study of Hanunoó, on the island of Mindoro, Philippines, documents not only that these shifting cultivators manage a large number of different crops but that they keep ninety-two varieties of rice. Neither Freeman's nor Conklin's pioneering monograph suggests an ecological framework for understanding why so many rice strains are maintained by swidden cultivators or what might happen to this diversity if intensification were to occur.

Ecological and social science researchers began more systematic description and analysis of Asian rice diversity in the late 1960s during the Green Revolution, when large portions of Asia's paddy rice were converted from local varieties to high yielding ones. Moerman (1968) observed that Lue farmers of Ban Ping in northern Thailand cultivated thirty types of wet-rice, distinguished mainly by whether the grain was hard or glutinous and by maturation date. Diversity was seen to play the functional roles of smoothing out labor demand and fitting microecological field conditions such as slightly deeper water. Moerman (1968) notes that rice types are unevenly distributed in the crop's population, subject to careful selection and related to yield, but his analysis of choice and change focuses on mechanization and fertilizer use rather than on changes in the rice populations of Ban Ping.

More detailed and ambitious analysis of rice diversity was undertaken by Lambert (1985), who studied the Pahang of Malaysia. Lambert's research is unusual because of his attention to the folk taxonomy of more than forty rice varieties kept by the Pahang. Lambert explains the logic of Pahang classification according to four steps for locating a particular variety. He discusses the cultivar's diversity in relation to environmental concerns, in particular variability in water depth, and finds that the Pahang are skilled in selecting and maintaining diversity as well as identifying new varieties. In a similar vein, Iskandar and Ellen (1999) discuss the human ecology of rice among the Baduy of West Java, Indonesia. These swidden cultivators maintain eighty-nine named varieties of rice that are distinguished by characteristics that have been found elsewhere: whether the rice is glutinous or hard, whether the seed is glabrous or hairy, the color of the seed, and maturation date. Iskandar and Ellen describe selection and planting practices, and they add ecological detail in showing the distribution of different rice types among households so that we have a picture of the population structure of the local crop. Lambert (1985) and Iskandar and Ellen (1999) exemplify progress in developing a human ecology of crop diversity.

An Andean Portal

The high altitudes of the Andes provide a good example of the complementary roles of environmental limitation and opportunity in understanding cultural adaptation. Here, indigenous cultures developed agriculture and complex sociopolitical systems that support a larger human population at higher altitudes than anywhere else on Earth. Andean farmers regularly work fields at four thousand meters above sea level — the highest permanent agriculture in the world — where oxygen deficiency (hypoxia), high diurnal temperature fluctuation, low temperatures, frequent frost and hail, and thin soils limit human endeavor. Besides presenting the difficulties of an extreme environment, the Andes offer the opportunity of great environmental heterogeneity and ready availability of distinct habitats with different physical and biological resources. The ancestors of modern Andean people learned how to exploit the advantages of environmental heterogeneity in several ways — domesticating plants and animals adapted to different altitude niches, creating settlement patterns for access to different altitude zones, developing economic and political systems organized around the principle of "verticality" (the use of a maximum number of altitude zones [Murra 1972]). The layer-cake nature of the Andean environment, its vegetation, and its agriculture has been analyzed by naturalists, geographers, botanists, and anthropologists since Alexander von Humboldt observed it in the early nineteenth century (Humboldt 1817). Archaeologists, physical anthropologists, ethnohistorians, ethnographers, and geographers have effectively used the concept of adaptation to this mountain environment to describe and interpret Andean human physiology, social evolution, and organization of culture, society, politics, and economics from the village up to the state level (e.g., Masuda, Shimada, and Morris 1985; Baker and Little 1976; Zimmerer 1996; Mayer 2002).

Andean people domesticated a suite of crops and animals and brought others from other indigenous American agricultural traditions. Particularly important are the numerous tuber crops from the region, of which the potato (*Solanum* spp.) is the most prominent. Others include oca (*Oxalis tuberosa*), mashua or añu (*Tropaeolum tuberosum*), ullucu (*Ullucus tuberosum*), arracacha (*Arracacia aequatorialis*), maca (*Lepidium meyenii*), achira (*Canna edulis*), and yacón (*Polymnia sonchifolia*) (NRC 1989). Ascending an Andean valley is to walk through different climates and plant communities and through a human landscape formed with terraces, irrigation works, patchworks of fields, and settlements. The vertical layering of Andean agriculture is as regular as the march of wheat fields, olive orchards, and vineyards in Tuscany, or rice paddies and vegetable gardens in Yunnan. Within this landscape and at a closer scale, equally marvelous diversity can be observed in the multitude of different fruits, vegetables, cereals, tubers, community forests, pastures, and animals, all products of generations of human hands. The kaleidoscopic variation of the Andes is not confined to biophysical mosaics but also includes biological variation within and between a cultivated field, a *chacra*. Like the landscape kaleidoscope, the crop elements of chacras are repeated and reconfigured over and again along the length of the tropical Andes.

The processes of cultural change and evolution constitute the other dominant theme in cultural ecology's efforts to understand the relationship between human society and the environment. Even before Inca times, the Andean populace had successfully integrated regions and cultures that were separated by the corrugated Andean topography. European conquerors and colonists carried the process of integration further by adding new social processes—culturally based exploitation, capital accumulation, market penetration, and peasant resistance (Orlove 1978). Ecologically, the European conquest led to a new agriculture and social landscape in the Andes. New crop and animal species—wheat, barley, sugar cane, sheep, and cattle—were introduced and competed with local crops and animals. Andean people were subjected to exploitation, resettlement, and redistribution, and as a consequence the population of the region suffered a near collapse within a hundred years of the conquest. Agriculture was reorganized from subsistence to extraction (Orlove 1978). A hybrid agriculture and economic system involving native and Old World crops and household and market production emerged as the foundation of post-colonial Andean society. Throughout the region's tumultuous history, Andean crops retained their remarkable diversity and served as a key element in the survival of Andean culture in the narrow valleys and peasant communities that were refuges (Mallon 1983). Ironically, Andean crop diversity went largely unnoticed by anthropologists and geographers who otherwise celebrated the resistance and resilience of Andean people, with the notable exceptions of Gade's (1975) study of crops and land use in the Vilcanota Valley of Peru, and an obscure paper by LaBarre (1947) on Aymara potato nomenclature.

In 1977 I began a decade of research on potato diversity in Peru involving qualitative, ethnobotanical research on crop diversity and farm surveys from contrasting regions, villages, households, and plots. My objective was to evaluate the persistence and loss of Andean cultivars under the pressures of agricultural modernization. Anthropologists (e.g., Redfield 1941) and economists (e.g., Epstein 1962) had successfully used intervillage comparisons to model historical change. My strategy was to work in two contrasting regions that reflected the continuum of agricultural modernization. The *mestizo* villages of

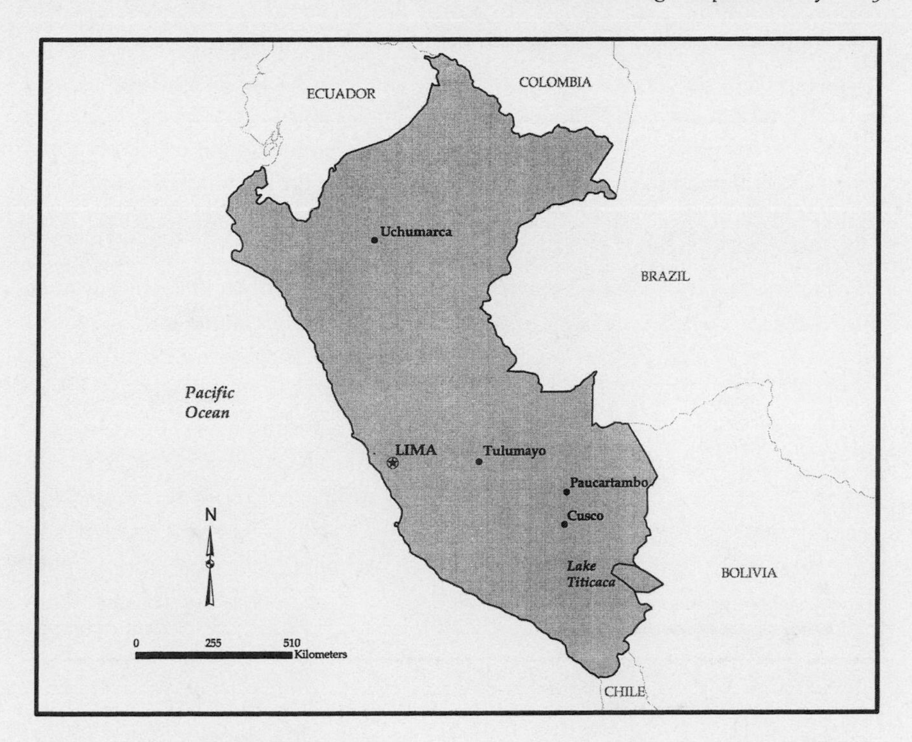

Figure 1.1 Study areas in Peru

the Tulumayo Valley in the central highlands east of Huancayo were fully integrated into regional and national markets. Modern potato varieties were a major cash crop. Quechua-speaking villages of Paucartambo in the Cuzco region of southern Peru used modern varieties and purchased inputs to a lesser degree. My purpose was essentially to compare households that varied according to the diversity of their potato fields and household characteristics such as farm size, education level, off-farm employment, and commercialization. In each region I surveyed six or more villages to capture differences in the ecological and social environments. I aimed for a minimum of one hundred household surveys in each region. Farm surveys enumerated information on the nature and management of each plot cultivated by the family, along with sundry data on social and agricultural aspects of the farm. The result was a very large number of plots that could be analyzed as separate management units.

Figure 1.1 shows the location of my various field studies in Peru. Potato diversity quickly led to research areas that were only distantly related to my initial research on the human ecology of Andean agriculture—crop ecology

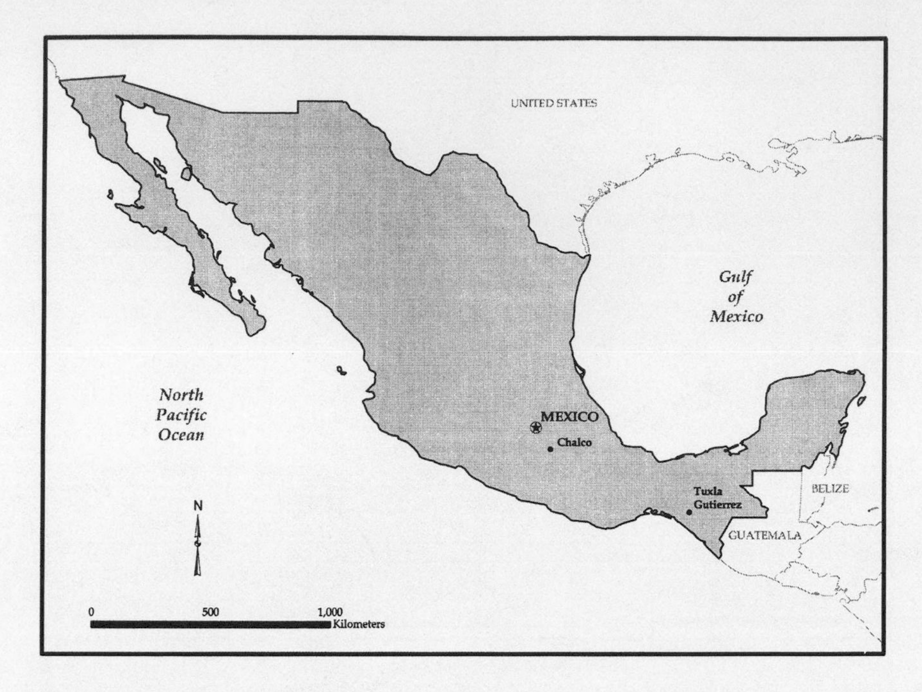

Figure 1.2 Study areas in Mexico

and crop evolution. Beginning in 1988, I began research on maize in its Mexican hearth, which has continued to the present. Between 1990 and 1994, I extended the same research program to wheat diversity in the western highlands and edge of the Anatolian plateau of Turkey. Figures 1.2 and 1.3 show locations of the field studies in Mexico (1988, 1997–1998, 1999–2002) and Turkey (between 1992 and 1994). Along the way, I began to investigate issues that reflect public concern for genetic resources and the farmers who are their stewards—how to conserve diversity on farms, and Farmers' Rights. Nevertheless, my Andean experience has continued to be a central reference point that has left its imprint on my research.

Methods

Methods for studying the human ecology of agricultural diversity include the collection and analysis of names and terminology associated with the crop, questionnaires applied to random samples of households among villages with contrasting social and agricultural environments, and the collection and characterization of crop samples. A key method is gathering data about each

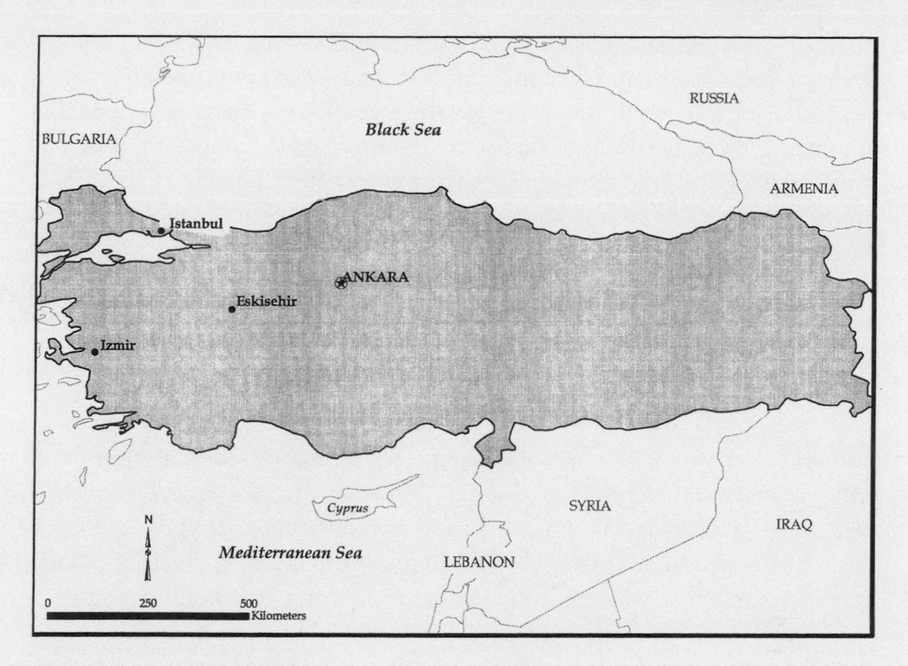

Figure 1.3 Study areas in Turkey

of the several fields that are cultivated by a single household. Field fragmentation allows the common pattern of maintaining several types of the same crop, and it also enables the researcher to study different cropping decisions by the same household. Fields may be environmentally distinct, have different tenure, be managed differently, and be used for different purposes. Collaboration is essential between anthropologists and each of three main groups: geographers, who specialize in spatial analysis; economists, who have quantitative tools to analyze farmer decision making; and crop scientists, whose interests are in population biology and genetics.

Site selection depends on research objectives, for instance whether to document traditional knowledge or to analyze competition between different crop types. In the Peruvian research, sites were chosen to investigate crop diversity across contrasting environments and social conditions and to evaluate the impact of agricultural modernization. Environmental contrasts can be captured by studying villages and fields across a heterogeneous landscape, for instance with rain-fed hillsides and irrigated valley bottoms. Access to market is a principal social feature that differentiates villages and can be captured by sampling villages and households on primary roads close to market centers and others on secondary roads or off-road altogether. The key to cross-sectional research studying the impact of modernization is to examine villages where both traditional and modern agricultural technologies are present. Competition between crop types for space and the attention of the farmer occurs in cropping systems where both traditional and purchased inputs are used and where agriculture serves both subsistence and commercial goals.

Farm surveys and cross-sectional comparisons among households and villages permit statistical analysis that relates variation in crop diversity on the one hand to social and environmental variation on the other. This method allows one to estimate the impact of factors such as increased use of modern varieties, commercialization, and off-farm employment on the diversity. It may show that certain types of farms and environments are more likely to keep traditional crops, and it allows the testing of hypotheses about the conditions of maintenance and loss of diversity. We hypothesized that selection of crop varieties was dependent on three major factors. First, environmental heterogeneity, such as field fragmentation and different soil types, will favor diversity because of adaptation by different crop varieties. Farmers are familiar with adaptation, which is measured by comparing yields of different varieties in the same environment. Performance, however, is modulated by management such as the use of fertilizers and fallowing, which in turn depends on social factors such as farm size, wealth, and education.

Second, risk is critical to variety selection. Crop types differ according to the variability of yields, and more variable yields are believed to pose higher risks (Scott 1976; Meng, Taylor, and Brush 1998). Diversity is often associated with stability, although the relationship is far from definite (Goodman 1975). Some households are able to live comfortably with the risk inherent in crop variability, while others avoid risk in order to guarantee survival (Scott 1976). Some villages provide means to cope with risk, such as off-farm employment, while others are deprived of these opportunities.

The third general factor in crop selection is consumption and other use and the ability of markets to meet consumer demand for diverse products. The history of agricultural development has emphasized the contrast between production for consumption or home use and production for the market, a contrast that might logically apply to crop variety choice. Heavy emphasis on a single commodity and long experience with that commodity might easily lead to a concern for quality that can only be satisfied by traditional crops. Households that produce their own food must be concerned with storage, but this concern may be irrelevant in modern crops that will be marketed, milled, and/or stored under more optimal conditions than in the back room or on the rafters of a farmhouse. Markets, however, can play an important role in decisions regarding consumption. If a farm family is confident that the type of crop they want to consume is available in the market, they may chose not to cultivate it themselves. So, the presence or absence of local markets (de Janvry, Fafchamps, and Sadoulet 1991) for specific crop types as well as for technical inputs becomes an important factor in crop selection.

Navigating the Issues of Crop Diversity

Three general themes have emerged from over a century of research on crop diversity, and human ecology has become a contributor to all three. The oldest theme is to connect crop diversity to domestication. Vavilov (1926, 1951) convincingly showed that crop diversity was one of the two criteria for determining center of origin, along with the presence of probable wild ancestors of the crop. Domestication, however, is an ongoing process in many crops because of a continued flow of genetic material from wild to cultivated plants. While genetics contribute most to this theme (e.g., Gepts 1998, Doebley 1990), human ecology is also relevant. Continued domestication depends partly on the management of crops that might favor gene flow and on selection of crops that might be produced by gene flow (e.g., Johns and Keen 1986; Elias, Rival, and McKee 2000).

A second theme is the relationship between crop diversity and crop evolution. This theme has been the purview of botanists and geneticists (e.g., Frankel, Brown, and Burdon 1995). The major questions that drive this theme have to do with population structure, population biology, and taxonomy of crops, issues tied closely to crop diversity. Issues in evolutionary biology raised by crop diversity include the interaction between genotype and environment, population processes such as competition and isolation, and the complementarity between natural and artificial selection. Following Darwin (1896), human, or "artificial," selection has been understood to be part of crop evolution (Donald and Hamlin 1984). Understandably, few crop scientists have attended to the role of traditional farmers in crop evolution. Human ecology's principal contribution here is to describe farmer perception and selection as they relate to crop population structure, and since 1985 many cropping systems in cradle areas have been described in these terms (e.g., Boster 1986, Bellon 1991, Zimmerer 1991, Teshome et al. 1999, Iskandar and Ellen 1999).

The third theme concerns threats to crop diversity and conservation measures to meet those threats. The Green Revolution that commenced in the mid-1960s affected several centers of diversity, such as the Philippines in the

Southeast Asian center of rice diversity, where nearly all lowland rice areas were planted to introduced, high yielding varieties by 1985 (Byerlee 1996). In twenty years, thousands of lowland rice varieties were lost. Comprehending and meeting the threat of genetic erosion emerged as a major theme of the science of agricultural diversity in the early 1970s (Frankel and Bennett 1970). Human ecology plays a central role in studying the patterns of agricultural change and their effect on traditional crops (e.g., Brush, Taylor, and Bellon 1992). On-farm conservation (in situ) has emerged as a viable and important means to preserve both diversity and the agro-ecosystem that created it (Maxted, Ford-Lloyd, and Hawkes 1997; Wood and Lenné 1999), and human ecology is again central to planning and implementing on-farm conservation. Contributions here include research to understand farmer valuation of diversity (e.g., Nazarea 1996), identifying optimal sites for in situ conservation initiatives (Meng, Taylor, and Brush 1998), and enhancing the participation of farmers in conservation (Soleri and Cleveland 2001).

2

A Naturalist's View of Crop Diversity

Human knowledge about plants reaches across time and culture. Collectively, human cultures recognize tens of thousands of species of plants, thousands of edible plant species, hundreds of crops, and hundreds of thousands of varieties of crop species. Arguably there is no collective realm of human knowledge about plants that is more elaborate than the last level in this sequence. This knowledge is the means whereby distinct varieties of crops are identified and described, and this knowledge is the basis for selection, property, and exchange among farmers, crop breeders, and consumers. Recognition and management of crop varieties are part of a single continuum of using and caring for plant species and plant communities, a continuum that is most apparent in the amazing diversity of crops. This diversity is renewed yearly during seed selection, a human routine that has guided crop evolution. However, after thousands of years as an integral part of agriculture, crop diversity appears threatened as farmers everywhere cope with economic changes, technological alternatives, and population pressure. Market and technological alternatives to diversity in the form of improved varieties, purchased inputs such as fertilizer and pesticides, and crop insurance can assume the agricultural roles that diversity previously played. Farming in North America and Europe has already seen a drastic reduction in the number of crop varieties, and this reduction has occurred in parts of the tropical world where diversity is even greater. Nevertheless, the availability of diversity to future generations will be no less important than it has been to previous ones.

The Search for Diversity—Exploration, Origins, Ecology

Engagement with crop diversity is ubiquitous in agricultural societies because farmers everywhere are aware of the potential of new crops or crop varieties to solve problems. Many different societies have organized the search for new crop varieties, and eventually this search has led to scientific crop improvement programs based on conscious selection, crosses among different varieties, and genetic manipulation. Engagement with crop diversity has also led to the study of crop evolution and the crop ecology that links biological, physical, and human elements.

Scientific engagement with crop diversity can conveniently be divided into three periods: (1) exploration, (2) search for the places of crop origins, and (3) conservation. Three themes weave in and out of these periods. The utility of crop diversity is the original theme, dominating the period of exploration but present also in subsequent periods. Crop evolution appears as a second theme during the search for crop origins. Finally, the ecology of crop diversity, or the study of the relationship of diversity to other biophysical and human elements in a single system, is the third theme, and the one that dominates the present period of conserving diversity.

EXPLORATION

It is not surprising that plant collectors and naturalists took passage on the ships of imperial navies and colonial governments—*Endeavour, Discovery, Beagle.* This passage was an efficient means to satisfy an age-old and widely felt passion to find nature's curiosities. Records of plant exploration and collection come from different societies and date far back into antiquity, and Woolley (1928) informs us that by 2500 B.C. the Sumerians searched outside of their Mesopotamian homeland for useful plants. Similar records testify to this practice in antiquity by the Egyptians and Chinese (Ryerson 1933). By the time of European expansion and colonization, these activities had grown into a well-organized and far-flung search for living material to fuel new industries and the extension of the agricultural frontier (Brockway 1979, Fowler 1994). However, farmers had established the practice of plant exploration and collection long before the organization of official programs.

Fairchild's (1898) suggestion that agricultural expansion is better accomplished with exotic plants than with endemic ones was confirmed by Jennings and Cock (1977), who demonstrated that crops almost always do better outside than inside their places of origin. The advantage of exotic over endemic species no doubt has been recognized for thousands of years and is behind both the search for seed and plants that can be pocketed or collected and the diffusion of crops outside of their native hearths. Anderson's (1952) description of agriculture as a "transported landscape" honors the thousands of generations of cultivators who have moved seeds and transformed the world.

The passion to learn about, acquire, and use life's diversity is not confined to European cultures or imperial states, although these have clearly benefited from the acquisition of living goods and their natural products. The passion to collect and experiment with new plants and animals is evident among hunter/gatherers (Etkin 1994) and in the transport of crops and domesticated animals away from their native habitats wherever humans planted seeds or tended animals. This passion most likely propelled maize out of its Mesoamerican hearth to the far reaches of North and South America. In the Eastern Hemisphere, crops moved eastward and westward with migrants and traders. Sorghum spread from its sub-Saharan homeland to India, and citrus moved in the opposite direction from its hearth in the southern Himalayan foothills to the Mediterranean. Indeed, the diffusion and development of crops during prehistory obscured the fact that crops had distinct and relatively limited centers of domestication.

Organized plant exploration has been described both as a boon to humanity (Fairchild 1938) and as a bane of the tropics that were colonized for industrial crop production (Brockway 1979). Exploration has been first and foremost a search for useful plants, whether for a broad public good such as freely providing new crops to farmers (Fowler 1994) or for private and perhaps ignoble purposes such as the voyage of the *Bounty,* which brought the seedless breadfruit tree from Polynesia to the West Indies to feed sugar plantation slaves. Nevertheless, plant exploration and collection existed long before European expansion and the rise of colonial plantations. In 1495 B.C., Queen Hatshepsut of Egypt is reported to have received frankincense trees from a royal expedition to the land of Punt, perhaps on the Somali coast (Watson 1983). In the eleventh century B.C. the Assyrian king Tiglath-pileser boasted, "I took cedar, box-tree and Kanish oak from the lands over which I had gained dominion — such trees as no previous king among my forefathers had ever planted — and I planted them in the gardens of my land. I took rare orchard fruits not found in my land and filled the orchards of Assyria with them" (cited in Watson 1983, 88–89).

Whatever the goal of exploration, incalculable scientific benefit accompanies the search for useful plants. The seeds of modern biological science — including systematic biology, evolution, and ecology — were hidden passengers aboard the ships, caravans, and mule trains that transported plant collectors. These seeds took root in catalogs, farmers' almanacs, and herbals, and in botanic gardens and herbariums that were created to record, promote, and maintain the seeds, cuttings, and plants that returned with plant collectors. By the eighth century, Islamic gardens in Egypt and Spain were filled with rare and exotic plants from Syria and beyond. Watson (1983) credits the Islamic introduction of heat- and drought-tolerant crops to the Mediterranean region with giving birth to an agricultural revolution based on more intensive and frequent cropping.

In 1827, American consular officers were requested to ship home seeds and cuttings of potentially useful species. In 1898, this activity was elevated to a formal program of the United States Department of Agriculture (U.S.D.A.) by the creation of the Foreign Seed and Plant Introduction Section, with David Fairchild as its founding director (Fairchild 1938). The early twentieth century is known as the Grand Age (Cunningham 1984) or Bonanza Years (Klose 1950) of U.S. plant exploration, when collectors from the U.S.D.A. ranged across the world in search of novel plants. The first botanical garden in America was established in 1730 on the banks of the Schuylkill River above Philadelphia (Klose 1950), predating by thirty years the founding of the Royal Botanical Gardens at Kew.

European colonialism and agricultural development in the United States transformed the ubiquitous habit of searching out and experimenting with new plants to purposeful and organized plant exploration and introduction. The "Columbian Exchange" (Crosby 1972) that moved food crops between the Old and New Worlds, and thereby revolutionized agriculture and human diets, was largely accomplished indirectly and through curiosity or immigrants' pockets filled with familiar seeds. Nevertheless, by the end of the eighteenth century, organized plant exploration and transshipment had become an integral part of European colonialism as a means to exploit territories in the tropics and new markets. Plantation crops that were moved from their original hearths to create colonial industries include sugarcane, tea, coffee, bananas, cacao, and rubber. Brockway (1979) chronicles the supporting role played by the Royal Botanical Gardens at Kew in colonial, plant-based industries of the British Empire, such as cinchona (an anti-malarial), rubber, and sisal.

ORIGINS

The willingness of naturalists to sail or trek with the flag bearers of colonial expansion should not obscure the scientific goals that drove them. A desire to reconstruct the Garden of Eden may have been a motivation behind the origin of European gardens (Glacken 1967), but gardens led to distinctly secular outcomes. As Findlen (1994) shows, the European penchant for collecting not only sired possession but also attitudes toward nature and the discipline of natural history. Knowledge about the diversity and distribution of species, which accumulated in herbals, almanacs, and botanic gardens, was at once a tool that helped to dismantle the Great Chain of Being (Lovejoy 1936) and that provided one of the building blocks of modern biological science. One of the many questions that emerged from the accumulation of knowledge about plant diversity concerned where crops came from. To some, such as the Maya, human creation and the creation of the primary staple were one and the same (Recinos 1950). Harlan (1992) notes that agriculture is accepted as a divine gift in cultures around the world. Besides this most common of explanations for the origins of crops, several explanations have been put forward that recognize humans as creators of crops.

Humboldt (1807) had despaired of locating the places of crop domestication: "The origin, the first home of the plants most useful to man, and which have accompanied him from the remotest epochs, is a secret as impenetrable as the dwelling of all our domestic animals" (Humboldt 1807, 28). The challenge in Humboldt's conclusion was soon taken up by geographers, botanists, and naturalists. By the middle of the nineteenth century, when evolution was recognized for species and for human civilization, plant geographers began to speculate about the reasons for plant domestication and the places where it occurred. Both of these questions still attract biological, geographic, and anthropological research.

Harlan (1992) lists four other general explanatory models for crop origins besides divine intervention: (1) domestication for religious reasons such as sacrifice, (2) domestication caused by crowding onto oases during drought, (3) domestication as an accidental but fortuitous discovery, and (4) domestication as an extension of gathering. Harlan concludes that none of these models stands up to scrutiny with data from the various instances of domestication, and he opts for a "no-model model" approach. Besides momentous consequences for human life, two striking aspects of plant domestication continue to invite research. First, it occurred independently on five of the six continents occupied by humans, Australia being the one exception. Crops were domesticated in vastly different environments and conditions ranging from the semiarid savannas of the Sahel to the Amazon Basin. Second, in the time scale of human evolution, plant domestication occurred virtually simultaneously around the world, that is, within a few thousand years. Domestication continues today, but its first appearance in southwestern Asia is separated from domestication in South America by an interval that is insignificant in the time scale of other major steps of human evolution.

The search for the locations of crop domestication began in earnest with the work of the Swiss plant geographer Alphonse De Candolle between 1855 (*Phytogéographie rationale*) and 1882 (*L'origine des plantes cultivées*). In his book on crop origins (1914), he argues for a field-oriented approach as opposed to a laboratory approach involving botany, archaeology and paleontology, history, and philology. De Candolle organized his search for centers of domestication around the search for wild forms. "One of the most direct means of discovering the geographical origin of a cultivated species, is to seek in what country it grows spontaneously, and without the help of man" (De Candolle 1914, 8). Wholly dependent on published records—floras, travelers' accounts, classical histories, and dictionaries—De Candolle did remarkably well. For example, he correctly identified the New World as the hearth of maize, potato, sweet potato, arrowroot, chirimoya, pumpkin, and chili pepper. But De Candolle was constrained by the ad hoc nature of his sources, confusion about taxonomy, lack of systematic archaeology and techniques for dating, and weak methods for making comparisons between wild and domesticated plants. His greatest legacy is to have established a holistic, naturalist's approach to filling the scientific void that Humboldt (1807) had identified. The naturalist legacy of De Candolle is unmistakable in the next encounters with crop diversity by scientists credited with the creating the scientific basis for the modern analysis of diversity—Darwin, Vavilov, and the first generations of crop scientists who directly studied traditional cultivators.

Darwin was familiar with De Candolle's writing, but he directed attention to the issue of variation rather than to crop origins. In this way, he laid out the framework for understanding crop evolution that has persisted for over a century of progress in crop science. Indeed, the first chapter of *On the Origin of Species by Means of Natural Selection* (1859) is "Variation Under Domestication." Darwin's two-volume treatise on *The Variation of Animals and Plants Under Domestication* (original 1868, U.S. publication of 2nd edition 1896) was his first major publication after *Origin.* Given that naturalists are generally dismissive of studying crops and domesticated animals, Darwin's emphasis on domesticated species might seem curious. He notes that "botanists have generally neglected cultivated varieties, as beneath their notice" (Darwin 1896, 1:322). However, explaining variation in domesticated species with the framework of the *Origin* was an astute political move. In the *Origin,* he had relied on the analogy of crop and animal husbandry to explain the process of natural selection. In *Variation,* he refracts and multiplies the analogy by repeated examples of the processes of natural evolution at work in agriculture, with the added and familiar hand of the farmer complementing natural selection. He defines varieties as "incipient species." In selecting and promoting variants of crops and animals, humans imitate the process of natural selection that favors the variant through reproductive success.

Darwin wrote: "No doubt man selects varying individuals, sows their seeds, and again selects their varying offspring. But the initial variation on which man works, and without which he can do nothing, is caused by slight changes in the conditions of life, which must often have occurred under nature. Man, therefore, may be said to have been trying an experiment on a gigantic scale; and it is an experiment which nature during the long lapse of time has incessantly tried" (Darwin 1896, 1:3). Darwin's framework still stands as a model of what a description of crop evolution should emulate (Rindos 1984).

Vavilov Centers

Vavilov is widely honored as the patriarch of the modern scientific program to understand, use, and conserve the diversity of crops inherited by the modern world. Vavilov began his career as a plant breeder in prerevolutionary Russia. The whirlwind beginnings of scientific plant breeding that followed the discovery of Mendelian genetics pulled in Vavilov along with virtually an entire generation of botanists. In 1916, he made an early and momentous step to systematically collect wheat and other species in the crop's center of origin around Russia's central Asian borderlands into Iran and the Pamir Mountains (Loskutov 1999). Collecting crop diversity in centers of domestication became Vavilov's passion and the basis of several lasting contributions. For twenty years, after rising to head of the Department of Applied Botany and Plant Breeding in Petrograd (Leningrad) in 1920, Vavilov made fundamental contributions that became the building blocks of subsequent programs around the world to understand, work with, and conserve crop resources. Having fallen afoul of Stalin and Lysenko, he was plunged into the Soviet Gulag and executed in 1943 (Popovskii 1984).

Vavilov's contributions to our understanding of crop diversity are both theoretical and substantive. His two most important achievements are (1) his synthesis of De Candolle and Darwin in order to identify the centers of crop origins, and (2) building a national collection of crop types that remained the world's premier collection and model for many years.

Vavilov's stamp on understanding crop origins is best known by the world map he created showing centers of origin (Figure 2.1). Vavilov's maps remain icons of the modern understanding of crop origins and crop evolution. Most importantly, they reflect a synthesis of the different approaches of De Candolle and Darwin to show the basis of our approach to domestication. This synthesis combines an emphasis on locating wild ancestors with a convergence of

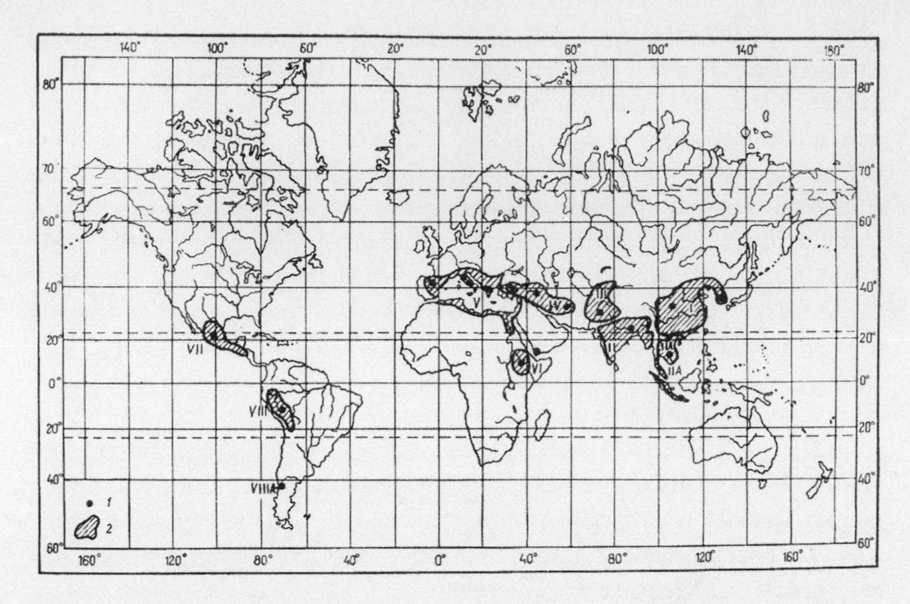

Figure 2.1 Vavilov centers (Vavilov 1992. Reprinted with permission of Cambridge University Press)

diversity of cultivated types. Vavilov (1951) was the first to propose specific evolutionary pathways that connected wild species with particular types of domesticates. Today, demonstrating these pathways defines the modern investigation of crop evolution (Harris 1989, Hillman and Davies 1990). Vavilov's synthesis rests on the recognition that potential wild ancestors may be widely dispersed and that crop diversity may be pronounced in "secondary centers" where variation has occurred after the diffusion of a domesticate. Indeed, diversity in some secondary centers may rival that of the original hearth of domestication. Ethiopia's diversity of barley and durum wheat is an example of this phenomenon (Asfaw 1999).

Vavilov assumed that domestication was a relatively discrete event rather than one that was dispersed in time and space. While not all crop domestication followed an identical pattern (Harlan 1992), the prevailing consensus is that many crops were domesticated in limited areas and over a relatively brief period. Thus, using the similarity between domesticated maize and its closest wild relative, teosinte (*Zea mexicana*), Doebley (1990) reports that the contemporary populations of teosinte from the Balsas River of Mexico drainage are the most likely to have been ancestral to maize. Hillman and Davies (1990) argue that self-pollinating crops such as wheat could have become completely domesticated within a few human generations, from two centuries to as quickly as twenty or thirty years.

Vavilov's second contribution was to build the great collection of crop types at the All Union Institute of Plant Industry, since renamed the Vavilov Institute in recognition of his contributions to Russia and the world. Collecting, crop taxonomy, genetics, ecogeography, and conservation at the Institute of Plant Industry at Leningrad became prototypes for agricultural research and gene banks. Vavilov led expeditions to over fifty countries, assembling collections of hundreds of species and hundreds of thousands of individual types (accessions). Adaptation and ecogeography of crops from around the world were studied through a network of 115 agricultural research centers dispersed throughout the Soviet Union but managed from the institute in Leningrad. Centralization enabled Vavilov and his colleagues to study broad patterns of crop adaptation and evolution before the formation of scientific networks linked together in professional societies with specialized journals and meetings. Although conservation of crop genetic resources and scientific investigation in the fields pioneered by Vavilov have been decentralized and reorganized with new research tools, his model, based on interdisciplinary crop biology and geography, remains vital.

Subsequent scholars have pointed out that there is plenty wrong with Vavilov's map and with the concept of "center" of domestication and diversity. Vavilov's map justifies his reputation as an indefatigable traveler, but he couldn't go everywhere and so he mistakenly omits large regions, such as the Amazon Basin and sub-Saharan Africa outside of Ethiopia. Both of these neglected regions are now recognized as centers of domestication of crops that diffused widely and became staples on other continents (Piperno and Pearsall 1998, Harlan 1992). "Center" is an ambiguous term, and the various geographic areas are configured differently, from highly focused cores to diffuse "noncentric" regions (Harlan 1992). For some crops, such as maize (Doebley 1990) and potatoes (Hawkes 1983), the idea of center for both domestication and diversity is appropriate. For other crops, such as sorghum (Harlan 1992), the idea is inappropriate. As we learn more about the diversity of crops and their relationship to wild ancestors such as wheat (Heun et al. 1997), a crop's center may take clearer shape. As Sauer (1994) observes, the heterogeneity of human motivations, environments, and species makes it naive to imagine that a single historical or geographic pattern pertains to all crop domestication. This complexity is compounded by the limits of our current understanding of the archaeology and evolutionary history of every crop.

Nevertheless, Vavilov's work in identifying key geographic regions where crop ancestors and crop diversity coincide has continued to attract adherents to the idea that agricultural origins might be associated with specific regions. The term *Vavilov center* is widely used to refer to concentrations of wild progenitors and genetic diversity. Hawkes (1983) attempted to resolve the controversy over whether centers of domestication and diversity are real or imaginary by distinguishing between origin and diversity in three combinations. Nuclear centers are well-defined regions where there is strong evidence for crop domestication. Hawkes (1983) identifies as nuclear centers northern China (the loess regions north of the Yellow River), the Fertile Crescent, southern Mexico, and central to southern Peru. Regions of diversity surround them, and outlying minor centers are more distantly related. Rather than offering a definitive list of nuclear centers, Hawkes's (1983) scheme reflects the state of crop biology and archaeology. As we learn more about tropical crops and the Neolithic in Africa and Amazonia, additional nuclear centers might be added.

For the remainder of this book, I will use the concept of Vavilov center in the way that Hawkes (1983) uses nuclear center. I am not proposing that all crops can be understood as having definitive centers of domestication and diversity. My use of Vavilov center certainly reflects the nature of the places where I have done research—Peru, Turkey, and Mexico. All three arguably fit the ideal of Vavilov centers of domestication and diversity.

ECOLOGY OF CROP DIVERSITY

It is logical that improved understanding of the origins of crops and crop evolution would open the possibility of connecting crop origins, diversity, and evolution to the farming practices of the contemporary inhabitants of the centers of crop origin and diversity. After all, evolution is a continuing phenomena, diversity was not fixed at the time of domestication, and crops are wards of the cultivator. A small group of intrepid crop scientists pioneered the no-man's-land between botany and anthropology after Vavilov and the beginning of agricultural genetics using the resources of traditional farmers. This early generation of crop ecologists went beyond Vavilov to probe the cultural practices of crop management that could inform research on the distribution of crop diversity and processes of crop evolution.

The contribution of this group of early ecologists is particularly apparent in research on maize in the Americas between 1940 and 1960. Scientists working in Mexico, Central America, Peru, and Brazil were initially absorbed in the systematic collection of maize, but they eventually were drawn to investigate the ongoing relationship between indigenous American cultures and maize. The first effort was the Mexican Maize Project, led by Edwin Wellhausen (Wellhausen et al. 1952). Efraim Hernández Xolocotzi was among the team members who collected maize throughout Mexico and went on to become a professor of botany at the Post Graduate College (Colégio de Postgraduados) outside of Mexico City and to lead a national undertaking to describe the

Table 2.1. Partial relation between maize diversity and specialized uses

Maize type (race)	Uses
Dulce de Jalisco, Dulcillo del Noroeste, Chulpi (*su su*)	Toasted; whole grain; toasted and ground for *pinole, chicha,* or *tesgüiño* (fermented beverage); and *ponteduro*
Elotes Occidentales, Elotes Conicos, Maíz de Ecuaro	Fresh ("corn on the cob"); *pozole*
Cacahuacintle, Blandito de Sonora, Harinoso de Ocho (all flour-type maize)	Toasted and ground for *pinole, pozolo-lero;* made into cakes and baked; made into *gordas* of uncooked maize
Bofo (a race grown only by Huicholes)	Toasted; *huacholes;* uncooked maize steeped in water to drink
Zapalote Chico (limited to the Isthmus of Tehuantepec)	*Totopos*
Reventador, Chapalote, Palomero Toluqueño, Apachito, Arocillo Amarillo	Toasted to pop for *palomitas;* toasted and dried with molasses for *ponteduro*
Dark blue races	Tamales and festive tortillas

Source: Hernández X. 1973a, author's translation

diversity of maize and its relationship to Mexico's geography and people. In writing on the ethnobotany of Mexican maize, Hernández X. (1972, 1985) suggests a three-way relationship between environment, culture, and maize. The role of culture is to create the context of specialized uses for the crop — for soup and stews, sweets, beverages, steamed cakes (tamales), and many forms of tortilla, the basic element of Mexican cuisine. In Table 2.1, Hernández X. (1973a) illustrates the connection between use and racial diversity in maize.

In addition to showing some of the connections between crop and culture, Hernández X. also pioneered in gathering information about the status of maize diversity over time. Working in the central state of Puebla and the southern state of Chiapas in the 1970s, he and his students made collections to compare with the ones that Hernández X. had helped to make in the 1940s. He thus helped to establish one of the longest time-series data sets on crop diversity. The accessions originally deposited by Hernández X. still dominate collections of Mexican maize.

Edgar Anderson (1952) of the Missouri Botanical Garden followed a similar path in Central America except that he looked at the practices and habits of indigenous cultivators of maize in Guatemala to understand the crop's evolution. "One cannot interpret population samples of maize efficiently without understanding as much as possible about the people who grew that maize. The long, dull hours spent in measuring the samples of maize yield a priceless harvest of understanding" (Anderson 1947, 435). Acknowledging the complexity of sorting out the dynamics of maize populations and evolution in its center of origin, Anderson (1952) described his research on maize as an "adventure in chaos." He and Cutler provided a definition of maize race that still is the prevailing unit of population analysis for the crop (Anderson and Cutler 1942). Likewise, they outlined methods for the study of geographic analysis of maize types in the Mesoamerican hearth before the availability of the large and systematic collections of the Mexican Maize Project (Wellhausen et al. 1952). Anderson noticed that Indian fields were uniform despite the diversity of maize in the region, an observation confirmed in later ethnobotanical studies in Mexico (e.g., Perales R., Brush, and Qualset 2003). Anderson (1947) initiated comparisons of maize populations in different regions, showing the strong effect of environment, especially altitude, and the effectiveness of selection in controlling the natural tendency of this out-crossing plant to diversify. After his fieldwork in Mesoamerica Anderson (1952, 215) regretted that he "did not sit longer in these sunny patios and gossip more with the old folks. I would know more about maize if I had." No doubt we all would have benefited.

Crop diversity is not merely an accident of history or geography but also the product of stewardship by specific societies and cultural traditions. The farming peoples who maintain the bounty of different species and crop types are far removed from the societies and cultures that domesticated plants, but they have sustained and nurtured diversity inherited from previous societies. Crops cannot be understood solely as biological entities apart from their human context, and critical questions about crops, including the survival of different forms, cannot be weighed without calling on social science. Crops are human artifacts that exist in human consciousness and as wards of human societies, unable to survive without the tiller's hand.

Social scientists have contributed to studying crop diversity in three areas: (1) ethnobotany, (2) economic botany, and (3) human ecology. These areas are not sharply divided by discipline or approach, but rather they occupy different locations on a continuum that runs between a genetic pole, such as molecular biology, and a social pole, such as econometrics. These three areas occupy the middle ground in this continuum and involve studies by both biologists and social scientists.

Ethnobotany

The study of knowledge systems has been a staple of anthropology almost since it became a scientific discipline. Because all languages manifest organized information about plants, anthropologists have focused on ethnobotany as a means to examine the relations between culture, language, and thought (e.g., Berlin, Breedlove, and Raven 1974). Since useful plants receive extensive attention in folk plant knowledge, nomenclatures of cultivated plants are especially rich. The varietal or infraspecific level is one of the six regular ethnobiological ranks, and this level is common to many ethnobotanical systems (Berlin 1992), and elaborate varietal classification of crops is conspicuous in many folk systems. For example, Andean farmers in Peru can name fifty or sixty different potato varieties, although few farmers grow that many.

Despite the acknowledged importance of landraces and their lexical prominence, the systematic analysis of the great infraspecific variation is often neglected, perhaps because of difficulties arising from the inconsistency and complexity of varietal names. While Darwin recognized varieties as incipient species and thus as models for natural evolution, varietal classification has been regarded as unfit for systematic botanical taxonomy (Anderson 1952). Cox and Wood (1999) remind us that crops are often "taxonomic nightmares," and Clayton (1983) conjectures that crop diversity may be too complex and our familiarity with crop complexity too great for crop diversity to be described or partitioned by conventional methods. Bulmer (1970) noted that a similar aversion seems to prevail among ethnobiologists, who may despair at the apparent confusion of nomenclature at the variety level. In their quest to describe universal building blocks of folk taxonomy, ethnobotanists seem to share the preference of systematicists for taxa at the species level and above. Reminiscent of Anderson's (1952) "adventure in chaos," Laughlin (1975) reveals his frustration when eliciting information about maize varieties in Zinacantan, Chiapas, Mexico. The Mayan-speaking Zinacantecos distinguished clearly between highland and lowland maize, but confusion seemed to reign in the naming of highland (local) maize, for which Laughlin elicited forty-seven names from five men for twenty maize varieties. Fortunately, exceptions to this frustration exist. At least a few ethnobotanists have delved into the richness of folk varietal names and the importance of varietal distinctions to cultivators (e.g., La Barre 1947, Boster 1985).

An important strand of cognitive anthropology has been to demonstrate the similarity between non-Western and Western knowledge systems (e.g., Atran

1987; Ellen, Parkes, and Bicker 2000). "Indigenous knowledge" has gained wide use (e.g., Brokensha, Warren, and Werner 1980; Brush and Stabinsky 1996), although its usage has been criticized (Agrawal 1995). The term *indigenous* delineates knowledge that is orally transmitted, informally generated and maintained, and culturally mediated, from knowledge that is formally generated and maintained according to written standards of specialized nomenclature. Distinguishing between indigenous knowledge and other knowledge systems has proven to be problematic (Agrawal 1995), but anthropologists and others have enumerated criteria that differentiate indigenous knowledge: (1) localness, (2) oral transmission, (3) origin in practical experience, (4) emphasis on the empirical rather than theoretical, (5) repetitiveness, (6) changeability, (7) wide sharing, (8) fragmentary distribution, (9) orientation to practical performance, and (10) holism (Ellen and Harris 2000). In ethnobotany of crops, indigenous knowledge consists of the folk names and associated knowledge used by gardeners and farmers as opposed to the systematic taxonomy derived from the Linnaean system of plant nomenclature.

As originally used in cognitive and ecological anthropology, the term *indigenous* was used interchangeably with *folk,* and glossed with *local,* or *informal.* The principal opposition implied here was between informal and Western scientific knowledge systems. This opposition is still an important theme, but a new, political theme has now entered the discourse. The current usage of the term implies notions of cultural and political resistance by tribal groups to domination by nation-states that represent other ethnic groups. In the current usage, indigenous takes on a more essentialist tone and is glossed with *autochthonous, native,* or *aboriginal.* When cognitive anthropologists first spoke of indigenous knowledge, they were not exclusively concerned with knowledge of tribal people or ethnic minorities (e.g., Brokensha, Warren, and Werner, 1980). While the current usage of indigenous suggests more precision, the term is often used uncritically, without examining the genealogy of the knowledge in question. Knowledge systems, as part of culture, tend to be naturally permeable to outside terms and information. Like crop species, both indigenous and scientific knowledge systems are open and fluid rather than hermetic and fixed. Thus Andean potato taxonomy freely combines Quechua, Aymara, and Spanish terms. The same is true for Mexico, where speakers of numerous native languages have pooled their knowledge about maize and joined European information systems to this American knowledge.

Boster (1985) elegantly demonstrates that perceptual distinctiveness is a primary reason for manioc diversity among the Aguaruna Jivaro of the Amazon Basin. These shifting cultivators identify cultivars on the basis of plant parts showing the greatest range of variation—leaf shape, petiole color, and stem color. Even if a manioc cultivar is superior in such qualities as yield or disease resistance, if it cannot be identified according to the visual criteria of the Jivaro, it will not be kept. This observation confirms that much of the diversity in domesticated plants cannot be directly linked to yield or other production factors (Harlan 1992, Bellon 1996). Like many other cultivars, color in manioc is a key to its selection by Amazonian horticulturists, but pigmentation is often the product of a single gene, not linked to other traits, and, therefore, not significantly related to the performance of the crop.

Economic Botany

Ethnobotany specializes in the plant nomenclature and classification, and while it may note the use of different species and types, its focus is taxonomy. Economic botany is a close cousin of ethnobotany, differing on the emphasis that it places on the description of use and on cultivated plants, their ecology, and their economic status. Economic botanists have asked two questions about crop diversity: (1) how does use of distinct types account for diversity? and (2) how does crop diversity serve specific farming systems? The first question addresses the issue of adaptation of specific crop varieties, and the second addresses the relation of diversity to the crop's production and ecology.

Varietal adaptation refers to the way individual cultivars are fitted to local resources and contexts. Adaptation to environmental conditions by landraces has long been identified with heterogeneous and marginal environments. Recall that Vavilov (1951) observed that crop diversity is strongly associated with mountainous environments. As a single farm or farming community confronts different, sometimes fickle, and problematic agricultural environments, it may find advantage in sowing specific cultivars adapted to particular conditions. Marginal environments such as acid or saline soils, drought, or flooding have also been directly linked to the maintenance of varietal diversity on farms and in farming communities. Bellon and Taylor (1993) demonstrate that the *Olotillo* maize landrace in southern Mexico is kept specifically because it yields better on poor soils than other varieties do. In the Middle East, mixed landraces of barley produce higher yields under the severe drought and cold conditions than single selections of landraces or improved barley lines (Ceccarelli, Acevado, and Grando 1991).

Where multiple cropping (two crops or more per year) is possible, varietal diversity allows households to maximize the use of scarce land resources. This is illustrated in the Asian rice cycles where extremely high population densities and seasonal rainfall patterns create pressure and opportunity to use land in different ways during the year. In the floodplain of Bangladesh, for instance, rice farmers manage three rice production systems during the year, each with specific cultivars (Ali 1987). The *Aus* paddy is short-stemmed, rain-fed rice sown on high land during the dry summer months (March–May). *Aman* paddy is long-stemmed, sown during the rainy season (June–August), and able to float on the seasonal floods and resist cold winter temperatures. *Boro* paddy is short-stemmed but irrigated, planted during winter months (December–February) in low-lying areas enriched by floodwaters. Besides the need for diverse cultivars to fit each of these agricultural niches, Bangladeshi farmers distinguish between broadcast or transplanted rice varieties and those that are local or improved. Bangladesh has contributed over five thousand types of rice to world collections (FAO 1998), a figure that is not surprising because its rice crop extends from the floodplain of the Ganges and Bramapurta Delta, where rice must survive yearly floods, to steep upland areas in the Himalayan foothills.

While varietal adaptation to physical factors has been emphasized, it also may be important in helping the farm household to confront social conditions such as labor shortage or the lack of credit. Labor availability is a constraint faced by many households, so it is advantageous to have several varieties that mature at different times or show tolerance to poorly timed weeding, irrigation, or fertilization. These varieties help families smooth out the demand for labor in managing crops. Labor scheduling is cited by Freeman (1970) as the reason why the Iban of Sarawak plant fifteen types of rice. In Chiapas, Mexico, the maize landrace *Rocamex* is valued because it is not overly demanding on the scheduling of fertilizer application and weeding, which can be a serious problem for households with limited labor (Bellon and Brush 1994). In Peru, Zimmerer (1991) reports that labor shortages among Andean households resulted in the loss of native potato and maize cultivars.

Besides the cumulative diversity among distinct crop varieties, economic botanists have also examined diversity at the population level. Although Anderson's (1947) observation that Guatemalan maize farmers plant monocultures has been confirmed for other areas and crops (e.g., Perales R., Brush, and Qualset 2003; Bellon, Pham, and Jackson 1997), diversity of a crop within a field may benefit the subsistence farmer. Population diversity within fields may confer an adaptive advantage that is different from the advantage of distinct varieties. Population adaptation refers to the advantage gained from interaction among different varieties that are planted together (Simmonds 1962). Two advantages from population interaction have been considered—yield gains and resistance to pests and diseases. At the end of the nineteenth century, agricultural scientists observed that varietal mixtures of self-pollinating crops such as wheat produced higher yields than did the individual components of the mixture (Simmonds 1962). One explanation was that competition between varieties led each of them to perform better than in pure stands. A century of research, however, has not confirmed that variety mixtures have a definitive yield advantage. Marshall (1977) reviews 328 comparative studies of the relative performance of mixtures and their components in eight crop species. Most studies (289) concluded that mixtures were equal to the mid-components, although 35 studies showed mixtures outperforming mid-components. In only 4 out of the 328 cases were mixtures inferior to the mid-components. Nevertheless, for farmers who lack yield trial data, planting a mixture might appear to be sensible because they have no means to predict the yield of the mid-component in their mixtures. It is important to note that these experimental studies do not address natural landrace mixtures in centers of diversity nor under farm conditions in marginal and heterogeneous environments.

Some researches have identified population diversity as a principal method whereby peasants or low input farmers manage risk (Clawson 1985; Cleveland, Soleri, and Smith 1994). Unfortunately, the relationship between diversity and successful risk management has rarely been empirically examined or tested against other explanations for diversity. Most of the discussion about diversity and risk is qualitative rather than quantitative, perhaps because empirical analysis requires complex, time-series data sets on crop production. As farmers and others weigh the contribution that diversity can make to risk management, two concepts are useful—vulnerability and stability. Vulnerability refers to the degree of exposure or threat faced by a particular crop or farming system. The concept of *genetic vulnerability* became prominent after 1970, when the corn crop in parts of the United States suffered severe losses from southern corn leaf blight (NRC 1972). The damage from this pathogen was directly related to the fact that virtually all commercial maize hybrids in the United States then carried a gene that made them susceptible to the blight-causing fungus. Vulnerability is also believed to exist because of reduced genetic variability in many crops. In the case of the southern corn leaf blight, vulnerability was traced to the presence of a male sterility gene in maize cultivars used by breeders in creating synthetic hybrids. From another perspective, vulnerability resulted from the replacement of diverse maize populations with different pedigrees by cultivars with common parents.

Perhaps the world's most famous agricultural catastrophe derived from genetic vulnerability was the Irish potato famine that occurred because of the small amount of diversity in the Irish crop owing to the bottleneck of the potato's trans-Atlantic journey from its Andean homeland. Irish potatoes never had much genetic diversity to begin with, because only a small portion of the Andean diversity of potatoes was brought to Europe. When the *Phytophthora infestans* fungus arrived in Ireland in 1845, it quickly overwhelmed the potato crop in a way never experienced in the more diverse Andean crop.

The principal advantage of mixed populations of landraces appears to be in offering a certain level of resistance or buffering to pests and diseases (Frankel, Brown, and Burdon 1995). Resistance from heterogeneity is an important and much debated issue in plant pathology, breeding strategies, and agricultural alternatives, and it relates closely to the concepts of risk and stability, which will be discussed below. The idea that variety mixtures provide disease resistance has received both theoretical and empirical support. Theoretically, variety mixtures in crops may contain different genes for disease resistance and may mimic the architecture of wild plant communities where infection is slower and not as devastating as in uniform crops (Browning 1981, Burdon and Jarosz 1990). The theory that fields of mixed varieties help farmers cope with unstable environments is supported by some research in less-developed countries (Clawson 1985) and by the spectacular, if periodic, failures of homogeneous crops, such as the 1845 Irish potato famine and the 1972 U.S. southern corn leaf blight (NRC 1972). Marshall (1977) and Wolfe (1985) review several positive cases where variety mixtures provide disease resistance. The importance of mixing varieties for disease resistance has been brought into industrial agriculture and crop breeding through such concepts as *multilines* (Wolfe 1985).

While crop diversity is often associated with heterogeneous and marginal environments and subsistence economies, it also has value in highly developed, input-intensive, and commercial agriculture. For example, peach farmers in California may have a dozen or more varieties in their orchards, each with a different harvest date, allowing them to spread the harvest over a longer period to capture price advantages and to avoid labor bottlenecks (Tadesse 2001) (see Chapter 11). In 1995, California fruit growers sold 161 varieties of plums and 165 varieties of nectarines. California orchards are highly advanced in terms of management and access to improved fruit varieties. Contrary to the notion that these characteristics lead to loss of diversity, there has been a dramatic increase in the number of plum, peach, and nectarine varieties in California. Between 1983 and 1995, the number of plum varieties that produced 1000 boxes of fruit or more increased from 29 to 56. Peach varieties rose from 34 to 80 and nectarines from 33 to 97 (CTFA 1996). Likewise, wheat agriculture in the United States has experienced an increase in the number of varieties grown, from 126 in 1919 to 469 in 1984 (Dalrymple 1988). This increase in wheat diversity is also reflected in a decline of uniformity at the field level (Cox, Murphy, and Ridgers 1986). In these cases of diversification in industrialized agriculture, the market value of diversity is evident.

Determining the contribution of diversity to stability in cropping and the impact of a loss of diversity is exceedingly difficult because of the noise created by constant, short-term fluctuations in yield and because instability has many different causes (Anderson and Hazell 1989). In agriculture, as in wild plant communities, the evidence of a positive association between diversity and stability is mixed (Tilman 1996, McCann 2000). Ceccarelli, Acevado, and Grando (1991) show that yields of individual barley landraces have lower variance than modern varieties in cold and dry conditions in the Middle East, but landrace mixtures are not specifically evaluated for variance. Instability caused by loss of genetic diversity in less-developed countries was suggested by increasing standard deviations of wheat and rice yields in India and Pakistan after the introduction of high yielding varieties (Barker, Gabler, and Winkelmann 1981; Mehra 1981). However, subsequent statistical analyses of country-level and time-series data challenged this initial finding and showed that instability in cereal production cannot be directly attributed to the use of high yielding varieties (Singh and Byerlee 1990). Research on wheat yields among fifty-seven countries between 1951 and 1986 showed that instability can be related to such factors as country size, moisture regime, and temperature but not to technological variables such as the level of adoption of modern, high yielding varieties or fertilizer use. In fact, despite a decline in genetic diversity in developing countries after the Green Revolution of the 1960s, Singh and Byerlee (1990) show that wheat yields are actually more stable.

Although criticized by May (1973) and Goodman (1975) as "folk ecology" rather than a scientific axiom, the premise of a positive relationship between diversity and stability proved resilient (Pimm 1986) and has recently regained credibility (McCann 2000). Tilman (1982, 1996) shows empirically that diversity helps stabilize production in wild plant communities. Agricultural instability at the local and national levels must be distinguished, and different factors must be sorted out in determining the role of genetic diversity in providing stability. Genetic vulnerability must be distinguished from other causes of yield instability such as co-variation at different spatial levels (synchronization). Determinants of yield unrelated to genetic vulnerability must be identified: price fluctuations, inadequate or poorly timed input supplies, and timing of crop management activities. While there is no single cause of instability, available evidence indicates that closer connection between localities and regions is most relevant (Hazell 1989). In particular, the agricultural synchronization between once isolated localities appears to play a large role in increasing the instability of cereal production at the country level (Hazell 1982). Synchronization might result from the widespread dependence on centralized input supplies such as electricity or credit and from increasing responsiveness to prices in national commodity markets. Theoretically, it is possible to have more stability at the farm level but elevated instability at the country level because of agricultural synchronization.

Although country-level data relating to the relationship between genetic diversity and risk does not confirm the idea that loss of diversity leads to instability, prudence provides several reasons for preserving diversity. First, there is evidence that world cereal production is becoming less stable as yield increases (Anderson and Hazell 1989). Second, there are numerous instances where the link between genetic vulnerability and instability or outright crop failure is well established (Lakhanpal 1989). Third, a statistical distinction between covariance and instability can be of little comfort to a farmer who does not have crop insurance or other safety nets when crops fail. Finally, the move to more uniform varieties began before we first gathered data on yield stability (ca. 1965–1970), so the real effect of the loss of diversity is almost impossible to assess.

Human Ecology of Crop Diversity

Ethnobotany and economic botany of plants provide methods for describing knowledge about crop diversity and the selection and maintenance of types of crops and crop populations. The goal of the human ecology of crop diversity is to understand crop diversity in a dynamic context of environmental and social change. In comparison with ethnobotany and economic botany, the human ecology of crops is relatively new, poorly developed, and limited in the number of studies and scientists who are attracted to it. This area of research is represented by such studies as Richards's (1985) research on swamp rice in Sierra Leone, the work of Bellon (1996) and Perales R. (1998) on maize in Mexico, and research on potatoes in the Andes by Brush (1992) and Zimmerer (1996), and on rice in Thailand by Dennis (1987). Because the social components that affect diversity change rapidly, they are the principal variables of the human ecology of crop diversity. These social components include cultural, social, economic, and technological factors that define the context of crop selection. These factors affect the agronomic, market, and cultural value of different types of crops. Operationally, they can be measured in an infinite number of ways, but the predominant measures include population growth and decline; farm size; the availability of labor; the availability of agricultural inputs such as fertilizers, pesticides, and irrigation; market infrastructure; consumer demand and commodity prices; and credit policies.

Selection and management of different types of crops and crop populations are the core of the human ecology of crop diversity. Selection refers to the choice that farmers make about which seeds to plant, and management refers to the process of seed selection and decisions about the quantity of specific crop types that are maintained. Three general factors determine how selection and management are carried out: (1) environmental and agronomic factors, (2) risk, and (3) use and markets. These factors also determine the likely outcome of selection in terms of the type of crop to be planted.

Environmental and agronomic factors delimit the specific production environments that farmers work with — altitude, soils, water availability. In addition, exposure to the hazards of crop production are included, such as wind, hail, frost, drought, disease, and pests. Farmers often work with different production environments on a single farm, especially in heterogeneous environments such as mountains and when the farm is subdivided into parcels. Specific adaptation is often assumed to be the basis of crop diversity (Harlan 1992). Because adaptation to specific environments by certain crop types is positively associated with yield, farmers will differentiate between varieties. As discussed above in the section on economic botany, varietal adaptation to distinct microenvironments is recognized by farmers and is readily seen in heterogeneous environments.

Production environments can be altered by the use of agricultural inputs such as fertilizer or irrigation, depending of course on the availability of inputs and the ability of the farmer to acquire them. Production environments may change dramatically, for instance, with the appearance of a new disease or pest such as late blight in Ireland. Likewise, a crop's adaptation can change, for instance, by the breakdown of disease resistance. Finally, adaptive advantage of one variety may be surpassed by another, either from natural crop evolution or from scientific crop breeding. A result is the turnover of crop varieties (Brennan and Byerlee 1991). The famous Composite Cross populations of barley, which were followed in situ in California over several decades after 1928, showed declines in diversity but the accumulation of certain alleles related to adaptedness (Allard 1988). David et al. (1997) observed the population structure of a population of wheat that was repeatedly sown, harvested, and resown over a seven-year period, and they found significant change in the composition of the population, the result of constantly new adaptations to the naturally dynamic environment. The dynamic nature of crop populations is shown graphically in comparisons of gene bank accessions collected at one time and new samples taken later from the same area (Soleri and Smith 1995; Tin, Berg, and Børnstad 2001).

Risk involves both instability of crop yields and the ability of farmers to cope with this instability. Risk is different from adaptation because it is measured by fluctuations in yield rather than by yield itself. Certain crop varieties may have a greater variation around their mean yields than others (Evans 1993). While we may assume that all farmers prefer a higher yield, some may be more affected by low yields and therefore averse to selecting the more variable type. Managing risk can be an equally or more important goal to farmers than yield, and poor farmers in marginal environments are especially sensitive to risk (Cleveland, Soleri, and Smith 1994). Scott (1976) describes the "moral economy" of peasants who create social and technological means to avoid the danger of falling below the level of production needed to sustain their families. Risk, therefore, is an important element in the context of crop selection and management. It may, for instance, lead a farmer to diversify seed to try to buffer the effect of yield fluctuations of one type. Clawson (1985) was told that the diversity of maize in Mexican peasants' fields is a buffering mechanism against unpredictable rainfall, although most research on this crop and region suggest a low level of diversity within maize fields.

The demand for various qualities of a crop and the availability of those qualities through market exchange shape selection just as adaptation (yield) and risk (stability) do. Selection depends on a set of qualities or utilities that are personally and socially determined (Bellon 1996). Selection is logically affected by wealth and access to resources such as land, labor, irrigation, markets, and credit to purchase agricultural inputs. In some instances, selection may be conditioned by gender. The need to produce the staple food or commodity augurs for an emphasis on yield, and yield also approximates the adaptation and fitness of a particular variety. Nevertheless, artificial selection tempers natural selection so that other qualities, including stability, are important or even prevailing. Qualities other than yield and stability are often glossed as "quality," but this term hardly captures the culturally determined aspects of selection. Taste is an obvious aspect, but taste itself is constituted in many different ways and judged according to an indefinable array of local standards. Andean farmers, for instance, measure taste according to the amount of dry matter (flour) in a potato, but they also seem to value the spectrum of subtle and unnamed differences in flavor and consistency in a bowl of mixed, native potatoes. Diversity has its own flavor. Numerous other qualities that affect selection are also apparent. In Mesoamerica, maize is woven into the fabric of ritual and cultural consciousness. Small infusions of colored and variegated maize are mixed into seed lots of white or yellow seed, and these variegated kernels are termed *Sangre de Cristo* (Christ's blood) (Anderson 1952), an agricultural example of the syncretism between European and Mesoamerican religion.

Another instance of religious influence on diversity of cultivars is found in Ethiopia, where the Ari people cultivate *ensete*, a relative of the banana (Shigeta 1990). Ensete leaves are used for wrapping, thatching, ground cover for sitting, containers, dresses, ornamentation at funerals and weddings, and shading from sun and rain. Fibers from the plant are used for rope, and the plant is ground for animal food. The most important use of ensete, however, is for the food from its pulp. Shigeta (1990) reports that his Ari informants recognized seventy-eight landraces. Diversity in the cultivated ensetes is promoted by designating sacred areas, called *kaiduma,* where wild ensete flourishes. The Ari believe that wild ensete was planted by God and that disturbing the kaiduma will be divinely punished. In fact, the Ari know that wild ensete has some of the same landraces as cultivated ensete; the ritual sanctuary of wild ensete creates a source of diversity that can be incorporated into cultivated forms through cross-pollination between the two forms.

Many more prosaic qualities shape selection, such as how well the seed stores between harvests, need for fodder or fuel, whether seed remains viable after storage, and the strictures of millers. Indeed, these other products or qualities of a single crop species may be more scarce or valuable than an additional increment of yield. Thus, Anatolian farmers grow some landraces of wheat because they are necessary to make asuré, a sweet compote of wheat and fruit that is served on holidays. In central Mexico, certain local types of the *Chalqueño* maize race are husked by workers who are paid in husks for wrapping tamales, rather than in cash. The husks and tall stalks of this landrace are as important as grain yield in the overall value of the crop. Other races are kept for use in soups (*pozoles*) or flour for tortillas (Perales R. 1998).

Bellon (1996) identifies five general types of concerns on which farmers base their evaluation of crops: environmental heterogeneity (e.g., soils, rainfall), pests and pathogens, risk management, culture and ritual, and diet. These general concerns are elaborated into specific criteria in crop evaluation, depending on the cultural, socioeconomic, and environmental context of the farmer. Using maize in Mexico as his case study, Bellon (1996) illustrates the specific ways in which the general concerns translate into specific criteria, and these are summarized in Table 2.2.

On-farm diversity is often required to provide various products a household wants to consume but cannot acquire through trade or market methods. While many qualities shape selection, markets have the capacity to release the farmer of the need to grow a particular variety to obtain the desired quality. Markets play a critical role in selection because they can substitute for direct production of specific qualities that are locally important. The profusion of diverse fruits and vegetables in the various markets of industrial countries testifies not only to the continued demand for special qualities but also to the ability of markets to supply those qualities. We may also expect markets everywhere to supply qualities that on-farm diversity otherwise might provide. However, market systems are often missing or unreliable, so households must themselves produce a desired product (de Janvry, Fafchamps, and Sadou-

Table 2.2. Farmers' concerns and ranking of three main maize varieties, Vicente Guerrero, Chiapas, Mexico

	Variables		Ranking of varieties		
Concerns	Variety traits	Gradient	Improved variety	Creolized variety[a]	Traditional variety
Midsummer drought	Growing cycle	Short/long	First	Second	Third
	Drought resistance	High/low	Third	Second	First
Strong winds (lodging)	Plant structure	Short/tall	First	Second	Third
	Stalk strength	Strong	First	Second	Third
Soil quality	Good for poor soils	Good/bad	Yes	No	No
	Needs good soil	Needs/does not need	Yes	No	No
Availability of labor for weeding	Ability to withstand delays in weeding	High/low	Third	Second	First
Availability of fertilizer	Quantity required to yield well	High/low	First	Second	Third
Availability of labor for fertilizer application	Ability to withstand delays in fertilizer application	High/low	Third	Second	First
Yield	Yield/volume	High/low	Third	Second	First
	Yield/weight	High/low	First	First	Second
Storage	Ability to withstand long-term storage	High/low	Third	First	Second
Use	Appropriate for self-consumption	More/less	Second	First	First
Taste	Appropriate for market	More/less	First	First	Second
	Taste	Better	Second	Second	First

[a]Creolized variety, an originally improved variety which has been mixing with local germ plasm for thirty years, see Bellon and Brush 1994.

Source: Bellon 1996 (Reprinted with permission from *Economic Botany.* © The New York Botanical Garden)

let 1991). Indeed, the ability of markets to become so advanced and proficient so as to eliminate on-farm diversity may be a rare phenomenon, found in particular cultures and societies. The more predictable situation is that markets will not be adequate to provide substitutes for the services and qualities that on-farm diversity offers. In Turkey, we found that an important reason for selecting local wheat instead of modern, higher-yielding varieties was the quality of local wheat for bread and the fact that no market was available to purchase wheat with the desired qualities (Meng, Taylor, and Brush 1998). A similar situation exists in Peru, where local potatoes are preserved for home consumption and for gifts (Brush 1992).

Apart from selection per se, seed management is also relevant to the human ecology of crop diversity. Seed selection, seed flow, and seed turnover are particularly relevant. Seed selection is often done at harvest time when seed is sorted from the rest of the harvest. Postharvest selection affects the choice of seed in two ways. First, it allows selection of only a few traits of the entire crop—those that are visible in the harvested mound of grain, ears, or tubers. Traits such as seed color or size may be linked to other traits, such as drought resistance, but relying on harvest traits to provide growth-cycle traits is uncertain. One reason why crop breeders are able to produce high-performing and competitive seed in a relatively short time is their attentiveness to preharvest performance. Preharvest selection, which can identify desirable traits that are not visible in the grain or tubers, is a laborious exercise that few farmers can afford. Wheat farmers in Turkey frequently told me that preharvest selection of wheat spikes for seed was done by their fathers or grandfathers, but no farmer who I interviewed practiced this type of selection. I did, however, observe this very same preharvest identification of wheat spikes in Abruzzo, Italy, in 2000. The result was a perfectly uniform stand of a traditional Italian landrace of bread wheat for homemade *pane*. Rouging, or discarding off-types (e.g., stressed or diseased plants) during growth is a compromise to preharvest identification of seed sources in the field, but rouging is rarely described in traditional farming systems. The second effect of postharvest seed selection is in constraining who does the selection. Harvests may not be attended by all of the people who may have a stake in the selection, such as women or other household members. For example, Mexican maize harvests are commonly a male activity, although women add to the seed pile while preparing tortillas (Rice et al. 1997).

Seed flow within and between farming communities and beyond is an important element of the human ecology of crop diversity. Seed flow is ubiquitous for several reasons. Seed stocks lose adaptive advantage as diseases or pests overcome their resistance. Drought, hail, frost, locusts, stray animals, and a host of other afflictions cause farmers to lose their seed. Almost all farmers are vigilant for new seeds, and most are eager to experiment. Neighbors find a better variety, or a new variety arrives with migrants or agricultural extension agents. Farmers may report that the seed they are planting was given to them by their parents, but what they may mean is that they are growing the same type of seed even though the actual seed has been renewed from some other source (Rice, Smale, and Blanco 1998).

Seed flow contributes to the mixing of crop populations and leads to seed turnover. Depending on the crop, even a small amount of seed flow can have a large effect over time, resulting in the replacement of one crop population with another. This is especially true for out-crossing crops such as maize, but even nominally self-fertilizing crops may have sufficient out-crossing to result in population replacement with a small influx of different seed. Louette (1999) has documented an active seed flow of maize landraces in Cuzalapa, Mexico. She concludes that the maize currently grown in the village is not the same maize as that grown a generation ago, even though the maize in question is a farmer-managed landrace that might otherwise be thought of as "traditional." In Peru, potato farmers routinely engage in long-distance seed flow networks (Brush, Carney, and Huamán 1981). These farmers distinguish their "native" varieties from "improved" ones partly because the latter require fresh seed every year to produce the expected crop. Nevertheless, trade in native seed over time results in dramatic changes in the populations of native potatoes. Valdivia et al. (1998) have described an international flow of seed for *oca* (*Oxalis tuberosa*) between Peru and Bolivia, carried out by subsistence farmers. Thus, modern farmers are still engaged in the same practice of transporting crops and varieties that rapidly diffuse agriculture beyond a handful of original hearths of domestication and spread single species far beyond their native habitats.

Summary

When Thomas Jefferson noted that "[t]he greatest service that can be rendered to any country is to add a useful plant to its culture" (Baron 1987), he was affirming the value of the biological diversity of plants. A century after Jefferson coined his aphorism and after naturalists had logged thousands of miles in search of exotic plants, the puzzle of diversity triggered the search for crop origins and for the nature of diversity. Why is there such extraordinary diversity of a few species in certain regions? Darwin, De Candolle, and then Vavilov framed this puzzle into a scientific enterprise that has intrigued countless other scientists and inquisitive laymen. A century of scientific inquiry about crop diversity has not provided a definitive or conclusive answer to the puzzle, but it has helped us to approach and understand the nature of diversity in ways unanticipated by Vavilov. Before now, the relation of crop diversity to cultural and social factors has been noted by naturalists, historians, and travelers, but there has been only limited organized research that connects diversity to the people who maintain it in the yearly rounds of cultivation. This blank area in the science of crop evolution is now being filled by ethnobotanists, economic botanists, and human ecologists.

3

The Measure of Crop Diversity

Careful seed selection has been an imperative of human survival for three hundred generations, since people first developed the arts of producing food rather than gathering it. The similarity of routines and goals of seed selection overlays a pair of scientific puzzles that have challenged us for over a hundred years. Why is there so much variation within crop species, and why is this variation distributed unevenly among farming regions? Specifically, why are there 125,000 distinct types of wheat, 90,000 rice types, and 30,000 potato types (Plucknett et al. 1987), and why is the diversity of each of these crops concentrated in a few regions? These puzzles have driven scientific progress in a number of disciplines, but they also have focused our attention on the importance of crop genetic resources and on their plight in the modern world. However, in order to understand crop diversity, an array of different definitions and measures of farmers, social scientists, and crop scientists must be recognized and reconciled.

Types of Crop Diversity

Crop genetic resources are multifaceted and can be viewed with either extreme reductionism or general holism. Thus, we may view diversity at the molecular level as base pair sequences or DNA fragments (e.g., Autrique et al.

1996), or at the level of ecosystems as interacting complexes of species, including human cultures and knowledge systems (e.g., Brookfield 2001). At the most fundamental level, genetic resources begin with the DNA codes that instruct plant development across the breadth of existence and expression of a crop species. Traits that define a species or classes within species are determined at different levels of genetic information: single genes, multiple genes, and gene complexes. These resources, however, must be defined at broader levels than gene or species alone for two reasons. First, crop species are not biological isolates from other species of plants; rather, they are members of relatively tight-knit families of domesticated plants, often distinguished by the presence of additional chromosomes or genomes (polyploidy). They are also closely related to noncultivated plants. Many crop species are able, in varying degrees, to exchange germplasm with these related cultivated and wild species. Second, crops are cultural as well as biological artifacts, extremely dependent on humans for survival. Recognition of the human element in crop resources and diversity has been gradual, but as ecological analysis and conservation have become prominent themes, the human element has also gained prominence.

When exploration and crop origins dominated interest in crop diversity, defining crop resources emphasized whole plants (genotypes) and, subsequently, plant populations, and genes (loci and their alleles). Two types of plant species were particularly significant in this period: (1) wild and weedy crop relatives and (2) landraces of crop species. While these two types are still central to any definition of crop diversity, additional types of information have become relevant. Interest in ecology and conservation of crop resources broadened the definition to include other components of the plant ecosystem: human knowledge and information systems, crop resources in gene banks and breeding programs, related plant species, environmental conditions, pollinators, pests, and pathogens. Studies of the ethnobotany, economic botany, and human ecology of crop diversity, discussed in the preceding chapter, reflect this broader definition of crop diversity and crop genetic resources.

WILD AND WEEDY CROP RELATIVES

Wild crop relatives share common ancestry with cultivated plants. In many crops, the processes of domestication and crop evolution isolated wild and domesticate species from one another, but the isolation varies according to several factors, for instance whether species are self-pollinating or outcrossers, and the conditions of cultivation and selection. Genetic exchange between wild and cultivated species has been observed in a number of crops and is probably a factor in the genetic composition of most crops (Ellstrand, Prentice, and Hancock 1999). Crosses from wild to cultivated species have been observed in maize, wheat, barley, sorghum, rice, potatoes, tomatoes, and radishes (Heiser 1973). Cultivated wheat is a relative of two large groups of grasses, *Aegilops* and *Triticum*. Wheat includes genetic material from both *Aegilops* and wild *Triticum*, but crosses between wheat and its wild relatives are inhibited by wheat's habit of self-pollination. In contrast, cultivated potatoes are outcrossers and able to reproduce with wild ancestors under natural conditions (Johns and Keen 1986, Rabinowitz et al. 1990). Exchanges between wild and cultivated potatoes are, however, complicated by the fact that the crop is managed through asexual propagation. In contrast to wheat and potatoes, gene flow between maize and wild teosinte (*Zea mexicana*) faces neither the obstacle of self-pollination nor asexual propagation. Nevertheless, there is sparse evidence of significant gene flow from teosinte to maize (Goodman 1988). In all crops, wild ancestors are important genetic resources because genetic material can be transferred from wild to domesticated species by crop breeding methods.

Weedy relatives occupy an intermediate status between wild and fully domesticated plants. Although members of crop families, they are often unwelcome guests, not tended by cultivators but dependent on agriculture to survive. They thrive in man-made habitats such as fields and hedgerows, but they retain qualities of wild species, such as shattering seed dispersal. Weedy relatives of some crops are biologically isolated—for instance, the *Aegilops* species that live around wheat fields in southwestern Asia but that do not cross-pollinate cultivated wheat. Others, such as some types of teosinte, act as potential bridges between cultivated and wild populations of related species.

LANDRACES

The most variable and abundant types of crop genetic resources are crop landraces, the thousands of locally adapted varieties that are named, selected, and maintained by farmers in different regions. Landraces are usually kept as populations rather than as carefully selected varieties or cultivars, and as populations landraces are characterized by great genetic and phenotypic variation within named types. Thus, wheat landraces within a single village in Turkey show traits such as white, black, or red grain; the presence and absence of awns; tightly and loosely packed seed heads; and different abilities to tolerate drought and poor soils. Figure 3.1 shows different wheat species and varieties that were collected in Kutahaya, Turkey, in a single field that the farmer identified as a having one landrace. The same type of listing can be made for wheat landraces maintained by other farming cultures in the Middle Eastern center of diversity and for other crops in their respective centers. In contrast, maize landraces of Mesoamerica show far less variation within a

Figure 3.1 Spikes from a wheat landrace in Turkey (courtesy of Calvin Qualset)

single named type, no doubt reflecting the fact that maize seed is selected from individual ears, while wheat seed in Turkey is usually selected in bulk from the pile of wheat on the threshing floor. The contrast between heterogeneous populations and more uniform morphological types, both referred to as "landraces," has been a source of some confusion in the definition of the term (Zeven 1998).

Harlan (1975, 618) provides the most extensive and often-used definition of landraces: "Land races have a certain genetic integrity. They are recognizable morphologically; farmers have names for them and different land races are understood to differ in adaptation to soil type, time of seeding, data of maturity, height, nutritive value, use, and other properties. Most important, they are genetically diverse. Such balanced populations—variable, in equilibrium with both environment and pathogens, and genetically dynamic—are our heritage from past generations of cultivators. They are the result of millennia of natural and artificial selections and are the basic resources upon which future plant breeding must depend."

Harlan's definition of landraces implies three assumptions. One is that variation among landraces is due primarily to adaptation to physical and environmental conditions. Although adaptation is undoubtedly important to the crop and its cultivators, other sources of diversity, such as curiosity, aesthetics, cultural identity, and market value, can also explain variation. The second assumption, related to the first, is that diversity among landraces is a product of direct and specific selection rather than of relaxed selection by farmers (Frankel, Brown, and Burdon 1995). Third, Harlan's definition states that landraces are balanced in relation to the environment and pathogens. This repeats the widely held belief that peasants and other "premodern" people live in stable harmony with their environment. The obverse of this assumption is that a departure from the practices of peasants, for instance a decline in diversity among crops (genetic erosion), will lead to instability because of vulnerability to pests and pathogens and susceptibility to the vagaries of climatic fluctuation.

Harlan's assumptions are plausible and widely subscribed to by agricultural scientists, but they have not been critically examined or tested with case studies of landraces in their native habitats. Indeed, testing these hypotheses is a daunting task requiring a wide array of biological, environmental, and social data, ideally gathered over time. There is little wonder, therefore, that the assumptions are repeated more often than tested. While landraces probably are indeed "traditional" and "ancestral" to contemporary crops, they undergo continual human and natural selection and thus are as "modern" as any other variety. The terms *traditional* or *ancestral* obscure the fact that landraces and the farming systems where they grow are dynamic and constantly changing.

Because of their large numbers, long history, wide distribution, and human importance, landraces have very different forms. Organizing and interpreting the diversity of crop plants hold the same importance to crop scientists as life's diversity holds for biology in general. The practical reasons for a systematic study of crop diversity include grasping the evolutionary history of the species, anticipating the crossability and compatibility of different types in breeding programs, locating useful agronomic traits, and estimating conservation needs and priorities. The taxonomy of landraces is complicated by the volume of slightly different types and the environment's influence on genetic expression. Since the discovery of the principles of inheritance, crop scientists have utilized various tools to measure similarity and difference among crop specimens. Plant morphology and agronomic characteristics were the initial bases for systematic studies of landraces, and they remain important because of their relevance to farmers and crop breeders. Biological markers, which are more discrete and less affected by environment, offer additional taxonomic tools. Measuring variation in isozymes and other proteins with electrophoresis added an important tool, and molecular markers have extended the ability to describe similarities and differences closer to the gene level.

Landrace taxonomy begins by reducing the variation within a crop species

Table 3.1. Classification of cultivated potatoes by chromosome number (ploidy groups)

Series	Chromosome number ($x = 12$)			
Solanum tuberosa	2*x*	3*x*	4*x*	5*x*
	S. × *ajanhuiri* *S. phureja* *S. stenotomum*	*S.* × *chaucha* *S.* × *juzepczukii*	*S. tuberosum* subspecies *tuberosum* *S. tuberosum* subspecies *andigena*	*S.* × *curtilobum*

Source: Hawkes 1990

into large-scale groups. Four distinct classes or groupings describe the population structure of landraces, depending on the crop species: (1) polyploid species, (2) subspecies, (3) races, and (4) varieties. Potato landraces are classified according to polyploid species, subspecies, and varieties; maize is classed by race, and wheat by polyploid species and variety. Polyploid species are closely related but differ in number of chromosomes. This difference can be associated with environmental adaptation, such as in Andean potatoes, and with the presence of certain traits, such as in wheat. Potato landraces, for instance, comprise varieties belonging to four cultivated *Solanum* groups with different chromosome numbers (24, 36, 48, and 60). Table 3.1 summarizes the polyploid grounds of cultivated potatoes in the Andes. Each of these chromosome groups has a relatively distinct place in the Andean agricultural environment (Brush, Carney, and Huamán 1981; Hawkes 1990) (see Chapter 4). The diploid *S. phureja*, with 24 chromosomes, is most often found at lower altitudes (<2,800 meters above sea level), while the pentaploid *S. curtilobum*, with 60 chromosomes, is found at higher, frost-prone altitudes (4,000 masl). The tetraploid potato species, with 48 chromosomes, is subdivided into two subspecies (*S. tuberosum* subspecies *andigena* and *S. tuberosum* subspecies *tuberosum*) according to whether it grows at tropical latitudes with short day length or at temperate latitudes with long day length.

Subspecies, races, and varieties are more relative groupings based on shared characteristics that include environmental adaptation, agronomic traits, morphological characters, plant protein patterns, and molecular markers. Some variation, such as seed color, is simple and is caused by differences at a single gene, while other variation, such as disease resistance or yield, is complex and is governed by multiple and interacting genes. Salaman (1926) attempted to describe the variation of the English potato using plant morphology. He observed that "we can account for all the known characters of the thousand and one varieties raised in this country by the interplay of characters derived originally from the two varieties described by Clusius and Gerard respectively, at the end of the sixteenth century" (Salaman 1985, 160). In 1926, Salaman described these two progenitors: "The distinctions between the two varieties are that Clusius's Potato is a very deeply coloured red and of the latest maturity, whilst Gerarde's is a white-skinned tuber of Maincrop or even earlier maturity. The similarity that is peculiar to the two, and differentiates them most from our present-day varieties, is the very irregular shape of the tubers disfigured by outgrowths, and the extremely deep eyes" (Salaman 1926, 5). While Salaman was overly optimistic in his confidence that all of the characteristics and varieties in England could be systematically related to two progenitors, he expressed a sentiment that still rings true — that systematic comparison of plant characteristics will reveal an underlying order in the vast array of different crop varieties.

Landraces are distributed according to many factors, but environment and economic development are the logical ones to investigate first. Landrace diversity appears to be more persistent in cradle areas of crop domestication, as illustrated by the contrast between maize types found in Mexican and U.S. agriculture. The diffusion of hybrid maize eliminated landraces in the American Corn Belt by 1950. In contrast, the vast majority of maize planted in Mexico is from seed produced by the farmer rather than from the seed company or some other centralized seed system (Morris and López-Pereira 1999). Nevertheless, closer inspection of Mexico shows a complex pattern of maintenance and loss of local maize types. Parts of Mexico have experienced similar changes to the American Corn Belt, such as the El Bajio region in the state of Guanajuato (Aguirre G., Bellon, and Smale 2000) and in the prime maize lands of Chiapas (Brush, Bellon, and Schmidt 1988). However, conservatism in selecting local maize is found in some surprising areas, such as in the farmlands around Mexico City where markets and agricultural services are well developed (Perales R., Brush, and Qualset 2003). This case challenges the stereotype that associates landraces with marginal agriculture, highly heterogeneous environments, and "traditional" farming systems characterized by small-scale farms and subsistence production (Cleveland, Soleri, and Smith 1994), poverty (Lipton and Longhurst 1985), and ethnic minorities (Nabhan 1989).

In sum, the idea of landrace embodies notions of crop and agricultural evolution in specific locations and environments. Long-term processes of adaptation and maintenance imply value for production under difficult conditions and for local cultural reasons. Because the term is applied to vastly different crops and farming systems, and often without definition, "landrace" means different things to different people—"primitive" varieties, varieties produced by "farmer-breeders," a store of genetic variability, or a resource for breeders looking for specific traits. After reviewing the numerous definitions of "landrace" since the term was first used in 1890, Zeven (1998) concludes that no single or all-embracing definition can be given. Landraces are not uniform varieties but rather populations that conform to a folk "ideotype" (Donald 1968) by morphological criteria such as height, grain color, and time to flowering. Variation within the population is common, reflecting seed mixtures that may be unintentional. Summarizing the important characteristics that emerge from over a century of research on this crop type, Zeven (1998) cites three distinguishing traits of landraces: high capacity to tolerate biotic and abiotic stress, high yield stability, and intermediate yield level under low input agriculture.

Measuring Crop Diversity

Research on crop diversity has proliferated in the seven decades since Vavilov (1926) published his theory linking the origins of agriculture, diversity, and the presence of wild ancestors (e.g., Almekinders and de Boef 2000; Wood and Lenné 1999; Brush 1999; Frankel, Brown, and Burdon 1995; Brown et al. 1990; Harris and Hillman 1989; Harlan 1995; Frankel and Hawkes 1975). This work, in turn, has produced its own abundance of ways to measure and understand crops. While a unified theory and unified methods of analyzing and describing diversity are imaginable, the science of diversity is, in fact, balkanized into different disciplines, technical languages, and popular understandings. Popular nomenclature used by farmers and based on plant morphology is but one way to identify biological diversity in crops. Other ways have been created by botanists, geneticists, and plant breeders to describe the biosystematics of plant species. More recently, ethnobiologists and ecologists have added their methods to describe and analyze crop species (see Chapter 2).

Diversity measurement varies according to whether it pertains to a species-specific, anthropological, or ecological study. However, as the science of crop diversity has proliferated and become more inclusive, the need increases to find a set of common parameters for measuring diversity. Plant diversity in an agricultural system exists at four different levels: (1) between species, (2) within a species, (3) within a population of a species, and (4) between populations of a species. The diversity at each of these levels and among them may be assessed differently by a molecular biologist, agronomist, or anthropologist, but recognizing these levels is a step toward creating a common framework.

Thus, diversity in an agricultural system includes not only plants in a farmer's field but also other varieties he/she can get from neighbors, a local market, or through public and private research institutions that keep and use genetic resources for crop breeding. Diversity can be measured at each of these levels, and change in diversity at one level may be offset by change at another. Although farmers could eliminate crop diversity from one season to the next by selecting a single variety or by switching crops altogether, diversity can persist at other levels. Farmers in North America, western Europe, the Punjab of India and Pakistan, and the Philippines have eliminated landraces and thereby have reduced diversity of crop populations on their farms. But on-farm diversity in these areas has been replaced by diversity from agricultural experiment stations and genetic conservation facilities. Smale (1997) shows that the pedigrees of improved crops have become increasingly complex because of breeders' use of a broad array of germplasm and sources. At an individual plant level, it is therefore possible that an improved crop plant will have more inherent diversity than a landrace plant that is presumed to have a local pedigree. Nevertheless, we must remember that numerous genotypes may constitute a single landrace as a population. Because of the success of crop breeding, diversity between plants and at the population level in farmers' fields has been replaced by diversity at the plant level by the combination of genes from many sources into single plants.

To a farmer, diversity might be the number of varieties of a landrace in a field or the number of crops in a kitchen garden. To a geneticist, diversity may be the number of alleles or the polymorphism found in the genome. An agronomist will focus on variability in traits such as yield, flowering time, plant height, or seed color. An ecologist or conservationist may concentrate on population structure, coevolution, patchy agricultural environments, biogeography, rarity, and abundance of different varieties or traits in relation to area. An anthropologist will look at folk nomenclature and taxonomy, the number of farmers who grow landraces, the number of varieties that are recognized, the link between knowledge and behavior in selection and management, and the exchange and market systems that move diversity between farms and regions.

These different groups carry on parallel discourses about the same phenomena using different terms and lexicons. Although a synthesis of these discourses might be possible, it seems unattainable at this time. There is simply too much invested in each of the specialized nomenclatures and interests to forge a new metadiscourse. Nevertheless, among these several discourses about crop genetic resources, three focal points stand out in viewing and discussing agricultural diversity: (1) folk measures, (2) genetic measures, and (3) ecological measures. Each way of viewing diversity is appropriate to different tasks and disciplines, none is more important than another to understanding a crop, and all are connected.

FOLK MEASURES

Folk measures of crop diversity are embedded in the names farmers give to crop varieties. Folk taxonomies are the systems of nomenclature and classification developed by people to describe and order the universe of plants that surrounds them. Folk taxonomies are utilitarian devices that help to order and retrieve popular knowledge about plants. Ethnobotanists (e.g., Berlin, Breedlove, and Raven 1974) have shown that folk taxonomies are both comprehensive and botanically accurate, but they differ from "scientific" taxonomies in their emphasis on utilitarian characteristics rather than on biosystematics. Folk taxonomies of crop species concentrate on the part of the plant that is used. Thus, the key to many folk classifications are the tubers, seeds, and fruits that are eaten or the bark that is used to dye cloth or produce fiber. These plant variations are generally confined to the infraspecific level, within rather than between crop species. Culinary qualities such as color, taste, and texture are important but may not reflect complex genetic differences. This may lead to underclassification, for instance when different wheat or potato species are lumped together. Conversely, infraspecific diversity is often of great importance to a farmer, cook, or merchant, while it may be far less significant to the biosystematics or ecology of a species. Consequently, folk classifications may appear to employ overclassification, for instance when different varieties of a single *Brassica* species (*Brassica oleracea*) are classed separately as a folk species — broccoli, kohlrabi, Brussels sprouts, cabbage, and kale.

Although most of the variations that are important to farmers are at the infraspecific level, they are biologically as well as culturally significant. Thus, Andean farmers' taxonomy of potatoes registers the difference between wild and domesticated species as well as the four polyploid groups of domesticated *Solanums* (Quiros et al. 1990). Turkish farmers' taxonomy of wheat notes the difference between the two primary wheat species. Mexican maize farmers distinguish races that are adapted to different environments and that are genetically distinct (Perales R., Brush, and Qualset 1998; Bellon and Brush 1994). Many key descriptors of crop species are exactly the same characteristics farmers observe, such as the shape and depth of potato eyes on the tuber or the size of maize kernels. The common warnings about the futility of using folk nomenclature in the systematic description of crop species should, therefore, be tempered with the appreciation that crops are as much cultural as biological artifacts. Harmony between biological and folk keys is logical since crops' separation into distinct biological groups has always been guided by human hands.

GENETIC MEASURES

The systematics of crop species have undergone tremendous change since De Candolle (1914) and Vavilov (1951) framed the puzzle of agricultural diversity for modern times. Biosystematics is faced with the daunting task of isolating genotypes from phenotypes, a job that requires detailed understanding of the interaction between genotype and the environment. Modern genetics and the development of genetic measures not subject to environmental or human selection have moved classification in the direction of "lumping," or reducing the number of biosystematic divisions in contrast to traditional botanists and natural historians who split categories based on environmentally sensitive characteristics. For example, Russian potato taxonomy of Vavilov's time recognized twenty-one species of cultivated potato based on the plant's morphology (Huamán and Spooner 2002). One standard taxonomy of cultivated potatoes (Hawkes 1978, see Table 3.1) reduces the number of species and subspecies of cultivated potato to seven, and Dodds (1962) proposed three species. Huamán and Spooner (2002) complete this trend by classifying all cultivated potatoes as a single species divided into eight cultivar groups. The reduction of species in wheat taxonomy is less dramatic than in potato, from eight domesticated species in De Candolle's (1914) time to four species today, distinguished, like potatoes, by chromosome number: einkorn (*Triticum monococcum,* 2n=10), emmer (*T. dicoccoides,* 2n=20), durum (*T. durum,* 2n=40), and common or bread wheat (*T. aestivum,* 2n=60) (Feldman, Lupton, and Miller 1995).

Ultimately, geneticists' measures of diversity are based on heterozygosity of alleles at given loci, and two measures of this heterozygosity are used: (1) quantitative and qualitative measures of plant characteristics and (2) genetic markers. These methods overlap, and each has its advantages and limitations. Qualitative characteristics are morphological or phenotypic features of a plant that indicate underlying genotypic similarity or difference. Diagnostic qualitative characteristics have been developed by observing crop species in different environments and over time to detect characters that have a low environmental effect, or a low gene-by-environment (G × E) interaction. Logically, the list

Table 3.2. Tuber descriptors for the cultivated potato

Characteristic	Variants
Predominant tuber skin color	White-cream, yellow, orange, brownish, pink, red, purplish-red, purple, dark purple-black.
Distribution of secondary tuber skin color	Absent, eyes only, eyebrows, splashed, scattered
Predominant tuber flesh color	White, cream, pale yellow, yellow, deep yellow, red, violet, purple, other
Distribution of secondary tuber flesh color	Absent, scattered spots, scattered areas, narrow vascular ring, broad vascular ring, vascular ring and medulla (pith), all flesh except medula (pith), other
General tuber shape	Compressed (oblate), round, ovate, obovate, elliptic, oblong, long-oblong, elongate
Unusual tuber shape	Absent, flattened, clavate, reniform, fusiform, falcate, coiled, digitate, concertina-shaped, tuberosed (lumpy).

Source: Huamán et al. 1977

of diagnostic qualitative characteristics of each crop species is unique, and these lists compose the descriptors of a given crop that key the systematic determination of similarity and difference. In the case of the potato, for example, fifty-four plant characteristics designate genetically linked markers or descriptors (Huamán et al. 1977). Table 3.2 summarizes a few of the qualitative characteristics of potatoes that make up the crop's descriptors, and these descriptors may have as many as a dozen different variants. Figures 3.2, 3.3, and 3.4 illustrate some of these variants.

Quantitative traits are measured on a population basis and reflect the action of many genes (loci) as well as a higher environmental effect on the genotype (a higher G × E interaction), such as plant height or yield. Quantitative characteristics are more difficult to use as diagnostic measures because of their multiple genes and environmental interaction, but these characteristics are often the most important ones in farmer selection because they are associated with yield, resistance to pests and pathogens, and tolerance of environmental problems. The difference between qualitative and quantitative characteristics

Distribution of Secondary Tuber Color

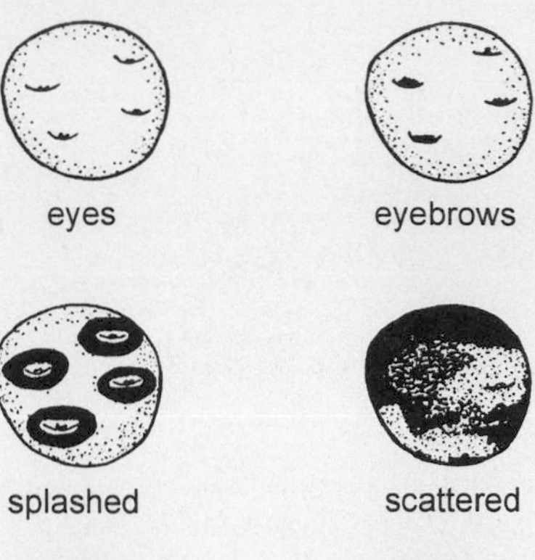

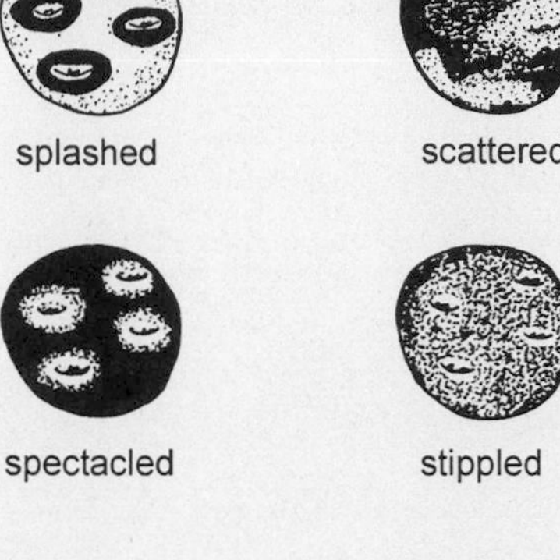

Distribution of Secondary Tuber Flesh Color

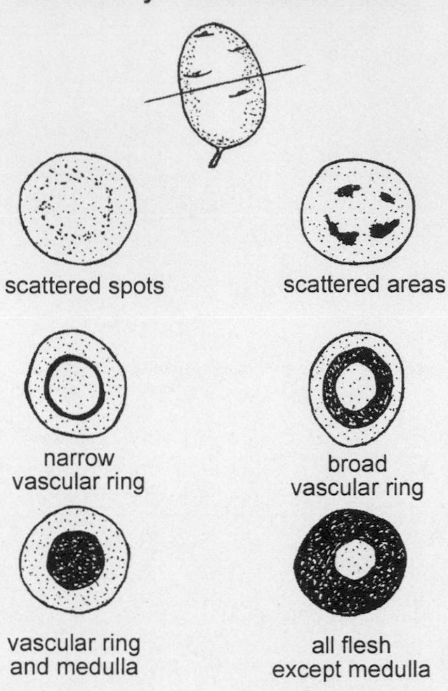

Tuber Shape

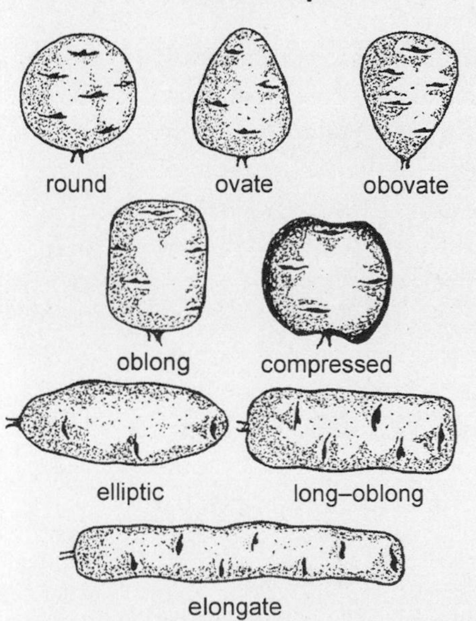

Unusual Tuber Forms

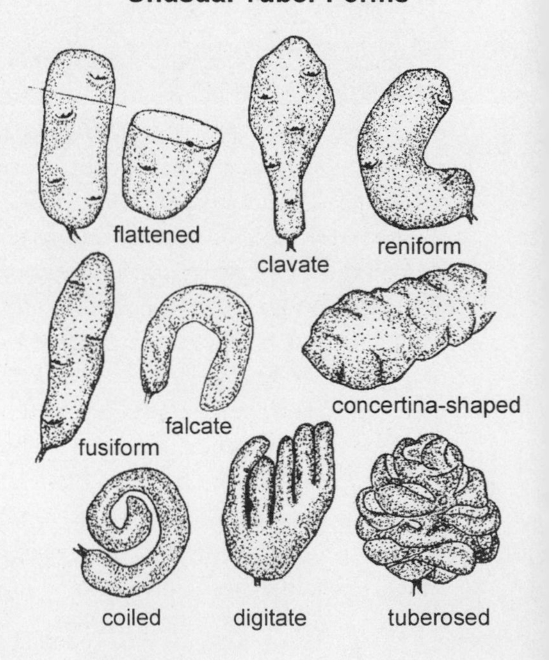

Figure 3.2 Tuber descriptors of the cultivated potato (After Huamán et al. 1977)

Figure 3.3 Yana Shokri, Tulumayo, Peru, 1978

Figure 3.4 Lumchipamundana, Tulumayo, Peru, 1978

opens a door to rather different assessments of diversity. Farmers may, for instance, tolerate variation of qualitative characteristics that are unrelated to yield, resistance, or lodging. A result, from a geneticist's viewpoint, is that farmers are underclassifying the true diversity that exists in a plant population that is variable and still segregating. On the other hand, farmers may see a geneticist's division of their landrace population as overclassification.

Genetic markers include the shapes of chromosomes, biochemicals in plant tissue such as isozymes, or storage proteins in seeds, and the order of base pairs in fragments of the plant's DNA. These genetic markers are not directly influenced by the environment or by human selection, so they are theoretically more true to the genotype of the organism. In addition, they may reflect simpler genetic sequences than those involved in plant morphological traits, so there are many more genetic markers to be measured than quantitative or qualitative characteristics. Genetic markers thus allow a rather abstract and quantitatively accurate measure of identity—a fingerprint, so to speak—as well as a measure of the relation between specific organisms. These markers are, however, invisible to people who might be selecting plant varieties, and they are not necessarily linked to morphological traits involved in natural selection. Thus, genetic markers may be only indirectly connected to the evolutionary processes affecting crop populations.

ECOLOGICAL MEASURES

Crop diversity is important not only as a store of potentially useful traits but also because it provides ecological services to farmers, such as the ability to exploit heterogeneous or marginal environments or as a buffer against disease. In addition, collection and conservation of the genetic resources embedded in crop diversity imply an understanding of crop ecology. Ecological measures of diversity are, therefore, necessary for assessing the role of crop variability and for conserving crop resources. Diversity has been a critical issue in ecology and consequently has been defined in many different ways according to the ecological problem, the environment, and the species in question (e.g., Soulé 1986). For crop ecology, two general types of diversity measures have been proposed. By far the most important are spatial measures of the amount of diversity and its distribution (Peet 1974). Secondarily, temporal diversity has been employed in a few cases (Brennan and Byerlee 1991). Spatial measures are primary because the need to access the functional role of diversity and to plan conservation involve spatial consideration.

Two data points have been widely used in spatial measures of crop diversity within a particular farming system: (1) the number of distinct crops types (e.g., species, landraces, cultivars) and (2) the distribution of different types within and among specific populations. To a lesser degree, the relation of pedigrees of crop types, such as improved varieties from crop improvement programs, has been used to calculate diversity (Brennan and Byerlee 1991). The simplest way to calculate crop diversity is to count the number of species, cultivars, or landraces present in a farm, village, or region. This count, of course, is often complicated by questions about what to count or how to measure individual elements of a crop inventory when counting components as ambiguous as landraces or farmers' varieties. Because of the problems of synonyms, and under- or overclassification, counts of crop varieties based on farmers' naming are useful at the level of an individual farm but questionable for assessing diversity among farms or at the regional level. Measures based on qualitative and/or quantitative characteristics, isozymes, or molecular markers can overcome these problems. An ethnobotanical alternative for village-level research is to create a representative sample collection of distinct morphotypes, using either actual crop samples or photographs, that can be used to assess the distribution of crop types among farmers.

Describing the distribution of crop diversity combines counts of crop types with data on how alleles, genotypes, or morphotypes of a crop species are partitioned over space—the evenness of distribution within and between different geographic levels, such as the field, farm, village, market region, province, or nation. In crop ecology, the starting point for population analysis is the farm, and the highest level is normally an environmentally distinct geographic unit such as a zone with given altitude or rainfall parameters. Political units such as states or departments are often substituted for the higher level (e.g., Spagnoletti Zeuli and Qualset 1987). Diversity is distributed within and between the populations of farms, villages, or regions. When diversity is found between geographic units, it indicates that populations within the sample are isolated or otherwise structured independently of each other. However, we frequently find that most of the crop diversity is found at a given geographic level, such as a farm or village; in other words, the largest portion of overall diversity is within populations rather than between them.

Besides assessing diversity within and between populations, the spatial distribution of diversity, or evenness, at different levels informs us about the ecological function and conservation of crop variability. Four data points are necessary to describe the distribution of crop diversity: (1) the number of distinct crop types (qualitative or quantitative traits, alleles, or cultivars), (2) the number of different populations in the sample, (3) the size of each population, and (4) the number of each crop type in each population. These data are processed in different ways to measure distribution. Three widely used measures are richness, evenness, and the Shannon-Weaver index. Richness is a

Table 3.3. Sample of diversity indices for crop population research

Index	What it measures	Formula	Description
Richness (Margalef)	Number of classes (e.g., races, cultivars)	$(S-1)/ln\ N$	S = number of classes N = total number of samples summed over all classes
Evenness (Pielou)	Distribution of classes in a population	H'/lnS	P_i is the proportion of the total number of samples in class i relative to the total number of samples
Shannon-Weaver	Combination of richness and evenness	$H' = \sum_{i=1}^{n} P_i\ ln\ P_i$	Shannon-Weaver value divided by the logarithm of the number of classes

Source: Peet 1974

measure of the number of distinct species, varieties, or morphotypes that are found in a population. Evenness is a measure of the distribution of distinct types in the population. This measure shows whether diversity is provided by the presence of a few rare individuals that are distinct or by comparable proportions of different types in a population. Evenness is relevant in two ways for assessing crop diversity. First, it relates to the agronomic benefit that might occur from the interaction of different crop varieties, for instance in affording disease resistance. Second, evenness is important to estimating the probability of extinction of different varieties that constitute a crop population. The Shannon-Weaver index combines measures of richness and evenness and is the most commonly used ecological measure of diversity in crop studies. The advantage to diversity indices, such as Shannon-Weaver, is to overcome problems caused by sample size (Rosenzweig 1995). Formulas for these indices are provided in Table 3.3.

Crop diversity on a field or farm lies somewhere between the theoretical poles of monoculture and polyculture of different species and crop varieties—between a population of similar or unique genotypes. Additional aspects of distribution are the relative abundance of a particular crop type (i.e., whether it is frequent or rare) and whether it is cosmopolitan or local. Figure 3.5 combines these aspects into a single and commonly used matrix. Distribution of crop types, whether they are evenly spread, frequent, or cosmopolitan, depends on many botanical, environmental, and human factors, including whether a crop is outcrossing or self-pollinated, the complexity of the environ-

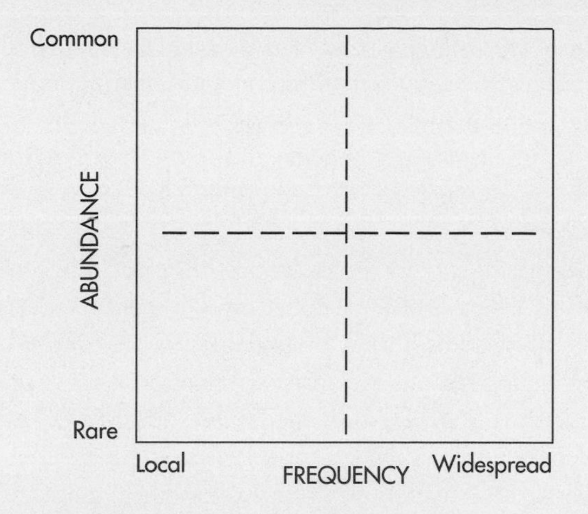

Figure 3.5 Frequency and abundance matrix

ment and the presence of strong selection pressures, and the isolation of the farm or its integration into larger market and technological systems. The degree of agricultural intensity, use of purchased inputs, production for home consumption or as a commodity, and the risk of hunger if a crop fails may be as important to the distribution of diversity as the number of soil types or risk of drought or disease. Obviously, there is no simple or single axiom about where and how diversity will be distributed in any farming system. As Harlan (1992) showed, crop evolution is an ongoing process of hybridization and diversification. This process continues today and in every farming system.

Explanations of Diversity in Agriculture

The concentration of crop diversity in certain areas was recognized long before the efforts of nineteenth-century naturalists to identify the geographic homes of crops (e.g., Theophrastus 1916, Watson 1983). Crop diversity is still concentrated in the places that Vavilov would show to be the hearths of domestication. The hills rolling southward from the Caucasus, across the headwaters of the Tigris and Euphrates rivers of eastern Anatolia to the Syrian Plain, are home to thousands of wheat varieties. Besides their staple cereal, Anatolian farmers plant many other native crops: lentils, chickpeas, parsley, cilantro, cherries, apples, pears, pistachio nuts, and almonds. In South America, Andean village markets in Peru or Bolivia are stocked with dozens of potato varieties, a collage of colors, shapes, and sizes. The same Andean market with multicolored potatoes is also stocked with dozens of other local crops and varieties: glistening piles of white, red, and purple oca, a half-dozen other tuber crops (*mashua, ollucu, arrachacha, maca, yacón*), red tree tomatoes, chirimoya in green armor, a dozen different types of chili pepper, and aromatic displays of spices and medicinals. In Mexico, farmers in a single village grow perhaps a dozen different kinds of maize in their milpas and house gardens, as well as fruits, beans, squashes, leafy vegetables, chili peppers, condiments, and medicinals (Alcorn 1984).

All farming systems harbor diversity within and between crops, but the amount and type of biological and cultural diversity vary greatly between farms and farming systems, just as biological diversity of trees varies between forests. Some farming regions are vast swaths of a single species or crop variety, for example the potato fields of the United States, where ten commercial varieties of a single potato subspecies (*Solanum tuberosum* subs. *tuberosum*) account for 80 percent of the crop's acreage (Quiros et al. 1990). Other regions combine many crop species and varieties into a single community, a complex mosaic that may include wild and semicultivated plants closely related to cultivars. These regions are agriculture's version of other reservoirs of genetic resources, such as tropical forests, albeit on a smaller and biologically less diverse scale.

We are immediately faced with two questions in accounting for the amount and distribution of diversity within crops. First, why is there so much interspecific and infraspecific diversity of crop species? Second, why is diversity distributed unevenly and concentrated in particular regions?

WHY ARE CROP SPECIES SO DIVERSE?

It is fair to assume that the domestication of plants created a bottleneck that reduced the diversity in crops in comparison with their wild ancestors (Cox 1998, Cox and Wood 1999). This bottleneck occurred because only a small portion of the biological diversity of a species was used by early farmers, who necessarily emphasized plants with a favorable mutation for the nonshattering trait that allowed cultivation. Because only a few individual plants of the species undergoing domestication had traits that were valuable to Neolithic farmers, the diversity of the species was not included in the populations that became crops. In addition, as several studies suggest, domestication in some crops, such as wheat and barley, may have occurred very abruptly, in perhaps as few as twenty or thirty years or within two farming generations (Hillman and Davies 1990). Such a time frame would logically confine the amount of wild germplasm that was originally incorporated in domesticated plants by limiting the number of generations of gene flow from wild to cultivated plants and by limiting domestication in space. Although many crops retain the potential of gene flow from wild species, farmer selection can be an effective constraint. Moreover, isolation from ancestral species and maintenance in the artificial environment of a field or garden may also restrict diversity by reducing selection to a few, anthropocentric criteria. A reduction of diversity between wild and cultivated species has been observed in beans (*Phaseolus vulgaris* L.) (Gepts and Bliss 1986), where gene flow is still possible.

Several factors explain an expansion of diversity following the bottleneck of domestication. Rindos (1989) lists four sources of increased diversity in his Darwinian model of crop domestication: (1) the gradual accumulation of favorable variants, (2) migration into new habitats with distinct selection pressures, (3) the growth of the human population, and (4) cultural change. The element of conscious selection may allow variation in crop species to accumulate faster than natural selection can limit variation. A result is that crop species resemble "species flocks" that appear in some habitats, for instance the fynbos of South Africa and the Kwongan of southwestern Australia (Rosenzweig 1995). The long evolutionary history in the presence of wild and weedy relatives provides opportunity for diversity to arise. Many crops have had five thousand to eight thousand years for farmers to recognize and accumulate diversity arising from hybridization, mutation, recombination, and introgression of germplasm from related plants. Conscious selection and protection of variation has encouraged diversity to accumulate by allowing less-fit genotypes to survive.

Habitat diversity logically plays a paramount role in crop diversity as it does in explaining biological diversity in general (Rosenzweig 1995). If domestication occurred most commonly in mountainous environments, as suggested by Vavilov, farmers since the beginnings of agriculture have been accustomed to exploiting different agricultural environments and finding crop varieties to allow this. The environmental heterogeneity of mountains provides not only numerous isolating factors but also a variety of selection pressures. Exposure to distinct environments in a single mountain valley must have prepared Neolithic farmers for rapid expansion out of the mountains and over continents.

Population growth may promote diversity by pressuring humans to exploit new habitats or to explore and exploit existing habitats more fully by subdividing them. If this scenario is valid, agricultural development that is induced by population pressures may actually lead to more diversity rather than to less, as is often assumed. Similarly, an expansion of cultural diversity may lead to crop diversity, assuming a positive association between cultural identity and the types of crops that are cultivated (Nazarrea-Sandoval 1995, Maffi 2001). Curiosity and conscious selection for phenotypic diversity seem to be natural allies of cultural expression and identity. A widespread human aesthetic for diversity must be significant in explaining diversity. Human diets around the world are dominated by starchy staples that provide most of the calories needed for subsistence. Plates of food comprised of grains or tubers slightly embellished with a few condiments have been the daily fare of most humans since the Neolithic. Only the wealthy can hide the staple under meat and vegetables. The human eye, mind, imagination, and palate everywhere seek diversity in staple foods, whether it be in an assortment of potatoes or in the shape of pasta. Quality and quantity are equally valued by all cultures. Humans, with their natural inquisitiveness and inventiveness, have had few realms for experimentation comparable to the plants that sustain them. Indeed, it would be remarkable if humans, with their ability to recognize and manage plant variation, did not create complex arrays of colors, tastes, shapes, and plant growth qualities.

Nevertheless, culture is only one element among the many forces in crop evolution that lead to increased diversity. The fact that cultural identity is often expressed in food preferences and cuisine lends credence to a positive relation between cultural and crop diversity. Unfortunately, this relationship has not been well researched and is, in all probability, complex and indirect. Anthropologists (e.g., Kroeber 1939) have demonstrated that cultures tend to sort themselves out into different environments and ecological niches, and this tendency obviously complicates the challenge of connecting cultural diversity to crop diversity.

WHY IS CROP DIVERSITY UNEVENLY DISTRIBUTED?

Diffusion, adaptation, human aesthetics, and selection may account for the inherent variability of crop species, but they do not explain why crop diversity is concentrated within specific regions. The exuberance of crop varieties in particular regions has been explained in two ways: (1) as a legacy of domestication and (2) as a product of farmer management.

The Legacy of Domestication

The legacy of domestication accounts for the concentration of diversity in Vavilov centers in four ways. First, continuous cultivation and selection in the homelands of crops provide these places with the longest time to accumulate and select for variation in crop populations. Assuming that a crop species has a given rate of mutation and recombination, it is possible that the additional time of existence has allowed more diversity to accrue in hearths of domestication than elsewhere. Longer time of adaptation in a single environment also allows for a crop species to fill all of the potential habitats and for farmers to recognize this. Second, Vavilov centers have not only crops but crop ancestors. The potential of gene flow, even at a small rate, between wild and cultivated types provides more raw material for crop variation than is available in places characterized by the absence of wild relatives. While isolation between domesticated and wild species is encouraged by self-pollination and farmer selection, more exacting measures of genetic relationships available with molecular markers indicate that a higher rate of gene flow between domesticated and wild relatives exists than has hitherto been demonstrated (Ellstrand, Prentice, and Hancock 1999).

Third, the crops that were taken by migrants or diffused out of the original Vavilov centers contained only a small sample of the diversity found in the original crop populations—the equivalent of a founder effect or genetic bottleneck in the diffusion of crops. Numerous instances of small samples of a crop forming the basis of a new crop population are reported. In the United States, for example, a single type of avocado and a single type of navel orange established the industry for these crops in California in the early twentieth century (Rosengarten 1991). Likewise, six varieties of peach formed the basis of that crop in California in the same period (Tadesse 2001). The wheat variety "Turkey Red" is credited with establishing the winter wheat agriculture that exists over tens of thousands of hectares in the north-central United State (Cox 1991). A few cases of exceptional crop diversity exist outside of Vavilov centers, for example barley in Ethiopia (Asfaw 1999) and maize diversity in the Andes (Grobman, Salhuana, and Sevilla 1961), but the story of crop migrants with little diversity is repeated around the world.

Finally, crop diversity in Vavilov centers is explained by the fact that they are particularly rich in the number of agricultural habitats. Habitat diversity may result from geographic features, such as a mountainous landscape. Such environments not only provide numerous niches for specific adaptation but also forceful isolating mechanisms for plants and people. However, the steep environmental gradient of mountains not only offers numerous opportunities for selecting distinct types, but it also pressures growers to select generalists that will produce when moved to different fields and will survive the fickle nature of weather in mountains, where frost, hail, drought, pests, and diseases can be fierce but localized. Indeed, the seemingly contradictory characteristics of being adapted to specific environments yet highly plastic in adaptation seem to be common to many crops. Diversity of agricultural habitat in Vavilov centers may also be a result of the accumulation of other species that are closely linked to the crop—symbionts, predators, pathogens—whose interactions with the crop create new habitats.

A Product of Management

The second general explanation of the distribution of diversity is that it reflects farmer management, implying active selection for diversity because of its functional value in agriculture. This explanation lacks the advantage of the first general explanation, the legacy of domestication, in explaining the particular concentration of diversity in Vavilov centers. Nevertheless, this explanation is important because it turns our attention to the active role of farmers in managing diversity and the role of diversity in certain farming systems.

Crop diversity at the farm level has been attributed to two aspects of management: (1) adaptation to different agricultural production conditions and (2) interaction effects of diversity that provide either higher yields or greater stability. A third aspect, production of desired products in conditions of missing markets, was added to these two in the previous chapter's discussion of the human ecology of crop diversity. The selection of different crop types for specific production environments has been observed in numerous farming systems and is the conventional explanation for the diversity of farmer varieties or landraces (Harlan 1992). Adaptation results in the crossover effect in which the yield lines of two varieties cross one another as we move from one environment to another (see Chapter 11). In Chiapas, Mexico, this crossover is observed between local and improved maize types as one goes from valley bottoms to hillsides and between soil classes (Bellon and Brush 1994; Brush, Bellon, and Schmidt 1988). The same situation pertains to wheat varieties in western Turkey (Meng, Taylor, and Brush 1998) and barley types in Southwest Asia (Ceccarelli and Grando 1999). The interaction effect of diversity at the farm level and the advantage of diversity in situations of missing markets have been discussed earlier (see Chapter 2). Suffice it to say here that there is sufficient evidence to suggest that management, as well as legacy, is an important basis for the uneven distribution of crop diversity. Thus, places with environmental heterogeneity, strong selection pressures, and missing markets should be places of crop diversity. In some cases, such as the Andes or the hill lands of southern Mexico, all three of these aspects are at play, while in other places one or two may drop out.

Summary

Diversity can be measured in different ways, and it exists at different levels, for instance within and between plants and within and between populations. Folk measures, which emphasize the useful part of the plant, are key to understanding crop species' distribution and evolution because it is folk knowledge that has guided the selection and maintenance of diversity over thousands of years. Genetic measures include both qualitative and quantitative traits and genetic markers. Ecological measures allow us to address population structure and distribution and to analyze the relation between crop diversity and changes in agricultural environments. All three of these measures are necessary to address the pressing question of the fate of agricultural diversity as human population, society, and technology undergo rapid and fundamental change away from the conditions that generated diversity in the first place.

Crop diversity has been both a key to identifying the places where domestication occurred and a puzzle. Domestication presented a bottleneck that restricted diversity, but migration into new environments, population increase, and cultural change have overcome this bottleneck. The aesthetics of diversity and its role as a cultural marker must certainly be counted as significant causal factors. The uneven distribution of crop diversity has been attributed to the legacy of domestication and to the advantages that diversity affords farmers in particular areas. While research has supported the association of these factors with crop diversity, the puzzle of crop diversity is far from solved. The more closely we examine the actual management of crops by farmers in very different types of agriculture, the more problematic the conventional explanations for diversity become, such as adaptation and interaction effects. This situation will be examined in more detail in relation to the problems of estimating genetic erosion or the loss of diversity as agricultural modernization occurs.

4

Crop and Society in Centers of Diversity

The development of agriculture is one of the defining moments of human experience. Within a blink of human history, people in Asia, Africa, and the Americas began to produce food rather than to hunt and gather, and we are today the inheritors of changes wrought at that threshold. Between six thousand and ten thousand years ago, people in vastly different environments — dry mountains of Southwest Asia, humid tropics of Southeast Asia, steep Andean valleys — crossed the dividing line separating food production from hunting and gathering. Agriculture prepared the foundations of modern life: permanent settlements, economic specialization, written records, and large human populations integrated by complex social and technological means. During the millennia after domestication, individual crops and cropping systems put their discernable stamps on people and places. Agricultural landscapes are as distinctive as regional cuisine built on key staples. Much of lowland Asia is landscaped by canals, bunds, and terraces to accommodate rice paddies, while Mesoamerica is sculpted by maguey (*Agave* spp.) fenced milpas where maize is intercropped with beans and squash. Because maize lacks gluten, Mesoamerican cuisine is based on tortillas and tamales rather than on maize bread, while gluten-rich wheat has stamped the cuisine of western Asia through many forms of bread.

The passage from food gathering to food producing is known as the "Neolithic Revolution," reflecting the appearance of new stone tools in archaeological strata. Axes for clearing forests and hoes for working soil presage dramatic changes in human ways. The term *revolution* is, however, a misnomer. The shift to agriculture took place over a hundred or more generations and makes sense only as an archaeological horizon, where human experience is compressed within buried strata. Rindos (1984) and Harris (1989) have cogently argued that a change so momentous as the shift to food production represented the culmination of social traits and cultural habits that evolved over long periods. Even if the specific act of domestication might occur relatively quickly (Hillman and Davies 1990), the transformation of wild plant to a crop would not be recognized or seized if the social and cultural ground were not already prepared by social evolution. The changes wrought by the invention of agriculture would have been impossible to detect in the short run, so that the material remains of societies appear similar immediately before and after domestication. The material culture left by the first farming peoples was virtually indistinguishable from the material remains of food gatherers. The passage was certainly not perceptible to the people involved, and the momentous consequences of the shift were apparent only long after it had been accomplished.

The eventual impact of the shift from food gathering to food production was indeed dramatic. Social systems and human landscapes became more dynamic than any that had existed in the previous five million years of human evolution. Settlements became large and permanent; material culture became varied and abundant; social organization became complex and dependent on webs of specialization and exchange that spread over ever larger regions; change, especially in technology, became a constant element in human life; and the imprint of humans on the landscape became more visible and enduring. Not all of the change wrought by domestication was positive, as malnutrition, social hierarchy, and ecological damage all followed.

Agriculture is the art of exploiting a disturbed environment. Farming is similar to riding a bicycle—sustainable only by the momentum of change. The invention of food production created a permanent imbalance between humans and the environment that could only be maintained by technological and social change. Farming demands that humans be just destructive enough of plants, animals, soils, and water systems so as to create appropriate conditions for the domesticated species that nourish and clothe humans without provoking environmental collapse. Human agricultural ingenuity makes food infinitely more abundant than in pre-agricultural times. Yet, the promise of abundance is forever yoked to the threats of environmental collapse and human privation.

Producing food relies on an integration of a common set of social, physical, and biological components that were discovered and invented independently in different places and at slightly different times. Physical and biological technology is easy to imagine—tools for tilling and harvesting, amendments to improve soils, crop varieties, and farm animals. Social technology, perhaps less visible but of no less importance, comprises human knowledge, culture, and social institutions that allocate labor and land and distribute the fruits of production. Technology thus embraces knowledge systems and the means to allocate resources: mechanical and chemical means to alter physical components of agriculture, and biological means to exploit or control a particular environment.

In all farming societies, the balance and momentum of agriculture is maintained by tinkering and adjusting technology—selecting seeds, experimenting with new techniques, improving tools, adopting crop varieties and tools from other farmers. Tinkering creates a momentum of its own, since it leads to discovery or adoption of new technology as farmers seek to work less, produce more, or achieve stable yields. This momentum can be observed in the expansion of agricultural frontiers, in new forms and amounts of energy to produce food, in the diffusion of crops and crop varieties, and in social organization to allocate land, labor, water, and harvests. The social results of agricultural momentum are clear: population increase, the ability for a few food producers to feed many people, and integrated social systems covering widely dispersed areas. Environmental results are equally apparent: ecological transformation such as deforestation, increased environmental disturbance such as soil erosion, and restructured populations of plant and animal communities.

Mountain Cradles

Generations of scholars have weighed questions about the location and processes of crop origins (De Candolle 1914, Darwin 1896, Vavilov 1951), but the boundaries of centers of origin of individual crops and crop complexes remain vague. Contemporary crop diversity, the presence of closely related plant species, and archaeological sequences showing domestication can be specified, but the intersection of these three lines of evidence of original domestication is elusive. In Mexico, evidence for the likely center of maize (*Zea mays* L.) domestication is found in all three vectors: the diversity of Mexican landraces of maize; the contemporary distribution of different wild relative ancestors, teosinte (*Zea mexicana*) (Doebley 1990, Iltis 1983) and gama grass (*Tripsacum* spp.) (Eubanks 2001); and archaeological remains (Benz 2001, Piperno and Flannery 2001). Ethiopia is one of the most diverse regions for barley, yet it lacks wild relatives of the crop (Asfaw 1999). Vavilov delineated centers of origin, but his map shows the regions of diversity of major crops rather than the limits of diversity or domestication. Harlan (1992) has argued eloquently for a "non-centered" view of agricultural origins—one that recognizes a fundamental human capacity to take the necessary steps toward independent invention of food production. Crops have been domesticated in a wide range of habitats, and domestication is an ongoing human practice (Johns and Keene 1986). These factors defy a simple characterization of "center" of domestication.

Mountain environments are often associated with crop domestication (Harlan 1992), although strongly contrasting habitats, such as tropical lowlands (Piperno and Pearsall 1998), are also recognized as cradles of crop domestication. Mountains provide ideal conditions for moving plants across ecological boundaries so that they may be isolated and prosper in a new environment. Furthermore, mountains provide abundant ways to isolate distinct crop populations as well as farming cultures. Historically and today, mountains pose barriers to social and technological change that has eroded diversity elsewhere. Mountains thus serve as repositories of social and biological diversity in agriculture. Crop domestication, diversity, and evolution undoubtedly have occurred in low altitudes and flat terrain, but today mountains are prominent places of crop diversity.

While the species that gave rise to crops might be cosmopolitan and found in different landscapes, a case can be made for domestication in mountains, so that the specific plant populations that were manipulated during domestication were mountain populations. Mountains offer favorable conditions for domestication of cosmopolitan species. Hillman and Davies (1990) note that domestication would be aided by annual shifts to virgin land using seed stocks harvested from last year's plots. Moving seeds across environmental boundaries (ecotones) would facilitate the process of isolating a crop species from its wild relatives. If ecotones played a role in domestication, then mountain areas are a logical site of domestication. The environmental heterogeneity of mountains provides another biological basis for crop diversity by creating a myriad of different niches and opportunities for local adaptation and isolation. Soils, moisture availability, drainage, exposure to sun and wind, temperature, and evapotranspiration rates are each highly variable in mountain landscapes, and in combination they present plants and farmers alike with numerous opportunities and limits.

Cultural and social factors must also be counted as possible explanations for the diversity that we find in mountains. While biological factors define the domestication process, cultural and social factors shape diversity during crop evolution and the development of agriculture. The biological explanation of highland diversity is that mountain crops have always been more diverse than the same species in flatlands. A social explanation is that flatlands and mountains may once have been equally diverse but have become different over time. In this divergence, mountain farmers add and maintain diversity over time, while farmers on flatlands lose diversity. The persistence of diversity in mountainous areas has several possible explanations. First, it is possible that mountains pose environmental and economic obstacles to the diffusion of varieties that are normally competitive and dominant. Perhaps the agricultural niches of mountains are too numerous and too narrow to permit the monoculture of highly competitive varieties that are found elsewhere. Agronomic limits, such as cool temperatures, higher risk of frost, lack of irrigation water, or poor soils, may impede the diffusion of otherwise competitive varieties. Second, mountains around the world are sites of cultural and ethnic heterogeneity and refuge (Aguirre Beltran 1979). Culture and crop variety type are logically associated through folk taxonomies, ritual use, and cuisine, thus increasing the probability that mountains will have a greater diversity of crop varieties than more socially uniform areas in the flatlands. Finally, mountains pose severe obstacles to the development of market systems that reduce social and agronomic relief. Urban markets are often far away and less accessible to highland farmers. Farmers in the flatlands are better served by providers of information and inputs, not to speak of the agronomic advantages of their situation. Mountains are notoriously neglected in infrastructure facilities, such as roads, telecommunication, and schools. Cultural diversity may retard the movement of information because of language barriers. Transport of inputs and commodities faces daunting obstacles that warp the economics of comparative advantage.

Potatoes and Their Andean Hearth

The Andes Mountains of South America rise abruptly from a narrow and bone-dry Pacific coastal plain to glaciated massifs over 5,500 meters high, march across high plateaus and tight valleys to the eyebrow of the jungle, and then plunge into the verdant, moist western Amazon Basin. In Ecuador, these transitions occur in a horizontal distance of two hundred kilometers, and in Bolivia, at their widest, the horizontal transect is only five hundred kilometers. Between the eastern and western foothills, the land has been corrugated by tectonic and geological processes, leaving a landscape unsurpassed in variety and complexity. Into a short longitudinal span the Andes pack most of the climates of an entire hemisphere, from tropical forests to arctic tundra.

Andean people have mastered the complexity of their landscape and have fashioned not only one of the great centers of plant domestication but also a home of great civilizations. Within the central Andean ranges, farmers and pastoralists depend on numerous ecological zones that are compressed into single valleys and dispersed longitudinally from the shores of the Pacific to the edge of the Amazon (Murra 1972). Altitude influences the critical factors in agriculture—temperature, precipitation, and humidity. A layer cake analogy for Andean environment is almost inescapable, but it oversimplifies internal complexity at single altitudes and the effects of longitude and latitude. Temperatures are approximately 6°C lower for every thousand meters in altitude gain, and thinner atmosphere at high altitudes causes a more dramatic diurnal fluctuation (Troll 1968). Precipitation is heaviest along the eastern slopes, but interior valleys in the rain-shadow of higher ranges are dry. Andean vegetation generally reflects the altitudinal and longitudinal moisture gradients, but the crumpled landscape provides numerous opportunities for distinct vegetation communities to exist. The botanist Weberbauer (1945) attempted to capture the profusion of Andean vegetation communities in a system of 160 "ecological floors."

At the center of Andean diets is a group of root and tuber crops dominated by the potato. Gade (1999) lists the potato as one of the twenty material symbols of *lo Andino*—the essence of Andean culture. As in other regions of domestication, a mixed cluster of crops surrounds the dominant crop. Cereals, legumes, tubers, fruits and vegetables, condiments, and narcotics were all domesticated in the Andes, but roots and tubers are notable and the potato is outstanding in its importance to Andean nutrition. Table 4.1 shows the wealth of domesticated species that trace their heritage to the Andes. In addition, movement of early domesticates between Mesoamerica and the Andes provided Andean people with imported crops to diversify long before the arrival of Francisco Pizarro in 1532. Prominent examples of Mesoamerican transplants to the Andes are maize (*Zea mays*), squashes (*Cucurbita* spp.), and avocado (*Persea americana*). From the opposite direction, tomatoes (*Lycopersicum* spp.) and tobacco (*Solanum nicotanea*) were brought from western South America to Mesoamerica.

Archaeological and genetic evidence indicate that the central Andes was the original hearth of the potato. Excavations in the highlands of central Peru date potato cultivation as far back as 5,800 years ago (Pickersgill and Heiser 1978), although other archaeologists suggest earlier dates (Engel 1970). At later archaeological horizons, there is ample evidence of the importance of potatoes in ceramics of pre-Inca and Inca cultures (Figures 4.1 and 4.2), and striking continuities exist between prehistoric and modern potato agriculture in the

Table 4.1. Selected Andean domesticates

Common name	Botanical classification	Quechua term	Aymara term
Tubers			
Achira	*Canna edulis*	Achira	Achira
Oca	*Oxalis tuberosa*	Oqa	Oqa
Mashua	*Tropaeolum tuberosum*	Añu	Isaño
Arracacha	*Arracacia xanthorrhiza*	Raccacha	Rakacha
Ulluco	*Ullucus tuberosus*	Ulluku	Colluku
Papa	*Solanum* species	Papa, Akshu	Papa
Yacón	*Polymnia sonchifolia*	Yakon	Yakuma
Maca	*Lepidium meyenii*	Maca	
Cereals			
Quinoa	*Chenopodium quinoa*	Kiñua	Hupa
Cañihua	*Chenopodium pallidicaule*	Qañawi	Qañawa
Kiwicha	*Amarnthus caudatus*	Quihuicha	Koyo
Legumes			
Popping bean	*Phaseolus vulgaris*	Ñuñas	
Lima bean	*Phaseolus lunatus*	Pallar	Pallar
Peanut	*Arachis hypogaea*	Inchis	Chokopa
Lupine	*Lupinus mutabilis*	Tarwi	Tauri
Fruits			
Chirimoya	*Annona cherimolia*	Masa	Yuructina
Guanabana	*Annona muricata*	Masasamba	
Pacae	*Inga feuillei*	Paqay	Paqaya
Passion fruit	*Passiflora linularis*	Ccjoto	Apinkoya
Tumbo	*Passiflora molissima*	Tumpaka	
Guava	*Psidium guajava*	Sawintu	Sawintu
Lucuma	*Lucuma obovata*	Ruqma	Lukuma
Pepino	*Solumum muricatum*	Kachan	Kachuma
Squashes and vegetables			
Zapallo	*Cucurbita maxima*	Sapallu	Tumuña
Achokcha	*Cyclanthera pedata*	Achokcha	
Condiments			
Paico	*Chenopodium ambrosioides*	Paiko	Paiko
Chili	*Capsicum pubescens*	Rocoto	Chinchi
	C. baccatum	Uchu	Waika
Huacatai	*Tagetes minuta*	Wakatay	Wakatay
Narcotics			
Coca	*Erythroxylon coca*	Coca	Coca

Sources: NRC 1989, Horkheimer 1973, Gade 1975

Figure 4.1 Mochica potato ceramic, Peru north coast (courtesy of Donald Ugent)

region (Figures 4.3 and 4.4). The richest pool of potato genetic diversity is found in the Andes and comprises different species, subspecies, and thousands of local varieties of domesticated tubers, along with wild and weedy relatives of the crop. Wild species closely related to domesticated species that are found in the Andes include *Solanum acaule* and *S. megistacrolobum* (Hawkes 1990).

The *Solanum* genus, to which cultivated potatoes belong, includes 176 potato species (Hawkes 1990). Like many other crops, cultivated potatoes can be divided into different groups (ploidy levels) according to chromosome number (see Table 3.1). The most common cultivated species in the Andes is the tetraploid *Solanum tuberosum* subspecies *andigena.* This tetraploid species accounts for two-thirds of the potato area in Peru and Bolivia and for most of the distinct varieties (Brush, Carney, and Huamán 1981). Also found regularly in Andean farmers' fields and larders are diploids *S. stenotomum, S. phureja,* and *S. goniocalyx,* the latter being a much prized delicacy resembling a boiled egg yolk. Among triploids, *S.* × *chaucha* is generally more rare but can be locally common. Three cultivated species flourish in high-altitude frost zones but contain enough alkaloids to be bitter and usually require processing by freeze-drying for consumption (Johns 1990). These frost-resistant species are the diploid *S. ajanhuiri,* found in southern Peru and northern Bolivia, a triploid *S.* × *juzepczukii,* and *S.* × *curtilobum,* a pentaploid with only two varieties.

Figure 4.2 Chimú potato ceramic, Peru north coast (courtesy of Christopher Donnon)

The pathway of potato domestication begins with the diploids. Hawkes (1990) identifies *S. stenotomum* as the most primitive species, based on resemblance between *S. stenotomum* tubers and tubers of wild diploid species, the most likely progenitors of all cultivated forms (Ugent 1970). Questions remain about which wild species—and whether more than one—contributed the original stock of domesticated potatoes (Hosaka 1995). Nevertheless, genetic diversity in the modern potato clearly is a result of several evolutionary processes occurring simultaneously over the entire history of Andean potato agriculture. These processes include successive domestication of diploids and the multiplication of diploid chromosomes into three other numbered sets (triploids, tetraploids, and pentaploids) by several mechanisms, and the continued flow, or introgression, of germplasm from wild species into cultivated stocks (Hosaka 1995, Rabinowitz et al. 1990, Johns and Keen 1986). Human action has also contributed to the long evolutionary process by moving potatoes out of their original habitat and to different altitudes beyond the native habitat of the earliest diploids.

Solanum phureja at low altitudes, *S. ajanhuiri* at high altitudes, and *S.*

Figure 4.3 Potato planting, Guaman Poma de Ayala, Cusco, 1587 (© Publications Scientifiques du Muséum national d'Histoire naturelle, Paris)

tuberosum ssp. *tuberosum* at temperate latitudes in Chile are all direct products of this centrifugal movement of cultivated potatoes and Andean cultivators. The accumulation of thousands of distinct morphotypes of potatoes is testimony to the plasticity of the species, especially of the tetraploid *S. tuberosum* ssp. *andigena,* and also to the identification, selection, multiplication, and maintenance of distinct varieties by Andean farmers. Some of this "artificial selection" has an obvious environmental and agronomic basis, for instance in the selection of frost-resistant types. Nevertheless, human selection for diversity in the potato cannot be understood solely as environmental adaptation but rather as a process involving the interaction of cultural aesthetics and production factors (see Chapter 5).

While potatoes possibly were not the first crop domesticated in the Andes, they became the lynchpin of Andean life, around which evolved the complex of crops and animals, settlement and trade that define Andean civilization.

Figure 4.4 Andean farming with foot plow *(chakitaklia)*

Scholars have often recognized that Andean societies are organized to utilize the diversity of ecological zones that are crammed into the dense topography of these mountains. In both prehistory and historical periods, most Andean people have lived at altitudes between 2,500 and 4,000 meters above sea level, and these altitudes have given rise to early and far-reaching civilizations: Chavin, Tiwanaku, Inca. Populations living at these altitudes focus their subsistence production around four major zones: (1) a low-altitude zone (<1,500 meters) for tropical fruits and, most importantly, coca (*Erythroxylon coca*), (2) a mid-altitude zone (1,500–2,800 meters) for maize, beans, squash, and some potatoes, (3) a higher-altitude zone (2,800–3,800 meters) for tubers and Andean cereals, and (4) a high-altitude zone (3,800 meters) for frost-resistant potatoes and pastures. In fact, most Andean people, whether in cities or villages, live around 3,000 meters above sea level, in the *kichwa* zone, and it is the potato crop that sustains this zone. The tuber-producing zone is actually subdivided into several more specific zones, according to altitude, the type of potato cultivated, and whether potatoes are planted for an early harvest (*maway tarpuy*) or the later and more major harvest (*hatun tarpuy*).

Diversity within and between Andean crops varies not only by location but also by whether production was destined for subsistence or other purposes such as paying taxes or commercialization (Zimmerer 1996). Emphasis on diversity for subsistence is expected to decrease as production is shifted toward other purposes. This pattern is seen across long stretches of Andean time, from the Inca period to the present (Zimmerer 1996). Centralization of agricultural management and selection criteria for the type of crop is thus a persistent counterweight to the momentum to diversify crops under highly individualized and isolated conditions. Centralization versus localized decision making is a tension that persists in contemporary Andean agriculture, where subsistence and commercial production coexist and the use of local and nonlocal production inputs are intermingled in most fields and villages.

Moreover, non-Andean crops and food have been present in the region since long before the Columbian exchange that brought European crops and animals to the region (Crosby 1972). The most notable prehistoric import was maize that arrived from its Mesoamerican home to South America long before local cultures coalesced into urban centers or states (Sevilla 1994). By the time of the European conquest, maize was widely diffused across the Andes and had evolved into distinctly Andean races. In contrast to the native Andean tubers, maize had acquired an elevated status in ritual life and political organization (Murra 1960). Indeed, the grand terraces of the sacred Vilcanota Valley of the Incas were built for maize — homage to the crop in the form of landscape architecture. Stratification by altitude resulted in a pattern that persists today. Maize is competitive and displaces certain types of potatoes, especially diploids at lower altitudes, but the cereal is less competitive and nondisplacing of the primary stock of tetraploid potatoes at higher altitudes (Gade 1975).

In Murra's (1972) terms, maize and potatoes form a complementary pair of crops as foods, ritual goods, and occupants of distinct agricultural environments. The Andean ability to absorb novel elements was exercised to an extreme after 1532 with the arrival of Europeans along with wheat, barley, broad beans, cabbage, sheep, cattle, pigs, horses, donkeys, and mules — now all blended into the Andean landscape in a way similar to maize. The ability to absorb such potentially dominant and destructive elements and to retain local crops and diversity is attributed not only to the great environmental heterogeneity of the Andes, but to the resilience of Andean people. This resilience is defined by the ability to find complementarity between the native and the exotic and to continually synthesize a distinctly Andean fabric out of new and old elements. Nowhere is this clearer than in Andean agriculture, where the synthesis of autochthonous and imported elements is almost as old as agriculture itself. Potatoes create a subterranean support to Andean life. A cultural superstructure celebrating the crop is relatively colorless in contrast to the elaborate rituals and art devoted to maize in Mesoamerica. Potato rituals in

Figure 4.5 Enjoying the harvest *(watia)*, Tulumayo, Peru, 1978

the Andes are never flamboyant, and most are private rather than public. But quiet potato rituals are the warp of Andean life — family rituals at planting and harvest, roasting freshly harvested tubers in the field in earthen ovens (*watia*), the humor and stories embedded in potato names, and gifts to friends, family, and workers (Figure 4.5).

Maize in the Mesoamerican Homeland

Maize (*Zea mays*) is one of the few crops that is so dominant in the regional culture and society of its origin that it might be perceived as having domesticated humans as much as humans domesticated it. Indeed, it is hard to overstate the position of the cereal in Mesoamerica's history, culture, and economy. Origin myths from different Mesoamerican cultures devote a preeminent place to maize in human affairs, for instance portraying the creation of humans from maize (Maya) or ascribing it as a gift to humans from Quetzalquotl (Aztec) (Markman and Markman 1992). The sacred book of the Maya, the *Popol Vuh,* chronicles how Xmucané, the grandmother and creator of humans, fashioned the first Mayas:

> And then grinding the yellow corn and the white corn, Xmucané made nine drinks, and from this food came the strength and the flesh, and with it they created the muscles and the strength of man. This the Forefathers did, Tepu and Gucumatz, as they were called.
>
> After that they began to talk about the creation and the making of our first mother and father; of yellow corn and of white corn they made their flesh; of corn meal dough they made the arms and the legs of man. Only dough of corn meal went into the flesh of our first fathers, the four men, who were created (Recinos 1950, 167).

Maize imagery continues to figure in the iconography of the Maya. For instance, at Palenque, the Temple of the Foliated Cross contains a tablet depicting blood sacrifice to maize that has human heads in the place of ears of maize (Schele 1974).

Later and in central Mexico, the same imagery reappears in the Mixtec pictographs of the Borgia Codex (Diaz and Rodgers 1993). The Náhuatl word for maize is *tonacáyotl*—"our support" (León-Portilla 1988). The *General History of the Things of New Spain* (the Florentine Codex), written originally in Náhuatl (ca. 1547) by Fray Bartolome de Sahagún (1979) and then translated into his native Spanish (ca. 1577), reveals the sacredness of maize to the Aztec people and the abundance of foods prepared with the grain (Figure 4.6).

For half a millennium after the flowering of Mesoamerican civilization, throughout the catastrophic Columbian exchange, and onto the modern transformation of Mexico, maize continued to permeate Mesoamerican life. A description of the sixteenth-century Maya still rings true in many villages and hamlets of southern Mexico and Guatemala: "If one looks closely he will find that everything [these Indians] did and talked about had to do with maize; in truth, they fell little short of making a god of it. And so much is the delight and gratification they got and still get out of their corn fields, that because of them they forget wife and children and every other pleasure, as if their corn fields were their final goal and ultimate happiness" (Vázquez 1937, translation by Morley 1946).

Five centuries later, maize continues to dominate the rural and cultural landscapes of Mesoamerica. Mexican peasants consider the plant to be kin to humans (Warman 1989). Morley (1946, 441) reports, "I am convinced that fully 75 per cent of all their (modern Maya) thoughts still center around this same important cereal." To the contemporary Náhuatl people of central Mexico, maize is their blood, and the future is prognosticated on a ritual table with

Foods that the lords ate [Maize-based]:

Hot, white, doubled tortillas; large tortillas; thick, coarse tortillas; folded tortillas of maize treated with lime, pleasing [to the taste]; tortillas formed in rolls; leaf-shaped tortillas;

White tamales with beans forming a sea shell on top; white tamales with maize grains thrown in; hard, white tamales with beans forming a sea shell on top; tamales made of dough of maize softened in lime, with beans forming a sea shell on top; tamales of maize softened in wood ashes; turkey pasty cooked in a pot, or sprinkled with seeds; tamales of meat cooked with maize and yellow chili;

Market food: white tortillas with a flour of uncooked beans; meat stewed with maize, red chili, tomatoes, and ground gourd seeds; birds with toasted maize; tender maize; green maize;

Tamales made of maize flowers with ground amaranth seed and cherries added; tortillas of green maize of or tender maize; tamales stuffed with amaranth greens; tortillas made with honey, or with tuna cactus fruit; tamales made with honey; tortillas shaped like hip guards; rabbit with toasted maize; green maize cooked in a pot and dried;

Hot maize gruel of many kinds; maize gruel with honey, with chili and honey, with yellow chili; white, thick gruel with a scattering of maize grains; sour, white maize gruel; sour, red maize gruel with fruit and chili; small green tomatoes with a maize gruel made with anonas; maize gruel made with amaranth and toasted maize; maize gruel with fish-amaranth seeds and honey; cold maize gruel; maize gruel with wrinkled chía, covered with green chilis or small, hot chilis; white maize gruel with chía, covered with yellow chilis; maize gruel with chía covered with squash seeds and with chili; maize gruel made of tortilla crumbs, and with ordinary and wrinkled chía, covered with small chilis;

All these foods came forth from within the house of the ruler.

And daily a man, the majordomo, set out for the rule with his food – two thousand kinds of various foods; hot tortillas, white tamales with beans forming a sea shell on top; red tamales; the main mean of roll-shaped tortillas and many [foods].

Sahagún 1979, 9: 37-39.

Figure 4.6 Maize in the Florentine Codex (Sahagún, ca. 1575) (Reprinted with permission of the School of American Research Press © 1981)

maize of colors in each corner (Sandstrom 1991). The Maya of Chiapas propitiate the gods with tortillas made of the grain, and maize is a ubiquitous element in the yearly ritual cycle (Vogt 1976, Teni 1989). Likewise in the national mestizo culture of Mexico, maize is a pervasive element—in cuisine, art, and folklore, as well as in the rural and national political economy of food. Mexican painter Diego Rivera used maize often in his murals depicting the birth and soul of modern Mexico. His homage to the agrarian foundation of Mexico in the vestibule of the Riveriana Chapel in Texcoco is decorated with a mural that is representative of the continued reverence toward maize in popular culture. Here, in Rivera's *The Blood of the Rural Martyrs: Emiliano Zapata y Otilio Montaño*, the two martyrs are buried beneath a field of maize, whose roots become an umbilical cord connected to the two graves.

The regional importance of maize is also reflected in long-term investment in science of the crop in Mexico and the United States. Within two decades after maize became the first commercial hybrid seed, Mexican and American scientists began the systematic collection and analysis of the crop in its Mesoamerican homeland (Wellhausen et al. 1952). This collection built upon botanical fieldwork (e.g., Anderson and Culter 1942) that was eventually enlarged to an unsurpassed scientific undertaking to comprehend a single crop. Only rice research in Asia rivals maize in attracting such a broad cross section of scholars, including artists, social scientists, botanists, agronomists, and geneticists. The economic importance of maize in Mexico and the United States is certainly an element in attracting so much scientific notice, but its place in the food system and culture of Mesoamerica transcends mere economic explanation.

The location of maize's cradle area in Mesoamerica is indisputable, and the consensus among geneticists and archaeologists is that central Mexico, lying in the heart of Mesoamerica, is the crop's heartland (Mangelsdorf 1974, Eubanks 2001, Doebley 1990, Piperno and Flannery 2001). Various measures of diversity, both morphological and biochemical, indicate that more diversity of cultivated maize exists in Mexico than in any other region. Also, Mexico provides one of the clearest archaeological records available for any crop, a sequence that strongly suggests the evolution of wild to domesticated maize during a 1,500-year period (Wilkes 1989, Goodman 1988, Mangelsdorf 1974). However, biological models of maize domestication propose a more rapid evolution based on genetic transmutations (Iltis 1983) and hybridization between *Tripsacum* and teosinte (Eubanks 2001).

Despite consensus on maize's geographic origin, there is little agreement on its biological ancestry. Cultivated maize's closest relative, teosinte—*Zea mexicana*—is common in parts of Mexico and is also found in Guatemala, but it is nonexistent elsewhere. Most teosinte is a weedy plant that grows in maize fields, indistinguishable from the crop in the early stages and fully capable of hybridizing with maize. Maize/teosinte hybrids are common occurrences in Mexican milpas where both species exist. Teosinte is a logical but disputed ancestor to cultivated maize. Figure 4.7 shows the continuum that exists from teosinte to maize, although to some (e.g., Goodman 1988) this depiction overstates the role in teosinte in the evolution of maize. Biochemical analysis (Doebley 1990) indicates that the *parviglumus* race of teosinte (*Zea mexicana* ssp. *parviglumus*) is the likely progenitor of cultivated maize. Nevertheless, several problems in citing teosinte as a direct ancestor can be raised (Wilkes 1989, Goodman 1988). No archaeological evidence has been found to verify teosinte's ancestry, and the earliest known maize, with eight rows, is distinct from teosinte. Because teosinte hybridizes with maize, it is very difficult to

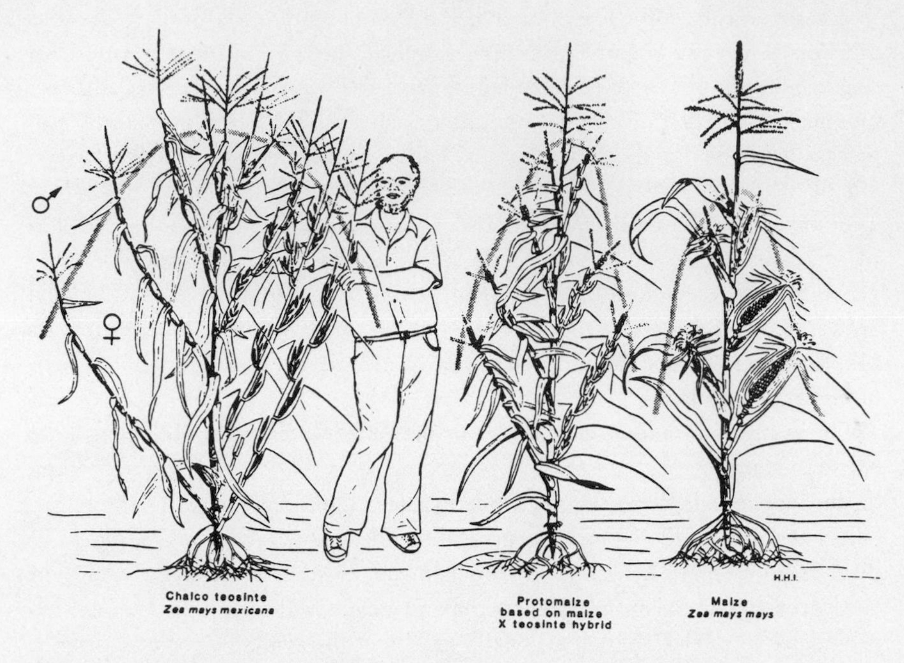

Figure 4.7 Transformation of teosinte to maize (Iltis 1983) (Reprinted with permission from *Science*. © 1983 American Association for the Advancement of Science)

domesticate maize through selection of teosinte populations (Wilkes 1989). Finally, the search for similarities between the chromosomes of teosinte and maize suggests that reciprocal introgression does not occur (Wilkes 1985), a view supported by isozyme analysis (Doebley 1990).

Maize is a prodigiously outcrossing crop, more dynamic in this regard than any of the other major crops of the world. Unlike potatoes or wheat, which each have several cultivated species, maize is a single species, and the same number of chromosomes ($2n = 20$) is common to all maize. Like all crops, the range of variation in maize populations is continuous, but interpreting the population structure and distribution of maize is made especially difficult by its facility for outcrossing. Dedicated collection and comparison of Mexico's maizes resulted in the first comprehensive analysis of the species in the landmark volume *Races of Maize in Mexico* (Wellhausen et al. 1952). For sixty years, the concept of race, promulgated by Anderson and Cutler in 1942, has been applied to Mesoamerican maize and has survived into the modern era of molecular biology: "A maize race is a group of related maize plants with enough characteristics in common to permit their recognition as a group. . . . From the standpoint of genetics a race is a group of individuals with a significant number of genes in common, major races having a smaller number in common than do subraces" (Anderson and Cutler 1942).

A maize race is, thus, a population of the species that has a certain morphological integrity and geographic identity developed through adaptation and human selection, involving the physical, biological, and social environment. Although more powerful taxonomic tools and elaborate data sets have become available, Anderson and Cutler's concept remains valid and is still the most widely used category for describing this species. The outcrossing nature of maize means that its populations are not described by distinct and stable genotypes but rather by swarms of genotypes with local reference but without sharp boundaries. Indeed, Wellhausen et al. (1952) concluded that modern races are the result of previous hybridization and that maize evolution continues through the same process. Wellhausen et al. (1952) identified twenty-five races and three subraces, and continued collection and analysis have, inevitably, augmented that number. In a comprehensive survey, Sanchez G. (1994) lists sixty-six races of Mexican maize.

Maize races are most clearly discernable at the regional level, where they can be delineated both morphologically and ecologically (Sanchez G. and Goodman 1992). Figure 4.8 is a dendogram of fifty Mexican races studied by Sanchez G., Goodman, and Stuber (2000). These races were distinguished by studying twenty-one enzyme systems (isozymes) and forty-seven morphological characters (agronomic, vegetative, tassel, spiklet, ear, kernel, and cupule). The morphological characters were gathered from maize collections that were grown in multiple locations between 1982 and 1984. This mass of data allows Sanchez G., Goodman, and Stuber (2000) to estimate the genetic distance (correlation coefficient) among the fifty races.

Comparison of maize samples, assembled from countrywide collections in Mexico, reveals a dual population structure. At the broad, environmental level, maize races can be grouped into a limited number of ecogeographic clusters based on similar genotype by environment interaction, such as response to moisture and temperature. Sanchez G. and Goodman (1992) describe four major ecogeographic clusters for Mexico with subdivisions into six areas: (1) low to mid-elevations of the Pacific coast, (2) low to mid-elevations of the Gulf coast, (3) low to mid-elevations of interior regions with late maturity, (4) mid-elevations of western Mexico, (5) high elevations of northwestern Mexico, and (6) high elevations of central Mexico.

Maize's broad distribution in Mexico is determined by biophysical factors —altitude, moisture, latitude—rather than social factors, although there is recognition that the diversity of maize at a more specific scale is vaguely associated with the distribution of Mexico's many cultural groups (Hernández

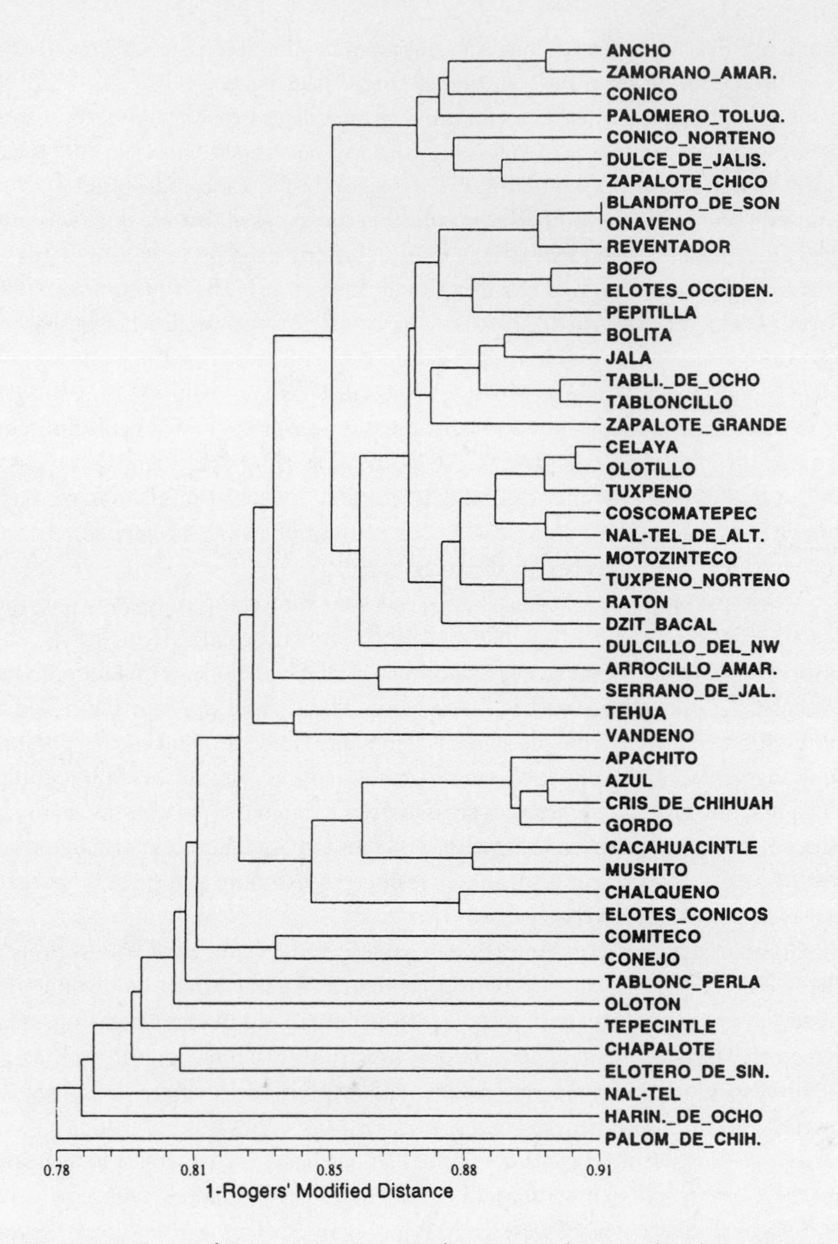

Figure 4.8 Races of maize in Mexico (Sanchez G., Goodman, and Stuber 2000) (Reprinted with permission from Economic Botany. © 2000 The New York Botanical Garden)

1985). At a narrower geographical level than the broad scheme of Sanchez G. and Goodman (1992), maize races can be clustered according to morphological characteristics that have low genotype by environment interaction—number of kernel rows, cob diameter, kernel width, and so forth (Sanchez G. and Goodman 1992; Sanchez G., Goodman, and Stuber 2000). Many of these characters are observed by farmers and serve as selection criteria, but several isozyme loci can also be included here. In other words, the specific geographic structure of maize populations, which is shown by neutral alleles of isozymes, reflects farmer selection of traits that may not be related solely to environmental fit. It is at the level of morphological race that we can begin to appreciate the crop as both a biological and cultural artifact.

Several morphological races can be grouped into each of Sanchez G. and Goodman's (1992) ecogeographic clusters, although there are some anomalous cases of morphological races that are geographically close but belong to different ecogeographic clusters. Morphological race as an analytical category for describing maize populations persists for several reasons. A linkage exists between morphological traits such as cob and kernel characteristics and isozymes (Paterniani 1969, Sanchez G. and Goodman 1992). Most importantly, farmers, consumers, and crop breeders identify and select maize according to the traits that form the basis of racial identification.

At the village level, below the broad ecogeographic distribution of maize races, the sorting of maize into morphological races and ecogeographic groups begins to break down under the relentless pressure for hybridization caused by the species' outcrossing nature and by the frequent exchange of maize between farms, villages, and regions. Comprehensive collections at the village level reveal that a single farming community often has numerous varieties of maize, some belonging to separate morphological races (Bellon and Brush 1994). Case studies of maize at the village level reveal the general success of local varieties in competition with improved maize that has been bred for high yield. These studies demonstrate that maize evolution continues to be dynamic and significantly determined by the farm folk of Mesoamerica, as it has been for hundreds of generations.

Brush, Bellon, and Schmidt (1988) and Bellon (1990, 1991) studied the mid-elevation zones of the Grijalva Valley in the southern state of Chiapas, where traditional maize varieties compete and complement each other by adaptive advantage in different farm environments. Approximately a dozen different maize varieties exist in a single community, although households keep a smaller number. Brush, Bellon, and Schmidt (1988) report that households plant an average of 2.7 varieties. Two Mexican races (*Olotillo* and *Tuxpeño*) and the exotic (*Argentino*) dominate the local farming system.

Maize races are adapted to environments at both the landscape and local level. In farming communities that have both valley and hill land, improved varieties of *Tuxpeño* dominate the valley land, and local, mixed populations of *Olotillo* and *Argentino* are important in the hill land. Bellon (1991) worked in one community in the Grijalva Valley and describes a dynamic situation of introduction of maize races, competition, and specific adaptation. Bellon found five Mexican and one exotic race in the community. The five Mexican races belong to a single genotype by environment group and three different morphology groups. Bellon shows that different races coexist in the same community because they perform differently in different soils and under different input regimes (Bellon and Taylor 1993, Bellon and Brush 1994).

A diverse and dynamic maize system is described for the low elevations of Jalisco state in central Mexico. In the valley of Cuzalapa, farmers keep twenty-six maize varieties (Louette 1999, Louette, Charrier, and Berthaud 1997) of both local and nonlocal farmer varieties, and advanced generations of improved varieties. One local variety is dominant, accounting for 51 percent of the maize area. Of the twenty-six varieties, three account for 71 percent of the maize area. The mean number of rain-fed varieties per household is 2.6, and 2.4 for irrigated maize. The maize system here is divided into four subsystems, depending on moisture and crop season length: irrigated/long, irrigated/short, rain-fed/long, and rain-fed/short. Notably, there is no strong selection pressure for distinct varieties to this agronomic gradient. The same varieties are grown in each of the four subsystems, albeit in slightly different proportions. This suggests broad adaptation of different maize varieties, in contrast to the findings of Brush, Bellon, and Schmidt (1988) and Bellon and Taylor (1993) in Chiapas, where farmers selected different maize varieties for different subsystems (valley bottom vs. hillside) and soils.

The higher altitudes of Mexico's central highlands supported the largest population and centralized state in pre-Hispanic times and continue to be the heartland of Mexican peasant society. These highlands, however, represent an environmental extreme for maize because of cool temperatures, low humidity, and high evaporation and transpiration rates (Eagles and Lothrop 1994). Maize populations in the central highlands are less diverse in terms of the number of races and varieties than those of the lower elevations of Chiapas and Jalisco, where conditions are more favorable. Moreover, the lowlands are hospitable not only to more diversity but also to modern maize from organized agricultural research (Bellon and Brush 1994).

The *Conico* (pyramidal) maize complex is adapted to the highlands—a group of closely related races are dominant above an altitude of 1,800 meters. Perales R., Brush, and Qualset (2003) found a single dominant type of white

Chalqueño maize south of the Valley of Mexico. This maize outperforms not only other local races but also improved varieties that have been bred to compete in the highland environment. An important finding of Perales R.'s (1998) study is that maize farmers who are fully integrated into the market, both in agricultural production and in other aspects of their lives, maintain a traditional maize variety despite the availability of improved and hybrid maize seed. This variety is grown across a wide region with different microenvironments and soils. In contrast with the mid-elevations of Chiapas (Bellon and Taylor 1993), the highland maize of the Chalco appears to be broadly rather than narrowly adapted to different conditions. Also contrasting with the specialized populations of mid-elevation Chiapas is the low-elevation maize of Jalisco, studied by Louette (1999). Like Perales R. (1998), Louette (1999) reports that seed exchange goes on within and between villages, requiring that maize be adapted to different conditions. Nevertheless, the Jalisco lowlands studied by Louette (1999) are similar to the mid-elevations of Chiapas studied by Bellon in their openness to new and exotic maize. In contrast, the *Conico* races of the highlands have been able to retain complete dominance, despite concentrated efforts to breed and introduce improved types.

Our picture of maize in its Mexican homeland, composed through biological and social research, is more complete than for any other crop. Nevertheless, this canvas also shows many blank spots. One poorly deciphered area is the relation between the cultural diversity of Mesoamerica and the biological diversity of maize. At what level does culture become relevant in determining the population structure of this crop? Another area is maize's relation to its closest relative, teosinte. How important is the flow of genetic material between these two species, and how is this relevant to the future evolution of the crop? A third area is how the rapid technological, economic, and cultural changes now affecting Mexico will impact its maize. Will the conservative maize populations of the highlands succumb to the pressure of modernization and integration into the world market from such agencies as the North American Free Trade Association and labor migration to the United States? What effect will the introduction of transgenic maize from the United States have on Mexican landraces (Quist and Chapela 2001)? Finally, how can Mexico protect this species that has been so critical to its biological and cultural heritage?

The Anatolian Homeland of Wheat

Anatolia provides a striking contrast in landscape, crop, and culture to the Andean and Mesoamerican valleys where the first potatoes and maize were cultivated. Travelers since antiquity have found the steppes of central

Anatolia a melancholy place, dry and treeless (Mitchell 1993). Anatolia shares the common historical distinction of being a hearth of domestication and center of diversity of a plant that became one of humankind's primary staples. Unlike the maize and potatoes whose seed and plants are individually handled, wheat—Anatolia's trophy—is managed as a population within a field. The concept of "indigenous culture," which is so pertinent in the Americas and relevant to understanding the contemporary agriculture and diversity of the region, is all but irrelevant in Anatolia, where migration and the dominant Turkish culture have obscured ethnic diversity on most of the plateau.

The Anatolian plateau has been home to many farming cultures, from the initial waves of Neolithic migrants to modern times. Early Indo-European cultures, Hittites, Celts, Hellenic and Byzantine cultures, Romans, and Turkish cultures have occupied Anatolia since the beginnings of agriculture there. On this steppe, at Çatal Hüyük, there is evidence of humankind's first steps toward urbanism, the social dividend of the Neolithic Revolution. It is possible that the quasi-urban settlement at Çatal Hüyük preceded cereal domestication in the Near East, and the material culture left here leaves no doubt of the close association of this region and the Neolithic Revolution (Hillman and Davies 1990).

Grain cultivation has been practiced in Anatolia for approximately ten thousand years. Anatolia proper is a high, dry plateau that is hardly distinguishable from the great Asian steppe that rolls eastward from Turkey. Geographically, the Anatolian plateau is not characterized by the dramatic variation and contrast of the mountains surrounding it on three sides in Turkey. Low rainfall amounts, a relatively severe winter with cold winds from the northern reaches of Asia, and few large rivers to cut through its deep, gray loam create a somber landscape. This dullness, however, is broken dramatically by the hills and mountains north, south, and west of the steppe. Between the Black Sea and Anatolia, a broken belt of uplands catches moisture flowing southward, producing cloud forests and supporting the hazelnut industry. To the south, the Taurus Mountains stand between Anatolia and the Mediterranean. In the west, the mountains that rise from the Aegean coast create a contrast of varied fields, forests, and landscapes that ends at the plateau.

Wheat has been a staple crop in the Anatolian region since prehistoric times. As with other crops in their cradle areas of crop domestication, the wheat of Anatolia is diverse both at the species and infraspecies levels. Like potatoes, wheat comprises several species that are distinguished in part by different numbers of chromosomes, from hulled diploids (2n=14) to bread wheat hexaploids (2n=42) (Figure 4.9). In the case of wheat, additional chromosomes add entirely new genomes to the wheat species. Farmers in the Pontic Moun-

Figure 4.9 Four species of domesticated wheat (After Schiemann 1948)

tains of northern Turkey still grow emmer *(Triticum dicoccum)* and einkorn *(Triticum monococcum),* two primitive species of hulled wheat that were the first wheat types to come into cultivation (Nesbitt 1995). Bread wheat *(Triticum aestivum)* dominates Anatolia, although durum wheat (tetraploid, *Triticum durum,* 2n=28) is also common. One of the few countrywide collections of wheat was done in 1948, when Harlan (1950) found 2,121 different types. Chapman (1985) reports 64,163 tetraploid and hexaploid Turkish accessions among various international collections.

Unlike the relatively common depiction of potatoes in Andean ceramics and maize in Mesoamerican ceramics and glyphs, wheat is accorded little attention in the prehistoric art of its hearth of domestication and diversity in Southwest Asia. In Anatolia, the Hittite grain god appears to hold a handful of wheat spikes (Figure 4.10). The lack of attention to the crop is not typical of areas adjacent to the Anatolian/Fertile Crescent zone of domestication. Wheat is depicted frequently in art of the Mediterranean area, for instance in Egyptian and Grecian art. In Greek mythology, wheat is seen as a gift of the goddess Demeter. Figure 4.11 depicts Demeter's gift of agriculture (grain) to humans through the hands of Triptolemus, whom she charged with sowing wheat throughout the inhabited world. The gift of wheat is reciprocity for Triptolemus for information about Hades' abduction of her daughter, Persephone.

The wheat of Anatolia proper has been deeply affected by the agricultural modernization that followed the creation of the Turkish Republic in 1923. Modernization included selection of Turkish varieties for better yields, importing high yielding varieties (e.g., *Bezostya* from Russia), and creating a national seed multiplication and distribution system. However, the resulting reductions in diversity caused by these development programs are unknown. Harlan made his collections after Turkey began to promote agricultural development by releasing selected and improved varieties. Since Harlan's collections were made, agricultural development in Anatolia has continued apace, with major impacts on wheat diversity from the introduction of new varieties. Nevertheless, on the rim of the great plateau, diversity persists in wheat. The presence of emmer and einkorn in northern Turkey is an example. Another is the hilly transitional zone of western Turkey, where farmers cultivate local and traditional wheat varieties, in some cases along with modern, improved varieties (Brush and Meng 1998, Meng 1997).

Following a pattern found in the Andes and Mesoamerica, hill and mountain villages of Turkey have more diversity than valley areas on the rim of the Anatolian plateau. In descending from mountains to the plateau, one follows a gradient from villages where local wheat varieties dominate to others where only modern varieties are found. Unlike potatoes and maize, wheat diversity

Figure 4.10 Hittite grain deity, Turkey (Hoffner 1979)

Figure 4.11 Demeter gives wheat to Triptolemus and humankind (Hoppin 1919)

in Anatolia is not the accumulation of different varieties cultivated by different households and on different fields. Rather, wheat diversity is within the populations of single types of wheat. The average inventory of wheat varieties of single farms here is small—one or two named varieties. Diversity between farms in a single village is somewhat larger but still limited to a handful of different types of wheat. Nevertheless, counting named varieties underestimates the true genetic diversity in the region. While farmers give a single variety designation to a particular field, the wheat of a traditional variety is, in reality, a composite population made up of numerous, distinct types (see Figure 3.1). A field may be dominated by a particular genotype but may also include other genotypes in various admixtures. Diversity within fields of traditional wheat is evident in both qualitative characters, such as awn type or kernel color, and quantitative characteristics, such as height or spike density (Meng 1997).

It is difficult to determine whether variety mixing is intentional. Turkish farmers recognize that diversity exists within their fields of traditional varieties but don't appear to do this purposively or to have an explanation as to why fields of local types are so heterogeneous. Cultivating a pure type as opposed to a mixture involves either much stronger seed selection, such as selecting spikes before harvest, or purchasing seed from a state or commercial source. Some mixture of bread and durum wheat is preferred because of the baking quality of flour from this mixture. Nevertheless, a single field shows diversity within both bread and durum wheat species, and proportions of different wheat species and varieties in a mixture were not specifically controlled. Mixing may occur through exchange, field rotation, or the use of common threshing floors where different varieties are brought together. Once mixing has occurred, recovering a pure type of wheat is exceedingly difficult, especially given the lack of a formal seed market for traditional varieties.

A Diversity of Patterns

The three cases of crop diversity introduced above—Andes, Mesoamerica, and Anatolia—are among a small number of hearths of plant domestication. Each of them bestowed a local crop that has become a world crop and staple for people in different continents and different environments. Each center is characterized by the presence of unusual diversity in the crop, in some cases the presence of different cultivated species. All three centers are also habitats for wild relatives of the cultivated species, a hallmark of a Vavilov center.

The singularity of these places dwindles somewhat when we turn our attention to human and geographic similarities. All three of them are mountainous, a landscape that undoubtedly is associated with crop origins and diversity. Here, environmental heterogeneity and the proximity of environmental boundaries (ecotopes) are pronounced. Farmers in all three regions cultivate numerous fields of the same crop, providing an agricultural context for diversity, past and present. Other mountainous regions are similarly associated with crop origins and diversity, for instance in Southeast Asia and Ethiopia. Nevertheless, mountainous terrain itself is not sufficient to explain either crop origins or the accumulation of diversity. Besides environmental heterogeneity associated with mountains, all three regions are characterized by sharply contrasting rainy and dry seasons. None are tropical in the sense of having continued high humidity or a monsoon season. But climate traits such as sharply contrasting rainy and dry seasons are common to many places that are neither cradles of crop origin or diversity, and these traits are difficult to connect to the persistence of diversity except that they define marginal conditions where industrial technologies are less competitive.

A trait shared by farmers of the Andes, Mesoamerica, and Anatolia is the persistence of household-based agriculture within a national context that is increasingly characterized by industrial production, market exchange, and urbanism. In all three cases, a positive correlation exists between the self-sufficiency of households in food production and the diversity of the staple crop, whether potatoes, maize, or wheat. Household production may be implicated in the persistence of diversity, but, like mountainous terrain, it is not sufficient to explain the accumulation of diversity in the first place. In the contemporary context, household production signals the relative absence of key markets for information, agricultural inputs, labor, and commodities. Diversity of crops can logically meet the demands of household food production in terms of fit to different soils by providing resistance to disease and pests or supplying commodities that are differentiated by use. Nevertheless, not all household economies have evolved diversity as a means to meet these demands, as evidenced by highland maize production in Mexico (Perales R. 1998).

In sum, no inherent similarity joins the Andes, Mesoamerica, and Anatolia other than their common history as centers of crop domestication and repositories of diversity. As many factors differentiate these centers of diversity from one another as draw them together into geographic, social, or environmental categories. In other words, crop diversity must be understood at a more intimate level and in terms of the local landscapes, societies, and cultures of each region.

5

The Ethnoecology of Crop Diversity in Andean Potato Agriculture

The profusion of species and genotypes in centers of crop origins plays a dual role in modern science and agriculture. Diversity is a resource that crop scientists and farmers use to improve yields and meet the demands of increased population and changing environments. Diversity is also a key to understanding the ecology and evolution of crops and societies. Crops are both cultural and biological entities that have been shaped by hundreds of generations of farmers. In turn, crop plants have shaped cultures and landscapes. Understanding how farming cultures nurture crop diversity is obviously important to success in interpreting crop ecology and crop evolution and to designing comprehensive conservation programs for crop genetic resources that are threatened by the social and environmental conditions of the modern world. Nevertheless, we have only a rudimentary understanding of how specific cultural traditions maintain and influence crop populations in centers of crop origin and evolution. Limited knowledge of the human contribution to agricultural diversity is a liability to both basic and applied science of crop population biology. A handful of studies describes the knowledge and management of crop diversity, but we are still in the preliminary stages of developing a systematic approach that can be applied across crops and cultures.

Among the social sciences, anthropologists and geographers have used the methods of ethnobotany and human ecology to describe and study agricultural diversity. Both of these specialized fields deal with the relation between humans and plants, and both have further specialization in the study of cultivated and noncultivated plants. Ethnobotany and human ecology are informal in that they lack definite theoretical boundaries and methods. Both are claimed as subfields by different disciplines, and they overlap in several instances. The relation between humans and plants has been examined on one side through the knowledge and information systems about plants and on the other through human behavior regarding plants. Ethnobotany describes the former—giving emphasis to the plant nomenclature and classification which are part of all cultures and languages. Ethnobotany is now included in the more comprehensive field of ethnobiology (Berlin 1992), but ethnobotany is both older and more replete than its more recent umbrella. Ethnobotany has drawn practitioners from both the biological and social sciences, and these backgrounds greatly influence the emphasis of particular studies toward plant systematics and botany or toward folk classification. Human ecology's focus on behavior is more comprehensive than naming and classification. Its research on agriculture includes land-use topics—energy flows, field rotation, intensification—as well as the economic botany of cultivated plants—classification, selection, management, and use of crop species. Human ecology seeks also to comprehend the decisions of different types of households regarding cultivated plants—which crops and varieties to plant, the distribution of distinct crops and varieties within farms, villages, and regions, and the impact of social trends such as commercialization, new technology, or increased population.

This chapter examines the case of potatoes (*Solanum* spp.) in the Andes of Peru, their cradle area of domestication and evolution. The emphasis here is on Andean ethnobotany of potato diversity. The Andean knowledge system of potatoes far exceeds a simple catalog of varieties and includes terminology for the crop's agricultural and social ecology. However, before delving into potato terminology and ecology, it is important to consider the issues of folk classification of infraspecific variation.

Folk Classification of Crop Varieties

Infraspecific variation is the raw material of evolution, and for crop plants it is the staple used by crop scientists and farmers to improve yield or other crop qualities. To taxonomists, however, infraspecific diversity is a problem—biological noise that obscures the underlying order among species.

The logical starting place to study the ethnobotany of crop diversity is the variety of names that abound in particular regions. On the surface these names are indicative of crop diversity and elaborate knowledge systems. However, the mere listing of names is insufficient as evidence of either biological diversity or agricultural knowledge, and botanists and agricultural scientists have suspected that the plethora of variety names is a result of overclassification rather than a reflection of significant biological diversity or subtle understanding of crop ecology. Often, crop scientists have invested more in analyzing relationships between crop species than those between the varieties or cultivars of a single domesticated species. Likewise, both botanists and ethnobotanists have been averse to analyzing the infraspecific level where most crop diversity is found. Burtt (1970) advises that the best way to treat varietal diversity is to ignore it: "[T]he best thing to do with a muck-heap is to leave it undisturbed so that it quietly rots down. In course of time the *Code of Nomenclature* will no doubt accept it as disposable refuse" (Burtt 1970, 238). Ethnobiologists have avoided such outspoken disdain, but few have invested themselves in sorting through the complex and often chaotic domain of folk names for crop varieties. Berlin (1992) devotes more attention to the levels of folk taxonomy where crop diversity is found, noting that specific and subspecific classification reflect the cultural importance of plants and are associated with cultivation. Botanists interested in crop evolution are inclined toward the relationship between cultivated species and wild relatives rather than the variation within a single species. With a few important exceptions (e.g., Sanchez G. and Goodman 1992), there is a paucity of botanical studies of the distribution of diversity within crop species and its relation to crop evolution after domestication.

There are, however, good reasons for grappling with the ethnobotany of plant varieties, especially of crops. The infraspecific level is common to many ethnobotanical systems (Atran 1987, Berlin 1976, Dougherty 1978), and elaborate varietal classification is conspicuous in some folk systems. The variety level is a primary unit in the management of agricultural ecosystems around the world, and "cultural memory" of crops is most highly developed at the variety level (Nazarrea-Sandoval 1995). Variety identification becomes more significant with industrialization and implementation of intellectual property rights.

Several case studies reveal the merits of drawing a more explicit connection between ethnobotany and farmer decision making in agriculture (e.g., Johnson 1974, Richards 1985). These studies reflect Bulmer's (1974) point that one of ethnobiology's principal goals is to understand how the determination of biological species relates humans to the biological dimension, a point that has been reiterated in the recent emphasis on "indigenous knowledge systems" (Brokensha, Warren, and Werner 1980) and ethnobiology (Berlin 1992). Perhaps most importantly, infraspecific variation is the primary locus of crop adaptation and evolution. It is only natural that the scientific analysis of crop ecology and evolution must grapple with the messy variation within and among different populations of a single crop. However, only after several generations of crop research by biological and social scientists can infraspecific variation now be appreciated and analyzed.

Potatoes and Knowledge of Potatoes in the Andes

Isbell (1978) drew the apt analogy of a kaleidoscope to describe the Andean world—bright fragments that are tumbled into different configurations in the tumultuous landscape. Potatoes and potato husbandry are important parts of the Andean kaleidoscope: species and subspecies at different altitudes; recombination of colors and shapes among tubers, fields, and field patterns; men and women toiling with foot plows, clod breakers, and harvest tools; smoke trails from earthen ovens at harvest; tubers set out to freeze on the cloudless nights of the Andean winter or skewered on sticks to dry in the brilliant sun; steaming clay pots filled with potatoes; guinea pigs eager for the peels; the warmth of a full stomach.

Both ethnobotany and human ecology have influenced our understanding of the kaleidoscopic diversity of potatoes in the Andes. Ethnobotany has focused on the indigenous knowledge of Andean people, especially the identification and use of local crops and production zones, emphasizing the diversity of biological resources used (Franquemont et al. 1990, Gade 1975, Tapia Núñez and Flores Ochoa 1984). Human ecology in the Andes has been influenced by Murra's (1972) concept that Andean social units, from the household to the state, show a similar pattern of adaptation to their mountainous terrain in prehistoric, colonial, and contemporary eras. Murra's approach emphasizes complementarity in a vertical landscape as one of the organizing principles of Andean society. The complementarity principle refers to the control and use of ecologically distinct, spatially separated production zones by single ethnic groups. This idea was originally articulated by Murra (1972) as "verticality" and was used to describe the organization of prehistoric Andean economies and states. In the contemporary Andes, three variants of vertical integration have been described—community control of distinct production zones in single valleys, migration between locations at different altitudes, and market integration of production zones (Brush 1977). Thomas's (1973) work on energy flow showed that multiple zones were better able to provide sufficient energy than single zones, and Golte's (1980) research suggests that multiple zone use smoothes labor demand, thus making labor more efficient and productive than could be possible within a single zone. A prerequisite of complementary land use is an inventory of crops that are suited to its different physical conditions: soils, temperatures, moisture, and evapotranspiration regimes.

Gade (1975) found thirty-six species of Andean domesticates in the Vilcanota Valley of southern Peru (see Table 4.1). The single most important Andean cultigen is the potato, and diversity within this crop in the Andes is greater than that within any other crop there. Originally domesticated by Andean pastoralists from tuber-bearing members of the *Solanaceae* family, the potato and Andean culture have co-evolved in their mountainous homeland for at least six thousand years (Pickersgill and Heiser 1978). Potatoes are grown throughout most of the agricultural zones of the Andes, but they predominate in the upper zones, between three thousand and four thousand meters above sea level. In some areas they provide up to 70 percent of the calories in people's diets (Ferroni 1979). Seven different species among four polyploid groups are cultivated in the region (see Table 3.1). Some of these species (e.g., tetraploid, *Solanum tuberosum* ssp. *andigena*) are cosmopolitan, while others (e.g., *S. ajanhuiri*) are very localized in their distribution. Some five thousand morphologically distinct varieties have been identified out of more than thirteen thousand Andean accessions held at the International Potato Center (Huamán 1986). Over one hundred varieties may be found in a single valley, and a dozen or more distinct varieties are kept by a typical farming household.

Spanish dictionaries of native languages provide an entry point into the unrecorded world of indigenous Andean people. Bertonio's (1612) dictionary of the Aymara language mentions some of the many names applied to the vast diversity of potatoes found in the central Andes. Both terms and the organization of knowledge found in contemporary nomenclature are also evident in Bertonio's dictionary. While it is unlikely that similarly named varieties grown today are biologically the same as those named four hundred years ago, the persistence of varietal names for fifteen to twenty generations of farmers is remarkable.

Among contemporary scholars, Andean potato nomenclature was first described by La Barre (1947) for the Aymara of Bolivia, and folk taxonomies have since been described for Quechua- and Spanish-speaking peasants of Peru (Brunel 1976; Brush, Carney, and Huamán 1981; Zimmerer 1996). Following La Barre, recent descriptions of Andean folk classification find three or four taxonomic levels for potatoes: genus, species, variety, and subvariety. Three criteria are important in potato classification: ecology (cultivated/wild/weedy, production zone), use (edible, for boiling, for freeze-drying), plant and tuber phenotype, and degree of polytypy (number of subclasses). The sim-

Figure 5.1 Potato identification, Paucartambo, 1978

ilarities of potato nomenclature across languages and types of production systems are notable. Whether in the seventeenth or the twenty-first century, Andean potato farmers distinguish potatoes by tuber phenotype, ecology, and use (Figure 5.1).

Figure 5.2 presents a schematic diagram of a folk taxonomy of potatoes from the central and southern highlands of Peru. At the genus level, the folk system distinguishes *Solanum* tubers from other Andean tubers such as oca (*Oxalis tuberosa*) and mashua or añu (*Trapoleum tuberosum*). At the species level, domesticated, wild, and weedy types (*mikhuna papa, atoq papa, araq papa*) are demarcated, and frost-resistant, high-altitude, bitter types with high glycoalkaloid content (*haya papa*) are differentiated from mid-altitude types without bitter compounds (*miski papa*). Table 5.1 compares the different folk species according to ecology, use, phenotype, and polytypy.

Varieties are distinguished primarily according to tuber characteristics, such as tuber shape (oval, spherical, flat, long), the configuration of the tuber's "eyes" (depth, number, location, color), skin color and pattern (white to deep purple, solid color, multicolored), and flesh color and color pattern (solid,

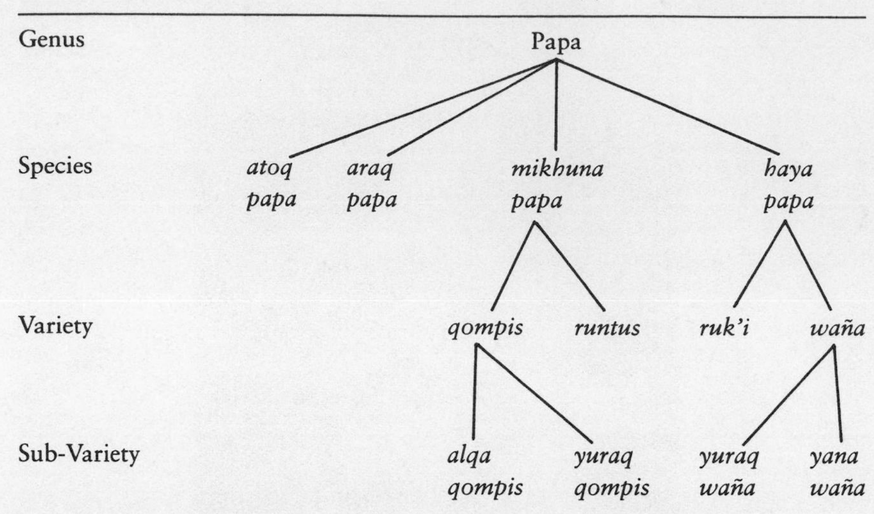

Figure 5.2 Schematic diagram of Andean taxonomy of potatoes (Cusco Quechua) (Brush 1992)

Table 5.1. Characteristics of folk species of Andean potatoes

Folk species	Ecology	Use	Phenotype	Polytypy
mikhuna papa	Broad adaptability; mid-altitudes (2500–3700 m)	Boiling; soups; frying	Nonbitter tubers; highly variable	Very high
haya papa	Frost resistant; high altitude (3,700–4,100 m)	Processing by freeze-drying; *chuño*	Bitter tubers	Low
araq papa	Weedy species; low–medium altitudes (2,500–3,200 m)	Boiling; soups	Nonbitter	Low
atoq papa	Wild species; all altitudes	Not used	Small tubers	None

Source: Brush 1992

ringed, white to deep purple) (see Figures 3.2–3.4). In rare cases, nontuber characteristics such as stem or flower color distinguish varieties. Tuber characteristics are qualitative and subject to somatic variation and environmental influences, and the relation between tuber characteristics and plant genotype is not well understood. Nevertheless, tuber characteristics are used by potato taxonomists as one of the means to determine biological similarity (Ochoa 1990). Because tuber characteristics are the criteria for conscious selection, they could also be partly linked to quantitative characteristics such as yield, adaptation, or disease resistance. This is suggested by the linkage between qualitative tuber characteristics and neutral and invisible traits such as isozyme variation (Quiros et al. 1990).

The final level of Andean potato taxonomy is the subvariety, where the only contrast is between tuber colors. Black (*yana*) and white (*yuraq*) subvarieties are frequent, and variegated skin color is labeled *alqa*. Skin coloration varies continuously and is transitory in some varieties, and the subvariety label is often understood to be unstable. In other cases, however, subvarieties are stable and biologically distinct. *Yuraq waña* and *yana waña* are the two stable variants of a single pentaploid species, *S.* × *curtilobum*.

Potatoes in the two lowest taxonomic levels of Andean folk taxonomy are grouped into named categories of intermediate ranking rather than into separate taxa. Table 5.2 presents a description of five intermediate ranks common in the central Andes. These intermediate ranks are labeled and cluster several varieties and subvarieties according to a single criteria, such as use or ecology. One grouping distinguishes potatoes with high water content (*unu papa* or *kal'wi papa*), suitable for soups or frying, from varieties with high dry matter (*wayk'u papa*), which are preferred for boiling or steaming. Farmers also contrast modern potato varieties that have been introduced since 1950 with local or "native" varieties. Modern varieties (Spanish: *papa mejorada*) are light skinned, white fleshed, smooth, and generally larger than local or "native" varieties, which are described as *chalo*. Chalo is used by both Quechua and Aymara speakers to describe mixed collections of potatoes with many colors and shapes. The word appears in Bertonio's (1612) Aymara dictionary (*cchalu*), suggesting that Andean farmers distinguished mixed collections long before the appearance of modern varieties.

The rich Andean nomenclature for potatoes is prima facie evidence for great diversity, and diversity at the species and infraspecific levels has been well documented for the Andes (e.g., Hawkes and Hjerting 1989). However, very little is known about the actual distribution of diversity either within or between regions or how diversity is affected by changes in agriculture. The measurement of genetic diversity and its distribution in Andean potato agriculture

Table 5.2. Intermediate folk categories at variety level in Andean potato classification

Category	Distinguishing criteria	Description	Contrast
wayk'u papa	Use	Dry/mealy potatoes for boiling or roasting	To *unu papa*
unu papa	Use	Watery potatoes for frying and soups	To *wayk'u papa*
k'usi papa	Use	Nonbitter tubers for freeze-drying (*chuño*)	To *wayk'u papa*
miska papa	Ecology	Fast growing; for short season (*maway tarpuy*)	To unnamed category for long season (*hatun tarpuy*)
chaqro/chalo	Phenotype	Mixed colors and shapes	To unnamed category for modern varieties (white potatoes)

Source: Brush 1992

is confounded by the complexity within the group of cultivated *Solanums* and by the great number of phenotypes and genotypes at the variety level. Somatic variation, introgression between cultivated and wild species, and hybridization within cultivated species also pose problems for measuring diversity.

While variety naming is a centerpiece of the Andean folk classification of potatoes, geneticists have long believed that, for two reasons, this system is not a reliable gauge of diversity. First, the folk system is based on tuber characteristics that are only partially relevant to the biological systematics of the crop. Second, Andean farmers are believed to overclassify diversity (Hawkes 1947), a practice that is exemplified by the use of several names in a single community for a single type of potato, and by the habit of changing names for such purposes as marketing. There is no evidence of a single, master list of names that farmers know or agree on, although they are aware of synonymy.

I became aware of the transitory nature of the farmer naming system one day in 1978 in the town of Pazos, south of the Mantaro Valley in Peru. I had been interviewing farmers about potato diversity and names. Most farmers, and especially the women of farm families, confidently named the different tubers in their household potato stores, but I also frequently encountered tubers unknown to the farm families. In our field notes and field mapping exercise these were dubbed "UPOs"—unidentified planted objects. Farmers were eager to help me sort out the complicated nomenclature, but they also showed some amusement at my concern to get just the right name for a specific tuber. One farmer finally decided that he should acknowledge my interest in names and in potatoes by bestowing the name "papa brush" on one of his tubers — a play on words to be sure. Within a few months, plant scientists from the International Potato Center station in nearby Huancayo began to find "papa brush" on other farms south of the Mantaro (P. Schmiediche, personal communication). Neither the other scientists nor I could determine whether "papa brush" was a name applied to a previously unnamed tuber or simply a new synonym for a potato known to farmers by a different name.

The individualistic, localized, and transitory nature of potato nomenclature would thus seem to limit names as a general tool for measuring diversity. Geneticists who work with the crop have preferred to work at the subspecies or species (ploidy) levels rather than at the variety level (Hawkes and Hjerting 1989). However, biochemical characterization of potatoes (Quiros et al. 1990) may help overcome some of the obstacles to biological assessment of diversity. These measures rely on isozymes and other neutral markers, including polymorphism in the plant genome — characteristics that are far less variable or environmentally determined than plant descriptors such as tuber shape. Quiros et al. (1990) found a clear correspondence between farmer segregation and identification of tubers and biochemical (isozyme) profiles of tubers that reflect genotype differences. The isozyme analysis is particularly relevant here, since one would not expect any degree of correspondence between a folk taxonomy largely based on one criteria (tuber characteristics), and biochemical identity based on characteristics that are invisible to Andean farmers.

Our study involved sampling ten fields in the province of Paucartambo, Peru — near the Inca capital, Cusco (Quiros et al. 1990). Two hundred plants in each field were randomly sampled and a single tuber taken from each plant. These were then bulked, and we asked the farmer to sort them into like-named varieties. When a single variety had more than thirty tubers, a representative sample of ten to twenty-five tubers was taken. The tubers from each variety group were assayed by starch gel electrophoresis for twelve loci known to be heterogeneous and useful in distinguishing between potato genotypes.

A comparison between farmer variety identification and electrophoresis identification resulted in several conclusions. First, a high degree of overall correspondence existed between farmer identification and electrophoresis phenotype. The most common incongruity between farmer and electrophoresis identification was a tendency by farmers to underestimate the actual biological diversity. This contrasts with the belief among many potato specialists that peasants overclassify their populations of native potatoes — finding more diversity than is indeed the case (Huamán 1986, Hawkes 1947). Most importantly, we must recognize that there is no absolute standard of identifying potato varieties, so comparisons between laboratory measures and farmers' identification are arbitrary. With this caveat in mind, we can see that farmers are generally accurate in recognizing types that can also be distinguished by isozyme analysis. There was, however, difference in the "accuracy" of farmers in recognizing biological diversity, as the number of varieties reported by the farmer and the number of genotypes found by laboratory analysis differed between 3 and 22 percent. Sambatti, Martins, and Ando (2001) also found inconsistency among farmers in their identification of cassava varieties in Brazil. In Peru, the "consistency" of names by a single farmer was higher than that among farmers (Quiros et al.1990). Most varieties found on different farms showed at least some congruence in electrophoresis profiles and names across farms, but this declined with the increase in distance between farms. For the thirty-two electrophoresis phenotypes found on more than one farm, eleven had a different name on each farm, while seven had the same name on all farms, and fourteen showed at least some congruence in names (Quiros et al. 1990). The data may understate the consistency between farmer name and biological identity, however, because they do not take synonymy among different names into account.

Human Ecology of Potato Agriculture

Isbell (1978) reports that new couples setting up household in Ayacucho, Peru, initially receive gifts of seed potatoes from their parents, after which husband and wife are generally on their own in selecting and maintaining varieties. This pattern pertains north and south of the Ayacucho area. Additional varieties are acquired through trade, purchase, gifts, and wages in kind. Women play an especially important role in the identification and selection of varieties, and women are involved in every stage of potato production: seed selection, production, harvest, storage, processing, and cooking. The key role of women has been described in Ecuador (Weismantel 1988), Peru (Allen 1988), and Bolivia (Johnsson 1986). Women's superiority in plant knowledge is acknowledged, and men often defer to women when questions arise about potato identification.

Both cosmopolitan and rare varieties are found in every village, and virtually every household cultivates cosmopolitan varieties, which often account for a significant portion of potato fields and area. At the other extreme are numerous rare varieties—a few plants cultivated by a few households. Cosmopolitan varieties are known to everyone—farmers, merchants, and consumers—and are widely traded. Well before the advent of scientifically improved potatoes promoted by agricultural development programs, Andean farmers had identified native varieties of unusual value, often because of taste but also for reasons of yield and adaptability. These varieties have large and robust markets—the *huayro* in central and southern Peru, the *qompis* in southern Peru, and the *imilla* in southern Peru and Bolivia. Approximately half of the total varieties in household inventories are these common varieties, but only a small percentage of a region's varieties are common (Zimmerer 1996).

Cosmopolitan varieties include both native and modern ones kept for different purposes—the native ones because of their culinary and commercial value, and the modern ones because of their yield and acceptance in the market. Improved and native commercial varieties are often grown as monocrops in single fields or as blocks within fields, and they may account for 70 to 90 percent of the potato area in many places (Mayer 1979). The prevalence of certain native and improved varieties means that most of the diversity can be maintained in only a small portion of the farm, where modern and selected varieties are not grown. This pattern is facilitated by the fragmentation of Andean landscapes, by complementary land use by single households and villages, and by the practice of cultivating numerous fields in the same year.

The ethnographic literature provides strong evidence that consumption is critically important in maintaining diversity. Virtually every study of potato selection refers to the importance of subtle yet elaborate contrast in taste, color, and texture of Andean tubers (Johns 1990; Johnsson 1986). Carter and Mamani (1982) note that certain varieties are prized for special meals, the most favored also being the most delicate and least productive. Brush (1977) describes certain varieties that are saved for gifts. Johnsson's (1986) study in Bolivia discusses the importance of potatoes and potato diversity to the cultural identity of the Aymara. His emphasis reflects Carter and Mamani's (1982, 98) account of the Mauca family's pride and prestige in possessing seed of many rare potato varieties. The contribution of potato diversity to Andean identity and prestige is echoed by Weismantel's (1988) study of Zumbagua, Ecuador. Potatoes are not a primary staple for most Zumbaguan families, but they retain prestige. Serving meals without European introductions, such as barley and fava beans, is a privilege of more affluent families. This contradicts the popular and Eurocentric notion that potatoes are always judged to be inferior food to cereals.

Intuitive logic asserts that diversity also exists because it is adaptive, leading to more stable production in the face of great environmental heterogeneity and abundant pests and pathogens. Brush (1977) and Carter and Mamani (1982) report that farmers recognize specific agronomic characteristics in certain varieties, such as resistance to disease or insects. While diversity may endow an adaptive or ecological advantage to subsistence farmers without other means to control disease or limit the effects of poor weather, this advantage is not particularly evident in potato names at the variety level. One exception occurs within the bitter species (haya papa), where more frost-resistant varieties (*ruki—S. juzepczuki*) are contrasted with less frost-resistant varieties (*waña—S. × curtilobum*). However, within the nonbitter folk species (miski papa), where diversity is greatest, there are exceptionally few widely shared names that refer to tubers with special resistance to insects, disease, or poor weather.

One-to-one relationships between particular varieties and environmental conditions such as soils, insect predators, or disease, have rarely been reported or evaluated (Hawkes and Hjerting 1989). Strong resistance to disease, insects, and drought is rare in domesticated potatoes, although some resistance is found within the predominant subspecies of the Andean region (*andigena*) (Hawkes and Hjerting 1989). While the ethnobiology of insects and crop diseases of the Andes has not been specifically studied, the observations of agronomists indicate a folk classification of six insect species that attack potatoes (Universidad Nacional San Cristobal de Huamanga 1983). Plant disease taxonomy is the least developed of the folk systems. Andean farmers gloss several diseases under the Spanish term *rancha* (late blight), but they do not recognize several major pathogens, such as nematodes and viruses. Knowledge of soils has been documented (Zimmerer 1994), but no single matrix seems to exist that maps potato varieties on to soil types.

The widely practiced sectoral rotation of fields (Orlove and Godoy 1986) is especially problematic to an ecological interpretation of diversity. In this system, potato fields are used only for one or two years before different crops and fallow are introduced. The ideal cycle is to return to a particular field once every seven years. This practice results in the yearly movement of the potato crop between fields in different parts of the community's territory. If ecological advantage of potato diversity results from fitting genotype to location, then this advantage would seem to be eliminated by the practice of frequent field rotation and the practice of growing diverse collections of potatoes together, instead of placing each variety in its special niche and keeping it there. Andean farmers do not emphasize site-specific adaptation in their nomenclature or management of potatoes, and agronomic trials suggest that most individual potato varieties perform equally well over a broad range of altitudes (Zimmerer 1998). While the contrast at the folk species level between nonbitter, mid-altitude types (miski papa) and bitter, high-altitude types (haya papa) is salient on several axes (production zone, relative hardiness, processing, and consumption), the contrast between varieties is based primarily on tuber phenotype. Potato farmers report that the most diverse native varieties are also the least resistant to disease, insects, and the effects of poor weather.

A diverse collection of potatoes may perform better than one without diversity because of an "interaction effect" in disease resistance (Frankel, Brown, and Burdon 1995; Zhu et al. 2000). Nevertheless, Andean farmers often concentrate their most diverse potatoes (chalo) into one small field and plant the majority of fields in selected varieties of both native and crop breeding origin. Figure 5.3 is a map of the diversity of a chalo field in Chinchero above the Vilcanota Valley north of Cusco. If an interaction effect exists, it would be observed only in one of several fields. In addition, diverse collections may be advantageous in the extensive field rotation system. Diversity may help ensure a certain productivity over time as farmers move their collection over the heterogeneous landscape and plant them at different altitudes, with different soils, moisture, and exposure to wind and sun. By this logic, a few varieties might be better adapted to a particular location, and because a farmer expects to move his potatoes each year, he/she does not force selection to limit diversity. Rather, diversity might guarantee a certain minimum yield because of the likelihood that some of the varieties in a collection will be adapted to the particular field that happens to be in use. Again, we must reckon with the fact that diversity is not found in every field or even in most fields. A diverse inventory may be seen merely as a pool of seed tubers that provide farmers with options for new varieties or as a pool to select new seed for less diverse fields.

Finally, diversity may have a very long-term advantage that is not specifically recognized in Andean folk biology or immediately apparent in the short term. Long-term stability of potato production may be enhanced by having a large repertoire of genotypes, some of which have particular advantage as environmental conditions, pests, and pathogens change over time. Under different conditions, varieties that are now rare may become advantageous, and thus prevalent. A large repertoire may seem superfluous in the short run, but it allows farmers to adjust to new conditions, including market demand.

Loss of Diversity

Andean agriculture has never been a static system, and cultivators have long been able to accommodate new technology such as European crops and animals. However, the pace of change appears to have accelerated since 1900. Market penetration, migration, population growth, political reform, and new technology are now ubiquitous. Integration of local communities into larger

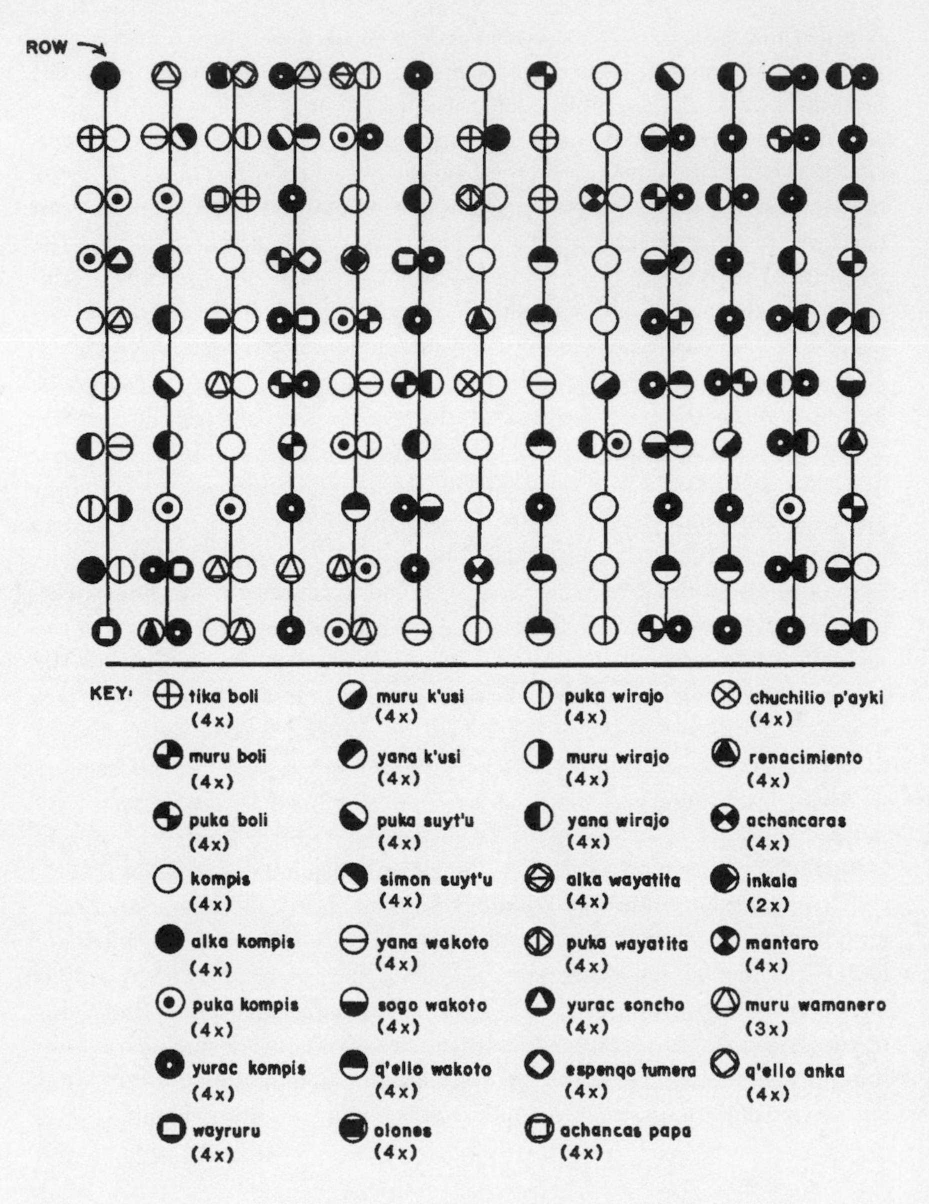

Figure 5.3 Quadrat of native varieties of potato in field near Chinchero, Cusco Dept., 3,820 meters above sea level (Brush, Carney, and Huamán 1981) (Reprinted with permission from *Economic Botany.* © 1981 The New York Botanical Garden)

political and economic systems has been present since pre-Hispanic times, but with the spread of capitalism this integration changed both qualitatively and quantitatively in the twentieth century through increased population, expansion of state power, roads, and mass communication. The Peruvian population nearly tripled between 1950 and 1990, from 7.6 million to 21.9 million (Urban and Trueblood 1990). Rural areas have experienced far less growth because of immigration to urban areas, but demographic pressures are felt everywhere, as market systems have expanded and rural hinterlands have been more closely incorporated into national economies. The presence of centralized state power has increased through such factors as agrarian reform, education, and regional development. Andean production is now characterized as much by commoditization and the acquisition of new technology as by complementarity and community regulation. Virtually every household now uses not only Old World crops and animals but also agricultural chemicals and modern Andean crop varieties.

Modern potato varieties were first released in Peru in the early 1950s, and they are now found in virtually every village in the highlands. These varieties were specifically bred to be higher yielding, better able to utilize fertilizer, and resistant to specific stresses such as disease or drought. Their adoption is directly encouraged by agricultural extension and credit policies of the government and indirectly by such factors as population increase or a farmer's wish to produce a larger surplus for the market. Two impacts of the diffusion of modern crop varieties in the Andes have been reported: increased productivity (Horton 1984) and loss of potato diversity (Ochoa 1975).

The concept of genetic erosion in farming systems is based on somewhat simplistic biogeography (see Chapters 7 and 8). Adopting modern crop varieties decreases the area that is planted to the traditional and more diverse varieties. Shrinking the area devoted to native crops should logically reduce diversity, just as the size of islands is directly related to biological diversity (McArthur and Wilson 1967). The basic flaws with this logic are to assume that the relationship between diversity and area remains unchanged and that farmers don't adjust their management of traditional crops to compensate for lost area. The biogeographic view of genetic erosion does not account for cultural, economic, and environmental buffers in agricultural systems that might protect diversity. Environmental heterogeneity, agronomic risk, market factors, and cultural factors are likely to limit the substitution of one or two varieties for the dozens that have evolved locally. The remainder of this chapter will examine whether this conservation has occurred in Andean potato agriculture in two highland valleys of Peru.

TWO VALLEYS

The impact on traditional crop diversity of the adoption of new varieties and agricultural intensification ideally should be studied in a historic framework. By all accounts, modern potatoes spread rapidly throughout the highlands after 1950 (Ochoa 1975). Unfortunately, we have neither biological nor socioeconomic benchmarks from a period before the diffusion of these varieties. The oldest systematic and preserved collections of native potatoes date only to the early 1970s, and our information on agricultural practices before 1950 is scanty and superficial. Without these historic benchmarks, comparison between regions, villages, and households is a valuable way to estimate the impact on diversity of such factors as commercialization of agriculture or the adoption of modern potato varieties. Comparison between two valleys in eastern Peru illustrates this research approach. These valleys were chosen because of their similarities and differences.

The Tulumayo Valley, fifty kilometers east of Huancayo in central Peru, is representative of areas of great potato diversity that have undergone extensive modernization and commercialization. The Paucartambo Valley, 50 kilometers east of Cusco in southern Peru, is representative of diverse farming systems that have experienced less modernization and commercialization. The Tulumayo and Paucartambo valleys are like many others along the eastern Andean escarpment. They share the traditions of complementary land use and community control mentioned above. Rolling upland pastures descend into steep and narrow valleys that have been landscaped into terracelike fields by generations of farmers. In each valley, peasant communities or corporate villages control agropastoral production over a large altitude range with different production zones. Each valley has some communities that existed independently and others that were part of haciendas before the agrarian reform of 1969–70. The community provides the modern framework of complementary land use, but households make most of the in-field decisions, such as crop and variety. Production is destined both for home consumption and for the market, and each valley is four hours from a major urban center, Huancayo or Cusco, by an all-weather road.

The Tulumayo and Paucartambo valleys are alike in their emphasis on potato production as well as in the organization of potato production. At the time of the study, potatoes were the predominant crop, accounting for 54 percent of the cultivated land in Tulumayo and 47 percent in Paucartambo. In both valleys, households cultivated numerous small plots in different production zones differentiated according to altitude, crops, agricultural calendar, intensity of land use, and degree of community control. Each valley produced both bitter and nonbitter varieties; each divided mid-altitude production into a lower, short cycle (maway tarpuy) and a higher, long cycle (hatun tarpuy); each relied on sectoral fallow and the simultaneous cultivation of different plots by the same household. Finally, both the Tulumayo and the Paucartambo valleys share regional fame as places where particularly high levels of diversity are found. These two valleys show a pattern that is familiar to agricultural systems in cradle areas of crop evolution and diversity. Small islands of traditional agriculture remain in a sea of more uniform, commercial agriculture based on the use of modern crop varieties and high energy inputs.

The major differences between the two valleys have to do with ethnicity, degree of commercialization of potato agriculture, and the history of adoption of modern potato varieties. A generation ago, the people of the Tulumayo were Quechua speakers, but today Spanish is the most widely spoken language. Labor migration from this valley to the mining industry in central Peru was a significant force in shaping local cultural identity during the twentieth century (Long and Roberts 1984; Mallon 1983). Tulumayo people identify themselves as mestizos, and they are explicit about the cultural differences between themselves and Quechua-speaking people who live south of the Mantaro Valley. Quechua is the predominant language of Paucartambo, where ethnic identity is Indian (Allen 1988). Paucartambo has not experienced such a singular integrating force as massive labor migration to mines. Its integration into the regional economy of southern Peru has been through the hacienda system and through periodic migration to Cusco and to commercial farms in the lowlands. This integration has not produced the fundamental shift in ethnic identity experienced in the central highlands, although commercial potato production by haciendas around Cusco has certainly favored particular local varieties, as documented in Figure 5.4.

Impact of Markets and Agricultural Modernization

A comparison between the two valleys and three types of potatoes (improved, native commercial, and mixed native—chalo) reveals a number of predictable contrasts but some intriguing surprises. Table 5.3 shows the distribution of different potato types according to the weight of seed as reported by farmers in the survey. The important contrast here is between improved and mixed types in the two production zones. Predictably, improved varieties were grown primarily in the lower zones, which are farmed more intensively throughout the central Andes for commercial purposes (e.g., Mayer and Fonseca 1979). Native types, especially mixed varieties, were concentrated in the high zone, where agriculture is less intensive and community control is more direct, but a striking contrast between the two valleys is the location of native

Figure 5.4 Potato harvest, Cusco, Peru, ca. 1935 (Martin Chambi, reprinted with permission of Teo Allain Chambi)

commercial varieties. Native commercial types were less important in Paucartambo than in Tulumayo. In the Tulumayo Valley, these were primarily grown in the high zone (hatun tarpuy) while in Paucartambo, they were grown mostly in the lower zone (maway tarpuy). In both valleys, the lower zone was dedicated to commercial production of the high-yielding improved and high-value native commercial varieties to take advantage of high, off-season prices. The difference in location of native commercial varieties reflected the fact that Paucartambo farmers plant another commercial crop, barley, in the upper zone (Zimmerer 1996).

Table 5.4 presents data on the percentage of potato area planted by type and on the use of the entire potato crop for sale and consumption for each valley. These figures refer to the total seed planted and harvest weight of potatoes as estimated by farmers. Eighty-nine percent of the Tulumayo Valley's potato area was planted to the more commercial types, in comparison with only 39 percent of the Paucartambo Valley. Improved types constituted the bulk of potatoes sold in each valley. In Paucartambo the consumption of mixed native potatoes was considerably higher, reflecting their larger production area.

Table 5.5 presents data on potato use (sold, consumed, saved for seed) of each of the three types of potatoes. These data show that improved potatoes are grown primarily for sale in both valleys, but they also reveal that relatively

Table 5.3. Potato distribution in the Tulumayo and Paucartambo valleys (%)

	Tulumayo Valley (*n* = 154)			Paucartambo Valley (*n* = 204)		
	Improved	Native commercial	Native mixed	Improved	Native commercial	Native mixed
Hatun Tarpuy (Long-season zone)	23	75	89	28	10	98
Maway Tarpuy (Short-season zone)	77	25	11	72	90	2

Source: Brush 1992

Table 5.4. Area and use of different potatoes in Tulumayo and Paucartambo valleys

	Tulumayo Valley (*n* = 154)			Paucartambo Valley (*n* = 204)		
	Improved	Native commercial	Native mixed	Improved	Native commercial	Native mixed
% of potato area	59	30	11	31	8	61
% of all potatoes sold	78	20	2	62	7	31
% of all potatoes consumed	41	37	22	30	8	62

Source: Brush 1992

high percentages of the mixed native varieties were also sold. The data presented in these three tables suggest that variation in use was continuous rather than discrete. Farmers in both valleys grew different varieties and mixes of varieties in separate fields, but the production from all fields was used for both consumption and sale. A comparison of Tables 5.4 and 5.5 indicates that no simple division can be made between production for sale of improved potatoes versus production for use of mixed native potatoes. The consumption of native potatoes was higher in Paucartambo, but so was their sale, since modern potatoes represented a smaller proportion of all potatoes produced.

The variation of diversity, management level, and commercialization between fields of different varieties and combinations of varieties was continuous. Tulumayo and Paucartambo farmers did not create, conceive of, or manage fields according to a matrix of discrete types: low input or high input, commercial or subsistence. The continuous gradation of management, selec-

Table 5.5. Use by potato type in Tulumayo and Paucartambo valleys

	Tulumayo Valley (*n* = 154)			Paucartambo Valley (*n* = 204)		
	Improved	Native commercial	Native mixed	Improved	Native commercial	Native mixed
% of type sold	80	62	45	64	50	25
% of type consumed	9	26	43	20	39	55
% of type saved for seed	11	12	12	16	11	20

Source: Brush 1992

tion, and use allowed ample opportunity for different mixes of local and outside inputs, enabling the conservation of traditional varieties and production technology throughout the system. Matrices of discrete field types (e.g., low input, high input, commercial, subsistence) reflect the views of outsiders such as advocates of "traditional" agriculture or those who disparage it in favor of "modern" production. In terms of managing the kaleidoscopic Andean landscape, these neat categories make little sense to Andean farmers.

The social context of Andean agriculture has changed fundamentally and recurrently since the European conquest: domination and exploitation based on ethnically defined castes, incorporation of local communities into larger regional systems (especially markets), demographic growth, development of economic and technological infrastructure, and political and social restructuring of Andean society through land reform. It has been common to assume that such changes bring about the rapid decline of diversity, as "traditional" agriculture is replaced by a "modern" system (Hawkes 1983).

Improved potatoes came to the Tulumayo Valley much sooner than to Paucartambo. At the time of the research (1985) Tulumayo farmers averaged almost twice as many years (13.59) as their Paucartambo counterparts (7.7) in producing modern varieties, and this longer period of adoption was reflected in the higher percentage of potato area planted with improved varieties in the Tulumayo Valley. Tulumayo suggests the path of agricultural change that Paucartambo may follow, as measured by agricultural intensification, commercialization, and adoption of new technology. If we accept the idea that the diversity of traditional crops will be adversely affected by the increased use of modern technology and increased integration into the market, then a comparison between the two valleys may give us some idea about the fate of native potatoes.

We may expect crop diversity to decrease as commercial and more intensive

Table 5.6. Technology adoption and potato diversity

	Tulumayo Valley	Paucartambo Valley
Average number of years using improved varieties[a]	13.9	7.8
Year of first introduction of improved varieties[a]	1950	1960
Farmers who have planted improved varieties (%)[b]	95.8	79.0
Cumulative number of improved varieties planted since introduction[b]	16	7
Farms using purchased fertilizer (%)[b]	97.9	85.8
Farms using pesticides (%)[b]	95.7	77.2
Average number of native varieties per farm[a]	12.8	9.6

[a]*n* = 87 for Tulumayo and 85 for Paucatambo
[b]*n* = 154 for Tulumayo and 204 for Paucartambo

Source: Brush 1992

agriculture relying on modern potato varieties take hold over a larger and larger portion of the two valleys. Table 5.5 shows that fields of mixed native varieties represented only 11 percent of the Tulumayo's potato area. These fields are small islands of diversity, surrounded by more biologically uniform fields. The size of these islands of traditional, mixed potatoes is critical to conserving diversity. In Paucartambo, mixed native fields constitute 61 percent of the potato area. Tulumayo should have less diversity than the southern valley. However, the average number of varieties per household in the Tulumayo Valley is 12.8, compared with 9.6 varieties per household in Paucartambo. Table 5.6 sets this comparison in the context of adoption of improved varieties, illustrating the idea that the farmers of Paucartambo are at an earlier stage in the adoption process of improved potato varieties. It also suggests that diversity of native potatoes remains, even after adoption becomes virtually complete, as long as some area is planted to native potatoes.

Higher potato diversity in the Tulumayo Valley, despite the loss of land to modern varieties, may reflect a history of greater diversity than Paucartambo before the introduction of modern varieties. However, the southern valley is regarded by many potato biologists to be within the region of greatest diversity in Peru (Hawkes 1983). Careful comparison between the two valleys (Brush, Taylor, and Bellon 1992) suggests a striking contrast. Paucartambo is in an earlier phase of adopting modern potatoes and clearly shows a negative impact of adoption on diversity. We used analysis of variance to compare diversity, measured by named varieties in a household's inventory, with socioeconomic factors and the use of different technologies (Brush, Taylor, and Bellon 1992). Our analysis indicated that when modern varieties replace traditional varieties on one hectare of the village's land, 5 traditional potato names drop out of the original inventory of 227 names in the village. The actual loss of diversity cannot be determined because of the ambiguous relation between names and biologically distinct varieties. The Tulumayo, on the other hand, is in a later phase of adopting modern potatoes and shows no loss of diversity, as more area is devoted to these types. The Tulumayo case suggests that modernization has not eliminated diversity because farmers adjust their management to maintain diversity on the small area that is not planted to modern types. It is possible that diversity was lost in the Tulumayo during the initial phases of modernization, but this loss does not necessarily proceed as modernization continues.

The Resilience of Diversity in Andean Agriculture

The loss of diversity from agriculture is neither simply described nor linear. Anthropologists argue that social reproduction in the Andes is a syncretic process whereby local and exogenous elements are continually combined (Allen 1988). Likewise, agricultural change in the Andes is not a dichotomous process of the replacement of older technology, but one whereby indigenous and imported technologies are combined into a single mosaic. Thus, fields of modern potato varieties are managed within the sectoral fallow system, and fields of mixed native potato varieties are rotated with European crops, e.g., barley and fava beans. Potato production in these two Andean valleys is not a dual system of production for use with native technology and production for sale with modern technology. Virtually every potato field has elements of both autochthonous and exotic technology, and production of all types of potatoes is used both for consumption and for sale.

Table 5.7 outlines the major factors that either favor or discourage native and improved potatoes in relation to consumption, commercialization, and production. This table shows the complexity of determining the advantages and disadvantages of producing either type. The framework is similar to Bellon's (1996) analysis of farmers' concerns (Table 2.2), but the substance of each case is determined by the crop, cultural identity, and environment. It also shows that there is no single axis on which to select the two types. Cultural identity, culinary quality, risk, yield, and commercial demand interact in the decision. Andean farmers are accustomed to wrestling with complexity of this nature, and they long ago learned that production and use decisions are not simple dichotomies.

Table 5.7. Selection criteria for native and improved potato varieties

	Factors favoring selection		Factors discouraging selection	
	Native varieties	Improved varieties	Native varieties	Improved varieties
Consumption factors	Good tasting; valued for gifts	Lower unit cost	Higher unit costs	Inferior taste; less suitable for usual cuisine; larger tubers require more fuel
Commercial factors	Higher market value; high exchange value	More profitable (good benefit/ cost ratio)	Low yield under traditional management	Low market value; limited local market
Production factors	Don't need new seed; seed readily available	Good short-term resistance to specific risks	Less resistance to specific risks	New seed required

Source: Brush 1992

Native potatoes are universally acknowledged to be culinary superior to modern varieties. This is not a significant criteria in maize selection in Chiapas (Bellon 1991) or in the central highlands around Mexico City (Perales R. 1998). The first measure in judging a potato's taste is how a particular variety tastes when cooked in the watia, a simple oven of rocks or sod constructed in the field at harvest time (see Figure 4.5). Varieties are also evaluated by how well they taste after boiling or steaming. Native potatoes are preferred as a class because they are drier than the improved varieties, which are thought of as insipid and watery (unu papa) and suitable for frying or for soups but not for standard boiling. Given the confusion and instability of farmer names for varieties, no single native variety represents a culinary paragon for other potato varieties. Native potatoes lack the standardization that accompanies commercial seed or a strongly regulated market. There is, therefore, no indigenous Andean equivalent of the certified North American "Russet Burbank" or European crops grown under sanctioned appellations. The Andean varieties that come closest to this status are the cosmopolitan, native commercial ones that are found in virtually every household, *huayro* in the central highlands or qompis in the south. I have been frequently told by informants that rare native varieties are equal or superior to these commercial native varieties.

Besides taste, cultural prestige also contributes to maintaining diversity and native crops. Ritual meals and celebrations and meals for guests emphasize native potatoes. Weismantel (1988) writes that potatoes occupy a primary place in the system of culinary signs and metaphors that comprise Quichua identity in Ecuador. She observes that "White" guests are served meals in which potatoes are minor complements to chicken and rice, while "Indian" guests are served guinea pig and potatoes. Native varieties are favored gift items and used to strengthen social ties, and some reports refer to them as "gift potatoes" (Spanish: *papas de regalo*) (Mayer 1979). Native potatoes are often expected as part of wages and during reciprocal labor exchange when meals are served. In the extremely tight labor market of Andean agriculture, the offer of *wayk'u papa* as partial payment is a good way to guarantee a supply of workers at critical times. These potatoes likewise are attractive to distant trading partners who bring meat and wool from the *puna* to exchange for potatoes.

There is a large and active regional and urban market for all varieties of potatoes, but the internal market for potatoes within the valleys is small and not well developed. Individual households do not trade, barter, or sell significant amounts of potatoes with their neighbors. Farmers speak of the desire to be self-sufficient in the different kinds of potatoes. Thus keeping mixed native potatoes is seen as an option that is preferable to specializing in one type and relying on a market to supply mixed types.

Although judged to be culinary superiors, native varieties are perceived as agronomic inferiors compared with the modern ones. Native varieties are lower yielding and more susceptible to the major diseases and environmental risks affecting potato production. Table 5.4 shows that in both the Tulumayo and Paucartambo valleys improved potatoes are concentrated in the lower, maway zone. This zone is less subject to frost than the higher zone, but it is more susceptible to some of the most severe threats in Andean potato production: aphids, viruses, drought, and especially late blight (*Phythphtora infestans*).

Modern potato varieties depend on regular supplies of fresh seed tubers, since farmers change the seed for these varieties after two or three years. The seed for native varieties is kept for many years and is renewed by rotation between fields at different altitudes. Seventy-two percent of the Tulumayo farmers and 79 percent of the Paucartambo farmers in the sample reported that they never changed native potato seed. Nevertheless, the fact that 20 percent or more of the farmers in each valley have changed their seed of native potatoes supports that idea that these crop populations are "open systems" (Zimmerer 1998) and thus difficult candidates for assessing genetic erosion (see Chapters 7 and 8). Sixteen percent of Tulumayo farmers and 49 percent of Paucartambo farmers said they never changed improved potato seed. Seed renovation in modern varieties is motivated by the wish to increase commercial yield with disease-free stock, a pressure that is less noticeable in native potatoes used at home. Larger, more commercial farms or farmers with more capital may thus be better able to plant modern varieties than capital-poor farmers. The seed of native potatoes is available from chalo fields at higher altitudes, where disease in less prevalent. In Cusco, a small seed industry has arisen to provide virus-free seed of the commercial qompis variety.

WHY SO MUCH DIVERSITY?

While Table 5.7 and the preceding discussion address the persistence of native potatoes, the question of why so much diversity remains unanswered. This puzzle cannot be solved by direct inquiry, because the question "Why do you grow so many types of potatoes?" is silly and nonsensical to Andean farmers. Informants were surprised and baffled when asked. From their point of view, diversity is natural and a given of the Andean kaleidoscope rather than something strange or unusual to be explained. They manage one of the most heterogeneous and complex agroecosystems in the world, and diversity within a single crop and field is a logical corollary of the variety of the world around them. From this perspective, the question of diversity may be asked only indirectly by examining why farmers don't eliminate diversity in favor of a single type of potato, improved or native. This question is less absurd to Andean farmers, but it asks them to speculate about something which they don't usually do.

Farmers in the Tulumayo and Paucartambo valleys have not eliminated diversity because they don't perceive any advantage to doing it and because there are advantages to keeping their mixed collections of native potatoes in a subsistence economy. Improved varieties don't taste very good, so they are not likely candidates to replace native varieties, and no single native variety meets everyone's criteria for best taste or best for exchange, gifts, or sale. Diversity is a pleasure in its own right when sitting down before a bowl of potatoes as the primary food at a meal. It is common for people to eat fifteen or more potatoes in a single meal, an occasion that is much more interesting if every other potato is a different variety. Diversity is akin to a condiment, like hot peppers, making meals more interesting. Naming often provides for word games that enliven meals. Some names are clearly evoked by the tuber's characteristics: pink, flat, and oval (cow's tongue, *wacapahallum*), cylindrical with eyes clustered at one end (cat's nose, *mishpasingu*), or a mottled, rounded oval (condor egg: *condor runtu*). Other names evoke places, perhaps where the variety originated (e.g., *Curimarca*). This nomenclature is rich in Andean wit, irony, and iconoclasm. *Lumchipamundana* "makes the young bride weep" because of its knobby form (see Figure 3.4). We find such folk varieties as "priest's ear" (*kurapalingling*), as seen through the confessional screen; another is "dog's vomit" or "dog's stomach" (*alcapapanzan*) in Quechua, glossed to "Peruvian flag" in Spanish.

When queried about replacing mixed native varieties with improved higher-yielding ones or selected native types, farmers point to two things that encourage diversity. First, there is no need to make such a simplifying replacement in the diverse Andean landscape. They cultivate potatoes in several fields each year, and in an Andean variant of agricultural involution (Geertz 1963), they always find space and time for a few native potatoes. Farmers of the Tulumayo Valley have reduced this to a very small portion of their fields, but they maintain a high amount of diversity on this parcel. Enough potatoes are produced to satisfy local needs, and the market is often saturated with potatoes during the main harvest. Farmers complain that they lose money when selling any category of potatoes (Mayer and Glave 2002), so that giving up prized native potatoes to produce more low-value modern potatoes may seem to be a treadmill to avoid. Thus, there is little incentive to eliminate the small fields of native potatoes that remain.

Second, the collections of mixed native potatoes are perceived to be a resource among farmers who are economically marginalized and disadvantaged. Mixed native potatoes are associated with traditional Andean agriculture and culture by all types of farmers (subsistence, commercial, Quechua, and mestizo) and urban consumers. Native potatoes are grown in the most marginal areas and on small farms. Their value, like those of other material elements of Andean culture, has been inverted in the Andean kaleidoscope (Isbell 1978) that at once depreciates and values items of traditional culture. Products of a humiliated group, native potatoes command premium prices in regional markets, such as Cusco and Huancayo, where they have been elevated to the status of an artisan crop. Within farming communities, native potatoes are also appreciated, perhaps as much for their cultural significance as for their superior flavor. They are favored gift items, and, in a rural economy that is increasingly short of labor, they are used as added incentives by landowners to attract workers.

Although mixed native potatoes are found in every Andean market, they are often a losing proposition despite their higher value. Modern varieties easily saturate the market, and, because native types fetch higher prices, producing them offers a way to spread the economic risk of poor prices of modern varieties. Nevertheless, low yields of native potatoes lead to dismal economics. In the village, native potatoes have been described as gift potatoes (papas de regalo) (Mayer and Glave 2002), but once in the market they become potatoes that in effect are given away (*papas regaladas*) because they are sold at a loss (E. Mayer, personal communication).

Merchants who purchase native potatoes have a narrow concept of diversity and prefer the one or two varieties that have won widespread appeal and recognition. However, the flow of diverse native potatoes to urban markets is sufficient to bring new native varieties to the attention of consumers and merchants alike. Periodic "booms" in demand for specific varieties are sufficiently common so as to be an incentive for farmers to keep diversity as a source of seed. The spread of the huayro variety from the central highlands to the Cusco region and the local appearance of *olones* (Franquemont et al. 1990) fit this pattern. The difficulty in multiplying seed rapidly and the ambiguity in folk taxonomy are reasons to prefer keeping one's own inventory of varieties rather than relying on exchange or the marketplace.

Summary

This chapter has explored the persistence of diversity in Andean potato agriculture. The ethnobiology of the potato crop emphasizes diversity at the infraspecific or variety level. Andean categories such as production zones (maway tarpuy and hatun tarpuy) and types of potato (dry—miski papa; watery—unu papa; boiling—wayk'u papa) all contribute to this emphasis. This case study is representative of several others on the maintenance of traditional crops in centers of agricultural origins in the face of economic and technological change in agriculture (Brush 1995). These studies document the resilience of traditional crops, like the cultures that have produced and nurtured them.

We can imagine two alternative futures for traditional Andean potatoes from the above analysis. On the one hand, there might be a gradual encroachment of improved and uniform native varieties under the inexorable pressures of population growth and incorporation into regional market systems. The impact of this encroachment would be to shrink the area devoted to mixed native varieties. This impact is evident in the comparison between the Tulumayo and Paucartambo valleys. The small area of mixed native potatoes with its tremendous diversity may be the last remnant of a waning agricultural system whose nemesis is already present. Ultimately, the area of native potatoes might shrink to nothing, thus completing the biological transformation of Andean agriculture that began with the European conquest five hundred years ago. Assuming that market incorporation, demographic growth, and technological innovation will continue and increase, the replacement hypothesis is plausible.

On the other hand, the continued presence of traditional potato area and diversity may be interpreted as biological evidence of the tenacity of Andean cultural elements in the technological polyculture that has existed since the European conquest. The persistence of diversity in the Tulumayo Valley, at an even higher level per household than in Paucartambo, suggests that the Andean tradition of diversity will survive. Though the disappearance of traditional crops and of diversity has been predicted by theorists of very different persuasions (e.g., Hawkes 1983; Fowler and Mooney 1990), the eclipse of diversity is confounded by the complexity of the tropical world and by the actions of the inheritors of ancient farming traditions. Native crops endure even in modern nations, and the persistence of crop diversity echoes the survival of Latin American peasantry in the face of major structural change (de Janvry, Sadoulet, and Young 1989; de Janvry, Sadoulet, and Gordillo de Anda 1995). Numerous factors dampen the predicted erosion of traditional potatoes in the two valleys described here. While adoption of modern varieties, market penetration, and population increase are significant, so too are cultural, economic, and environmental factors that buffer their impact. There is no single axis on which to chart the fate of these genetic resources as farming systems change. What seems to be predictable is that farmers will continue to be active agents in reconfiguring the material base of their Andean agricultural legacy.

6

The Farmer's Place in Crop Evolution: Selection and Management

Two broad periods divide the modern inquiry about crop evolution. The first period lasted a century, from the mid-nineteenth century, when Darwin cataloged the diversity of domesticate crops and animals and De Candolle sought to identify regions of crop origins, to the mid-1960s, when the success of the Green Revolution signaled the importance of crop genetic resources and darkened their future in centers of diversity. At the midpoint of the first period, around 1920, the beginning of genetics and concerted crop improvement provided a backdrop to inquiries about crop evolution. Two general themes dominated this initial period: (1) where, how, and why did domestication occur? and (2) how did domestication affect the species that became crops and domesticated animals? An expansive literature in anthropology, geography, genetics, and plant sciences exists for these two issues (e.g., Vavilov 1951, Ucko and Dimbleby 1969, Harlan 1992, Rindos 1984, Harris and Hillman 1989, Zohary and Hopf 2000).

The second period of crop evolution research was heralded in approximately 1970 by a series of conferences that addressed the loss of crop genetic resources, a loss resulting from the successful deployment of genetics and crop breeding around the world. During the second period, research on crop evolution has focused on conservation of genetic resources. As part of the conservation effort, researchers have studied the habitats of crop evolution and the crop diversity that exists in these habitats. Crop evolution research has steadily become more ecological and interdisciplinary, involving both biologists and social scientists. We now recognize that crop evolution itself, like organisms that have been domesticated, is subject to social forces. Research today addresses the effects of modern conditions such as population growth and technology on crop diversity as well as on crop evolution itself. Studies on crop evolution have coalesced around the need to understand contemporary farming systems that generate and maintain crop genetic diversity. Such understanding is essential to improving the use of genetic resources in agriculture and conserving genetic resources that have resulted from crop evolution.

The Structure of Crop Evolution

Our understanding of crop evolution is directly descended from the work of Darwin. Following the same pattern as general evolution, crop evolution embodies processes that differentiate and order life — variation and selection. In *On the Origin of Species* (1859), Darwin relied on knowledge of variation in domesticated species and selection by farmers and animal breeders, and he used this knowledge to describe the processes of general evolution. The 1868 publication of *The Variation of Animals and Plants Under Domestication* (1896)brought the importance of domestication and conscious manipulation of variation to the foreground of evolutionary theory. Although Darwin's idea of "pangenesis" put forward in *Variation* was Neo-Lamarckian and mistaken, his description of crop evolution has survived as the standard theoretical framework.

As in his framework for general evolution in the *Origin of Species,* Darwin's emphasis in the *Variation of Animals and Plants Under Domestication* was on the ordering of variation through selection rather than on the cause of variation. Variation is the product of a handful of processes affecting the composition of genes — mutation, recombination during reproduction, genetic drift, and gene flow among populations. These processes are unchanged by domestication, and during the vast majority of crop evolution since domestication farmers have done little to affect or manipulate the natural processes that produce variation. Since the rise of crop genetics and directed crop improvement, however, natural processes of variation, such as mutation and gene flow, have been accelerated and purposively manipulated. Very recently, gene flow has been expanded and broadened by the use of transgenic or recombinant DNA technology to move gene sequences between species or even between phyla. While knowledge of the genetic basis of inheritance has greatly augmented our understanding of the nature of variation and has complemented

Darwin's framework for evolution, it has not fundamentally challenged the Darwinian vision of evolution.

The novelty of the evolution of domesticated plants and animals is that natural selection is not the only type of selection that operates on variation. Rather, two additional kinds of selection — methodical and unconscious — also operate on domesticates and are often combined under the term "artificial selection" to distinguish them from natural selection. Darwin summarizes the three types of selection in crop evolution: "*Methodical selection* is that which guides a man who systematically endeavors to modify a breed according to some predetermined standard. *Unconscious selection* is that which follows from men naturally preserving the most valued and destroying the less valued individuals, without any thought of altering the breed; . . . *Natural selection* . . . implies that the individuals which are best fitted for the complex, and in the course of ages changing conditions to which they are exposed, generally survive and procreate their kind" (Darwin 1896, 177–78).

As in wild plant populations, natural selection operates on cultivated plants to winnow the products of variation. Crop plants that are more fit in specific environments, because of tolerance to stress, ability to capture scarce resources, or production of more viable seed, are more likely to be represented in greater numbers in subsequent generations. Thus, yield and fitness are naturally linked in a positive manner (Evans 1993), although yield is not a function of natural selection alone. Farmers' emphasis on yield therefore subsumes an element of natural selection in farmer choice of seed.

Methodical and unconscious selection are the unique aspects of evolution under domestication. Several scholars have reinterpreted Darwin's use of methodical and unconscious selection (Donald and Hamblin 1984), but his approach is still largely intact. Methodical — or conscious — selection is purposeful and direct for specific traits that farmers and consumers find desirable and against traits that are undesirable. Methodical selection is the kind we observe among farmers and gardeners who choose seed according to the type of plant they wish to produce with larger grain, stiffer stalks, deeper color, and any one of an almost infinite number of characteristics. The result of methodical selection is evident in the long record of crop evolution toward crop plants that allocate more and more energy to the parts of the plant that humans intend to use (Evans 1993). Indeed, domestication itself must have involved conscious selection for traits that distinguish cultivated from wild plants — for example, larger seeds or tubers with lower levels of bitter glycoalkaloids (Hillman and Davies 1990, Johns 1990). During thousands of generations of farming, and with the accumulation of knowledge about cultivated plants, the number of traits that are consciously selected for and against has logically increased.

Unconscious selection is more difficult to distinguish from natural selection, and its separation from conscious selection is complicated by the lack of information about selection in the distant past or among farmers who may not articulate selection criteria. An example is the practice of "rouging," or removing crop plants that appear to be diseased or otherwise undesirable. The recovery of nonshattering seeds during domestication may have been both conscious and unconscious. Darwin's definition of unconscious selection was drawn without the benefit of an understanding of genetics, which is necessary for any modern discussion of evolution. Nevertheless, an understanding of conscious and natural selection of crop populations presents us with opportunities to observe unconscious selection. One opportunity occurs when a trait that is consciously selected for is linked to traits that are not observed by the farmer; an example is the linkage between grain color and other traits in maize. Traditional maize farmers in Mexico and elsewhere are known to name and select maize according to kernel color. Paterniani (1969) demonstrated that selecting for color and starchy or sugary quality in maize kernels could rapidly establish isolated populations and that the days to flowering of the tassel and ear were affected by selection for kernel characteristics.

Methodical Selection

Methodical or conscious selection of variation within crops has been studied in a number of contexts. Researchers have used a general rational choice framework to approach farmer selection of crop diversity (Brush, Taylor, and Bellon 1992; Bellon 1996). A logical starting place was the microeconomic analysis of crop selection decisions under the assumption that selection is driven by a limited set of goals—maximization of production, optimal use of resources such as land or labor, or minimization of risk. Selection affecting crop diversity can be treated as a form of technology selection, a topic that has been extensively researched as part of the process of agricultural modernization (Feder, Just, and Zilberman 1985).

Two issues pertain to the study of conscious selection: (1) what criteria are employed, and (2) what is the process. The criteria issue concerns the objectives of selection (e.g., yield or quality), while the process issue concerns who, when, and how selection is carried out. Determining selection criteria has received most of the attention of researchers who have studied farmers' decision making relating to crop diversity (e.g., Richards 1985, Brush and Meng 1998). This reflects the assumption that farmers around the world operate within a common framework of interests and attitudes in making decisions regarding which crop type to select. This framework includes the maximization of production, optimal use of land and labor or other scare resources, and minimization of risk.

SELECTION CRITERIA

Yield is the obvious focus of rational choice approaches to selection. It is an observable variable among crop types, geographic regions, environments, and time periods. It is all but impossible to imagine a farmer who is not aware of yield as a measure of the returns to his/her labor or land because it is directly linked to food availability and income. And, while yield can be viewed as a principal element of artificial selection, it also reflects the impact of natural selection on crop populations. Consequently, yield has been the starting place for most studies of crop selection and crop evolution (e.g., Evans 1993).

Nonetheless, yield alone is not sufficient to comprehend farmer selection or crop diversity. Three objections can be raised to using yield as the sole selection criteria. First, yield is an ambiguous concept. Are we merely measuring the weight or volume of edible grain or tubers, or are other products such as straw or husks also to be counted? In parts of Mexico, the *totomoxtle* or outer husk of the maize ear, which is used to wrap tamales (steamed maize cakes), is nearly as valuable commercially as the maize grain itself (Perales R. 1998). How do farmers calculate yield when the quantities of grain and other products are not positively correlated, for instance when fodder, which is a function of plant height, is reduced to allow higher grain yield?

Second, quality may be equally as important as quantity in calculating the benefit of a particular crop variety. Quality is even more ambiguous than yield as a criteria, potentially masking many other relevant aspects of the harvest. Taste, processing and cooking qualities, resistance to pests and spoilage during storage, and market demand have all been shown to affect selection. Because quality may be negatively correlated with quantity, it is reasonable to treat yield and quality as though they are separate criteria in selection.

Third, perceived risk of crop failure or yield instability may be as important to farmers' decision making as consideration of quantity or quality of the harvest. We have long recognized the importance of "safety first" in the decisions of subsistence farmers (Lipton 1989, Scott 1976). A farmer may be confronted with different microenvironments among fields. More importantly, some crop types are better able to withstand unstable or threatening conditions such as drought, flooding, wind, weeds, insects or other predators, and disease. Farmers who are fortunate enough to have sufficient agricultural resources and supplies can control the effects of these conditions with irrigation, drainage, wind breaks, and herbicides, pesticides, fungicides, and other chemicals. However, many of these amendments to farming are recent inventions or have not been available to most farmers during the long evolution of crop plants. Rather, farmers have met these threats by selecting crop types that are resistant or tolerant to negative conditions, possibly at the expense of yield.

Fitting Crops to Environment

Crop breeders recognize differential performance of crop types in different environments as the "crossover effect" (Evans 1993; Singh, Ceccarelli, and Grando 2000). When grown under stressful conditions, such as poor soils or aridity, varieties that have high yields under optimal conditions may not do as well as varieties that yield poorly under optimal conditions. Farmers who face uncertain and marginal environments, such as those in hill lands or semiarid areas, may choose varieties that are potentially lower yielding but stable in unfavorable conditions. Other farmers may keep different varieties for specific conditions. This phenomenon has been observed in numerous crops and comparisons between farmers' varieties and improved varieties under different agronomic conditions. Indeed, Zeven's (1998) definition of landraces cites their ability to tolerate biotic and abiotic stress.

Improved varieties can outperform traditional varieties, but this advantage may hold only in environments with good soils and sufficient moisture. The yield lines of the two types have been shown to intersect as one passes to less-favorable conditions, such as from good rainfall to drought. For example, Ceccarelli and Grando (1999) show that crossover occurs between the yield of improved and local barley varieties in Syria as one shifts from stress (aridity) to nonstress environments (Table 6.1). Here, the advantage that one type enjoys in one environment disappears in the second environment.

In Chiapas, Mexico, farmers keep certain varieties for rocky hillsides that are tilled with the traditional *coa,* or digging stick, and use other varieties for valley bottoms that have good soil and are plowed with oxen or tractor (Brush, Bellon, and Schmidt 1988). Bellon and Taylor (1993) found that farmers in Chiapas use different maize races on specific soils within the farming system of a village. Traditional maize (maize race *Olotillo*) is planted in poor soils (*tierra amarilla*), while improved varieties (maize race *Tuxpeño*) are reserved for the best soils (*tierra negra*).

Bellon (1991, 1993) points out that crossover derives also from the social environment of agriculture. Maize farmers in Chiapas refer to the improved, short-stature maize (Tuxpeño race) as "rich man's maize" not only because of its high yields on good soils but also because of its debility when confronted

Table 6.1. Crossover in yields of improved and local barley varieties in Syria under stressed (YS) and nonstressed (YNS) conditions

		YS[b]		YNS[c]	
Type of germplasm	*N*[a]	Yield	Range	Yield	Range
Modern	155	488	0–893	3901	2310–4981
Landraces[d]	77	788	486–1076	3413	2398–4610
Best check		717		4147	

[a]Number of entries
[b]Average of two stress sites
[c]Average of three nonstress sites
[d]Pure lines obtained by pure line selection within landraces

Source: Ceccarelli and Grando 1999, 62

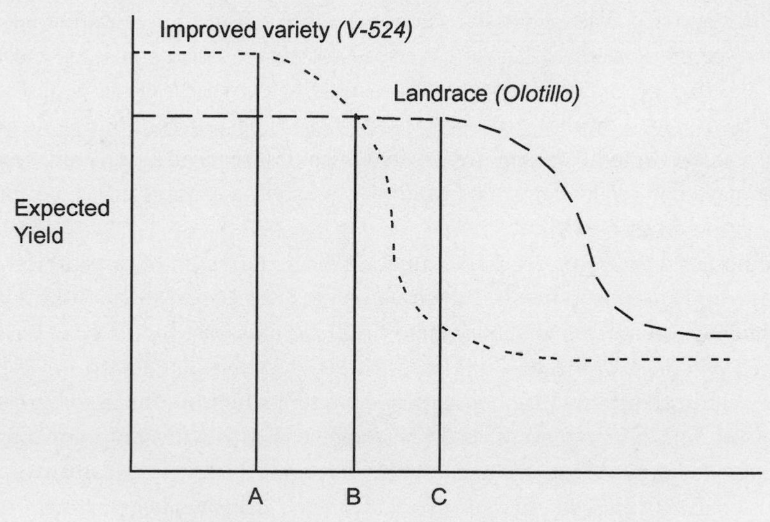

Figure 6.1 Socially mediated crossover of maize varieties in Chiapas, Mexico (Bellon 1990) (Reprinted with permission of M. Bellon)

with weeds and poorly timed fertilizer application. Rich families can afford to hire labor to weed and they don't necessarily depend on the inefficient loan programs of public agencies to purchase fertilizer. Poor farmers, who work as day laborers and who don't have good access to credit, value the ability of the Olotillo landrace to withstand (*aguantar*) difficult conditions along with them and to be forgiving of their poor timing in weeding and fertilization. Figure 6.1 depicts this socially induced crossover effect in lowland Chiapas maize production.

Variety Mixtures

Selection among varieties is one choice that farmers make, and whether to plant a field in a single variety or with a mixture of varieties and/or crops is another choice. Some farmers may have found that diversity itself provides protection against yield instability and threats to food security (Clawson 1985). Frankel, Brown, and Burton (1995) discuss "interaction effects" of heterogeneous crop stands and observe that plant disease is reduced by intercropping different genotypes. Field research on rice in China shows that yields are higher when disease-resistant genotypes are planted in the same field as disease-prone varieties (Zhu et al. 2000). Nevertheless, conscious selection and planting of diverse genotypes in a single field by farmers have not been well documented or analyzed. Perhaps stability is best understood in terms of covariance among fields that are planted with different varieties. Low covariance may dampen the normal instability of yield, but this effect is not an interaction effect per se.

Although they are often glossed under a single name, landraces are diverse mixtures or populations of different genotypes (Zeven 1998). In addition to conscious selection for diversity, many reasons exist for the mixture of genotypes in centers of origin—mixing during harvesting on common threshing floors, natural differentiation arising in crop reproduction, the mixing of seed lots over time. Awareness of crop heterogeneity varies by crop and farmer knowledge (Sambatti et al. 2001). Peruvian potato farmers are fully aware of the numerous types in their *chalo* fields, but Turkish wheat farmers gloss highly mixed fields under a single name. Most farmers I have interviewed and observed understand diversity in crops as separately named varieties that are selected for reasons of production, use, or risk. It is not understood as the mixture of genotypes that can be observed within a single field or landrace. The one exception in my experience is the Andean chalo field of potatoes that are purposively mixed.

Values of Diversity

Selection criteria may be embedded in the nomenclature and folk taxonomies of crop varieties, but it is rare for variety names to explicitly reveal selection criteria. Several approaches exist to consolidate yield, quality, and risk into a single framework for understanding selection. One approach is to view selection criteria as values that can vary by context, either among farmers or over time. Value, either social or individual, may be defined in different ways (Brown 1990, McNeely 1988). Brown (1990) identifies three general categories of values (option, indirect, and direct) and further refines them into specific types. Option values derive from future use of a biological resource (Brown 1990, Krutilla 1967); this category is the most relevant to the public and political concept of genetic resources. Krutilla (1967) identifies three types of option value: existence, bequest, and scientific. The last two are the most pertinent to arguments in favor of conserving genetic resources. As a bequest, crop genetic resources may have a high value to future generations of farmers and consumers, who may face different agricultural conditions than those confronting present generations. The importance of bequest is demonstrated by the high turnover rate for crop varieties, a rate that depends on the availability of genetic resources (Brennan and Byerlee 1991). Most importantly, genetic resources have value for science as essential components of research to improve public welfare. High returns to investment in agricultural science, especially to crop breeding, strongly suggest a high value for genetic resources (NRC 1993).

Indirect values refer to environmental services rendered by plants and animals and benefits resulting from biological resources that do not depend on harvest and consumption. One indirect value of crop diversity is the interaction effect mentioned by Frankel, Brown, and Burdon (1995), which is based on the observation of lower incidents of disease in diverse crop populations. This effect is attributed to buffering against pathogens and slowing their spread.

DIRECT VALUE OF WHEAT DIVERSITY IN TURKEY

Direct value of crop diversity refers to the harvest and use of different crop varieties as parts of noncommercial, commercial, and/or industrial production. The value to farmers of specific crop characteristics can be studied by ranking exercises in which farmers are asked to rate different varieties according to such elements as yield, quality, stability, disease, or drought resistance (Brush and Meng 1998). Another exercise is to ask farmers to rank varieties and then determine which characteristics give varieties higher ratings.

The selection of wheat varieties in Turkey illustrates how criteria can be described and used to analyze the variety choices. The cultivation of wheat in Turkey for more than eight thousand years has resulted in a large number of local wheat varieties in addition to the existing wild and semidomesticated wheat relatives. Modern varieties have been available in Turkey since the early part of the twentieth century, and semidwarf varieties were introduced from Mexico in 1966. However, the adoption of modern wheat in the country varies greatly from region to region. The International Maize and Wheat Improvement Center (referred to as CIMMYT) (1993) reports that, in 1990, only 31 percent of Turkey's wheat area was planted in modern varieties. Research conducted between 1990 and 1994 in the western Transitional Zone, located between the major wheat-producing Anatolian Plateau region and the western coastal plains, showed considerable variation between provinces and from village to village in the proportion cultivated in modern wheat varieties.

Data were collected in 1992 by a survey covering 287 households conducted in 24 villages in the provinces of Eskişehir, Kütahya, and Uşak. Villages varied with respect to three agroclimatic zones: valley land, hillside land, and mountain land. Valley land is more likely to have irrigation and to be connected with urban markets, while mountain land is most distant from markets and situated in or around forested zones. Hillside land shares attributes of both valley and mountain land. The socioeconomic survey covers a broad range of information regarding household characteristics, detailed production data, and consumption preferences. Households in the sample include those that cultivate only modern varieties, those that cultivate both modern and traditional varieties, and those that cultivate only traditional varieties. Differentiation among households also exists with respect to percentage of production output marketed. Table 6.2 summarizes several household characteristics by province and agroclimatic zone for the sample.

In the Turkish research, we found evidence that environmental heterogeneity, crossover, risk, and the high transactions costs of obtaining desired quality in the wheat marketing system contributed to the continued cultivation of landraces in specific areas of a nation that has successfully promoted modern varieties. An econometric analysis of the factors influencing variety selection confirms the importance of acknowledging multiple motivations for cultivating traditional varieties (Meng, Taylor, and Brush 1998). The probability of cultivation of traditional varieties in a given field significantly increases when the plot is situated in less-fertile soil or when distance to market and bad road quality increase a household's cost of accessing markets.

Our survey also reveals that a substantial percentage of households participate in some kind of market, but market integration is not merely a function of

Table 6.2. Characteristics of sampled households in western Turkey, 1992

	Sample number	Off-farm income (%)	MV only	TV only	Both MV/TV	Total land (ha)	Total irrigated land	Total plots	Mean plot size (ha)	Number of household wheat plots	Distance from market (km)
All households	285	26.5	68	167	50	12.3	2.6	12.4	1.3	7.0	16.6
Eskişehir	96	21.9	65	16	15	15.5	6.7	13.2	1.8	7.2	20.4
Kütahya	96	22.7	1	64	31	10.4	0.9	20.3	0.7	10.8	12.9
Uşak	93	35.1	2	87	4	10.8	0.1	8.5	1.4	2.8	16.6
Valley	100	17.0	46	33	21	15.0	6.7	15.8	1.7	7.3	10.6
Hillside	101	22.5	10	66	25	13.9	0.6	16.2	1.2	8.4	13.5
Mountain	84	42.4	12	68	4	6.9	0.02	9.5	1.0	4.8	24.0

Source: Brush and Meng 1998

road quality or distance to market. A relevant weakness of markets is that it is all but impossible to purchase a specific wheat variety in the formal market system. Private and public wheat buyers categorize the grain by color and hardness rather than by quality aspects that farmers recognize for specific varieties. This marketing policy results in the mixture of varieties in each market class. If households wish to consume a specific variety, cultivating it themselves is the most certain method to guarantee availability.

Traditional varictics arc oftcn considered to have more attractive quality characteristics than modern varieties. An examination of household rankings of specific attributes on a scale from 1 (best quality) to 5 (worst quality) sheds light on the association of these characteristics with individual varieties. Households ranked each variety cultivated with respect to taste, bread quality, milling quality, yield, disease resistance, and drought resistance. These characteristics reflect important considerations for the household on both the production and consumption side. Table 6.3 presents the average scores among all households for five characteristics pertaining to both traditional and modern varieties. In general, modern varieties rank higher for yield attributes, while traditional varieties are ranked higher in terms of taste and baking quality. Traditional varieties also appear to be associated with better drought-resistance attributes.

Additional information on the relative importance of varietal attributes comes from household responses listing reasons for discarding a previously cultivated variety. Table 6.4 presents this information for all households that have given up the cultivation of traditional varieties. Yield is the most frequent reason for their decisions. Quality-related reasons decrease sharply among households that cultivate modern varieties. These findings suggest that quality

Table 6.3. Wheat variety ranking by Turkish farmers

Household type	Sample number	Attribute (1=best quality, 5=worst quality) (SD)			
		Yield	Drought resistance	Disease resistance	Baking quality
All households	352	2.13	2.24	1.86	1.93
		(0.87)	(1.07)	(0.87)	(0.87)
Households cultivating modern varieties	133	1.73	2.72	1.97	2.22
		(0.67)	(1.02)	(0.81)	(0.92)
Households cultivating traditional varieties	219	2.37	1.95	1.79	1.75
		(0.89)	(1.00)	(0.90)	(0.79)

Source: Brush and Meng 1998

Table 6.4. Reasons for no longer cultivating traditional wheat varieties

Reason	All households responding (*N*=163/285)		Households cultivating modern varieties (*N*=88/118)	
	Number of responses	Percent	Number of responses	Percent
Yield	124	38.8	78	38.8
Cold susceptibility	38	11.9	20	10.0
Production-related	27	8.4	16	8.0
Quality problems	26	8.1	6	3.0
Drought susceptibility	24	7.5	20	10.0
Lodging	23	7.2	21	10.4
Other	18	5.6	14	7.0
Seed availability	16	5.0	10	5.0
Marketing problems	12	3.8	9	4.5
Disease susceptibility	5	1.6	1	0.5
Price	4	1.3	4	2.0
Climate adaptability	2	0.6	1	0.5
Soil adaptability	1	0.3	1	0.5
Total responses	320	100	201	100

Source: Brush and Meng 1998

Table 6.5. Reasons for no longer cultivating modern wheat varieties

	All households (*N*=100/285)		Households cultivating traditional varieties (*N*=54/217)	
Reason	Number of responses	Percent	Number of responses	Percent
Yield	43	25.4	22	25.9
Drought susceptibility	26	15.4	19	22.4
Production-related	20	11.8	7	8.2
Quality problems	14	8.3	10	11.8
Disease susceptibility	13	7.7	1	1.2
Seed availability	12	7.1	3	3.5
Cold susceptibility	10	5.9	10	11.8
Price	10	5.9	3	3.5
Other	9	5.3	4	4.7
Lodging	4	2.4	1	1.2
Climate adaptability	4	2.4	4	4.7
Soil adaptability	4	2.4	1	1.2
Marketing problems	0	0	0	0
Total responses	169	100	85	100

Source: Brush and Meng 1998

issues are no longer of great importance for those households that have given up traditional varieties. Table 6.5 presents reasons for opting against the cultivation of modern varieties. Again, yield remains the most important reason, but different motives may pertain here for households other than those that listed yield as the primary reason for giving up a traditional variety. For modern varieties, it is more likely that the yield response reflects a failure to attain expected yield. In addition, the percentage of households listing quality and drought resistance as reasons to give up modern varieties is highest among households that continue to grow traditional varieties. These results further suggest that both consumption demand and environmental factors affect variety choice. Moreover, they show the advantages associated with landraces for consumption quality and adaptability to difficult climates.

Interpreting Farmers' Criteria

A common approach to understanding selection criteria is through the analysis of farmers' choice of crop types. Bellon (1996) lays out a set of generic "farmer concerns" in the selection process—environmental heterogeneity (e.g., different soils in and among fields), pests and pathogens, risk management, culture and ritual, and diet (see Table 2.2). These concerns are hypothetically found in every farming system and form the basis for all crop choice. However, some concerns may be less important or relevant to one community or farmer than to another. For example, farmers in a community with a benign environment (e.g., good soils and adequate rainfall) and good market access (e.g., predictable supplies of agricultural inputs) may be expected to give less attention to environmental constraints than those in a community with a marginal environment and missing markets. Some communities and cultures place far more emphasis on use of different crops in cuisine or ritual. For instance, cultures that identify rice varieties with lineage give more emphasis to the ritual concern than do others that do not make this association.

A framework such as Bellon's (1996) list of farmers' concerns is useful to understanding the ethnographic context of crop selection, but ultimately the actual choice of which crop type to plant is the key to understanding crop evolution. Crop choices have been studied in numerous cases through farm surveys (Feder, Just, and Zilberman 1985), and a common focus is on the choice between local or traditional crops and modern or improved crops. Describing crop choice by using farm surveys gives the quantitative summary of numerous individual decisions, but this summary does not provide an explicit statement of why crop types are selected. Rather, it allows us to intuit the basis for selection—whether yield, quality, risk, or some other concern.

Household surveys on the choice between traditional and modern wheat varieties in Turkey illustrate the complexity of modeling selection. Table 6.6 presents data on thirteen variables that were used to model the choice of traditional and modern wheat (Meng 1997). This table shows that certain variables are more commonly associated with the choice of whether to plant traditional wheat. The most important repercussion is the negative effect of market development, shown by the fact that households in districts that sell more wheat in the market are less likely to plant traditional wheat. Likewise, larger farms and better-educated farmers are less likely to plant traditional wheat. Older farmers and farm families with more children are also more likely to plant traditional varieties, and these varieties are favored in places where the yield of traditional wheat approximates the yield of improved wheat.

Table 6.6. Traditional wheat variety choice in western Turkey

Variable	Change in the probability of cultivating traditional wheat variety
Yield difference: traditional/modern	0.42
Yield variance difference: traditional/modern	−0.04
Yield covariance difference: traditional/modern	0.06
Price difference: traditional/modern	0.02
Age (head of household)	0.21
Years schooling (head of household)	−0.14
Number in household >13 years old	0.17
Off-farm income	0.01
Household wealth	0.07
Total land (farm size)	−0.19
Total high quality (fertile) land	0.12
Number of livestock owned by household	0.10
District wheat production sold in market (%)	−1.48

Source: Meng 1997

SELECTION CONTEXT

Identifying and analyzing selection criteria has been the object of most studies of farmer involvement in crop evolution. A second concerns the context in which crop choice and selection are actually carried out. This issue relates to the effectiveness of selection in terms of the strength and speed of selection. While some concern has been expressed over the role of gender in the selection process, there is little systematic research on whether men and women make different types of selection decisions. There is little doubt that both men and women are involved in the selection process—but perhaps at different times and locales. Communication and joint interests within the family's household are likely to dampen the effects of gender on selection. Observing ranking trials of maize varieties by men and women in Oaxaca, Bellon et al. (2000) found that each sex ranked six out of ten varieties in "top ten" variety groups differently, although the study also found that men's and women's criteria were essentially identical and similarly ranked.

Using the farm household as the "unit of selection" is the logical way to avoid the potential problem of gender bias in analyzing selection. An example of the usefulness of seeing selection as a household activity is provided by a study of maize selection in Veracruz, Mexico (Rice, Smale, and Blanco 1998). Men play primary roles in the field operations of the milpa—the complex assemblage of maize, beans, squash, minor cultivars, and wild greens—including selecting maize ears for the next season's seed. However, Rice, Smale, and Blanco (1998) found that selection does not end at harvest. Rather, during the process of preparing maize for the daily tortillas of the household, women frequently added maize ears to the seed pile, thus complementing the men's choices and making selection a household activity.

Knowing how selection is made is critical to understanding farmers' roles in crop evolution. The common pattern among many farming systems is for selection to take place after the harvest, by choosing seed from the harvest pile according to criteria such as size, color, and healthy appearance of the grain or other seed material. In postharvest selection, farmers are directly selecting only the traits that are visible to them in the seed head or tuber, and they are indirectly—or unconsciously, in Darwin's terms—selecting only traits such as the height, leaf area, or disease resistance of a crop type. Farmers are likely to prefer higher yielding crop types, but this requires a choice among known types, for instance choosing between two different types grown in a single field or environment. Obviously, comparison and choice occur in all farming systems, but crop choice may not be a common feature of the year-to-year farm management in traditional farming systems. In fact, relatively low numbers of varieties per household are common in several centers of diversity, especially in cereal crop inventories of peasant households. Reviewing the literature on traditional lowland rice in Southeast Asia, Bellon, Pham, and Jackson (1997) show that fewer than two rice varieties are cultivated per household in villages studied in Cambodia, Malaysia, and Thailand. In Mexico, Perales R. (1998) sampled twelve villages in the central highland states of Mexico, and Morleos and reports an average of only 1.2 maize varieties per household. In Turkey, we found an average of 1.2 wheat varieties among twenty-seven villages in western Anatolia (Meng 1997). Studies of tuber crops have shown a similar pattern of low numbers of Andean potatoes (Quiros et al. 1990) and Amazonian manioc (Salick, Cellinese, and Knapp 1997), although in these cases one field per household or village often is highly diverse.

Yield may be directly selected for in some seed lots by choosing seed heads that show optimal size and grain filling. Nevertheless, contrary cases are common; for instance, small potatoes are often selected for seed in the Andes. In other words, traditional farmers in centers of crop diversity appear to exert strong selection pressure on some traits (e.g., color) but weak pressure on other plant characteristics that may be particularly relevant to the crop's performance. Because farmers do not make preharvest selection, some observers think that farmers play a minor or even nonexistent role in crop evolution. Wellhausen, Fuentes, and Hernández-Corzo (1957), writing about the evolution of maize in Central America, opined that traditional farmers added little to the natural forces of maize evolution.

This is an extreme position that has not been widely embraced by other researchers of crop evolution. Indeed, Darwin (1896, 185) rejected the assertion that "semi-civilised people" have not carried out methodological selection since ancient times. On the other hand, it is equally extreme to argue that farmers exercise a strong and directed control over the process of crop evolution in the same way that scientific breeders do, as the latter group manipulates gene flow and the pace of variation and extends selection to cover many more plant traits. This is not to say that farmers have not had the primary role in crop evolution, but rather that their role has been incremental and apparent only over many generations. Only a few cases of breederlike activities by farmers have been reported. In southern Peru, the selection of potato plants from seed beds of true potato seed is a technique that apparently was observed at a regional agricultural experiment station and later copied in the village of Chincheros (Franquemont et al. 1990). Again, Darwin (1896) notes that methodical selection has become more systematic and effective in modern times, but he acknowledges the accomplishments that have accumulated during previous generations of selection. The difference between farmers and crop breeders rests on this point.

Products of Crop Evolution

Since domestication, crop plants have passed through thousands of cycles of reproduction, variation, and selection. Crop evolution since domestication has produced three striking results across many different species. First, yields of crop plants have grown exponentially (Evans 1993). Second, crop species have become increasingly diversified. Third, a few crop plants have become dominant, thereby reducing the number of species that sustain the human population.

YIELD INCREASES

Yield increases are due in part to the evolution of farming technology, which has improved nutrition to crop plants and limited the ravages of pests and pathogens. But the yield potential of crop plants has also been increased through selection. In particular, selection has steadily changed the allocation of plant resources to favor the parts of crop plants that constitute yield, such as grain number and size (Jackson and Koch 1997). More favorable harvest

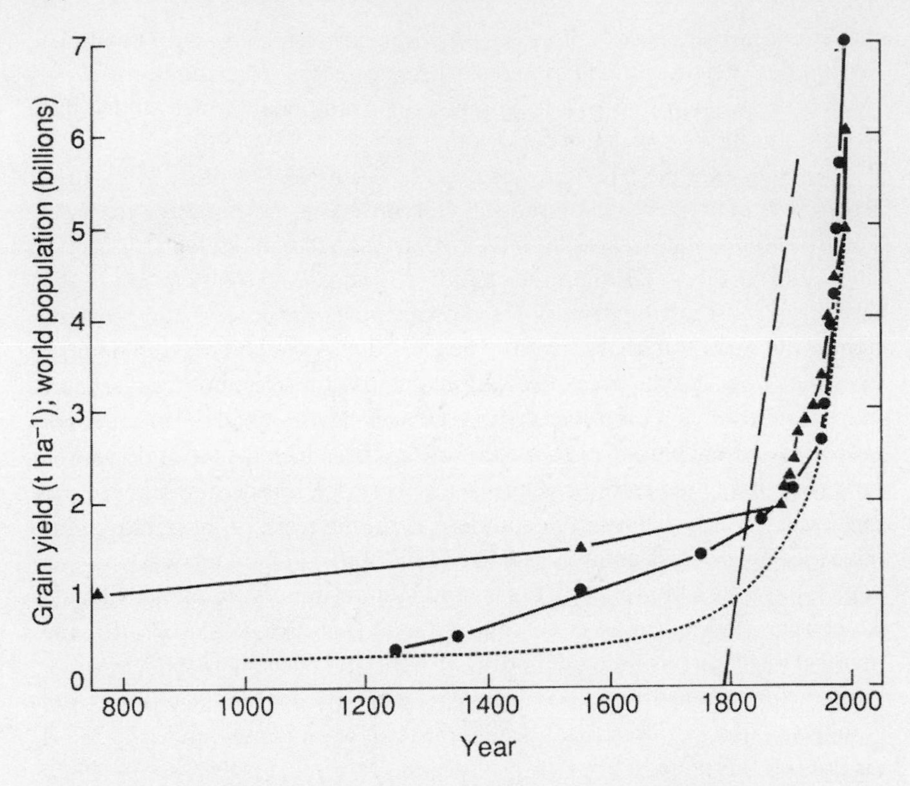

Figure 6.2 Evolution of wheat (●) and rice (▲) yields; [(••••) population] (Evans 1993) (Reprinted with permission of Cambridge University Press)

ratios, measuring the ratio of harvested to nonharvested material (e.g., grain to leaf area), reflect the success of selection for higher yields (Evans 1993). Yield gains since the development of industrialized agriculture with scientific crop breeding, mechanization, and chemical inputs have been impressive, but gains are apparent before industrialization. Figure 6.2 depicts the yields of wheat and rice over several centuries. The nearly vertical increase in yields since industrialization in the mid-nineteenth century extends a pattern of growth in yields that begins before industrialization. This is not to say that nonindustrial agriculture is capable of achieving or sustaining the same yields as those found in industrial societies, but descriptions of modern agriculture as "revolutionary" are also exaggerations.

DIVERSIFICATION

Diversification is also a product of crop evolution, as Darwin (1896) noted in commenting on the power of artificial selection to differentiate among

Table 6.7. Gene bank holdings of selected crops

Crop	Total accessions among genebanks	Largest holder	Percentage of world collection in largest collection
Wheat	784,500	CIMMYT	13
Rice	420,500	IRRI	19
Maize	277,000	Mexico	12
Tomato	78,000	United States	30
Potato	31,000	CIP	20

Source: FAO 1998

the useful parts of plants. Among domesticated plants, diversification is evident in the development of different species, polyploid groups, and crop morphotypes. Examples of speciation within crops include wheat (four species of *Triticum*), potatoes (seven species of *Solanum*), beans (three species of *Phaesolus*), and chili pepper (four species of *Capsicum*). Crop morphotypes that make up landraces reflect genetic differences, and these are counted in the thousands. These morphotypes constitute the vast bulk of the accessions that are stored in the hundreds of gene banks around the world (FAO 1998)—the result of plant collectors who took samples of crop types that appeared to be unique or unknown. Table 6.7 gives some indication of the numbers of morphotypes found in major gene bank collections. Landraces are especially prevalent in national and international inventories of crop genetic resources, accounting for nearly 60 percent of all gene bank accessions (FAO 1998).

A landrace's diversity is the store of slowly accumulated genetic variation nurtured by farmers through countless generations. Perhaps the most visible and impressive products of farmer-based crop evolution are diverse and fragmented populations of landraces, or named varieties, that are for the most part adapted to a limited area (Harlan 1975) and low inputs (Zeven 1998). Landrace populations are not static, but change is dampened by the decentralized nature of selection and limited by the chance appearance of beneficial mutations or other genetic alterations. A myriad of reasons exists for saving crop variants. Certainly, yield and adaptation to particular environments are important, but personal and cultural reasons may be equally as important. Curiosity is a universal human trait, and it is reasonable to extend this to fascination with and celebration of diversity among cultivars. Personal pride or the honor of naming a crop variety after a family or lineage may also play a role in the diversity of crops.

In part, diversification results from the diffusion of crop plants out of their original habitats and their adaptation to new habitats. Crop diffusion is as old as agriculture itself and has contributed to the modern distribution of the human population (Ammerman and Cavalli-Sforza 1984). Long before the modern era, several crops became truly cosmopolitan, having spread into regions that represent many of the Earth's habitats. Maize, for instance, spread thousands of miles and nearly 90° in latitude from its cradle area in central Mexico to the Great Lakes in North America and to the Southern Cone in South America. Diversification is also the product of stewardship that derives from both utility and curiosity. Not only do farmers value individual crop types, they also value diversity per se—as a source of future variation, as an object of pride or culinary pleasure, or for agricultural benefits such as disease resistance (Frankel, Brown, and Burton 1995; Bellon 1996). Andean fields of chalo potatoes reflect a long tradition of the purposeful cultivation of diversity. Another example is reported by Salick, Cellinese, and Knapp (1997), who studied the diversity of manioc among Amuesha cultivators in the Amazon Basin, where the crop was domesticated. While the average household only maintains 5.9 manioc varieties, the garden of the local shaman is a trove of diversity with 53 distinct types.

DOMINANCE

The third result of crop evolution, increasing dominance of a few major cultigens, grows out of the interaction of the first two results, increases in yields and diversification. Wilkes (1993) observes that "only about 5,000 species have historically fed the human population. This represents less than 1% of the flora of the world." Far fewer plant species have been cultivated; Wilkes (1993) estimates that 1,500 species have entered agriculture, and even fewer have been major sources of food. He suggests that 150 plant species provide calories to humans today. Hawkes (1983) gives similar numbers: 3,000 plant species used for food out of 200,000 possible, 200 domesticated plant species, 15 to 20 species of major importance in the modern world. Listing both food and nonfood crops, Harlan (1992) enumerates 348 species among the world's farming systems. For the Near Eastern Complex, he lists 57 food crops. However, the bulk of the human diet depends on far fewer species, and the number appears to be decreasing. Manglesdorf (cited in Harlan 1992) argued that 15 crops feed the world, a figure that is approximated by some (Harlan 1992, Wilkes 1993) but disputed by others (Prescott-Allen and Prescott-Allen 1990). A handful of mega-crops, such as wheat, rice, maize, manioc, and potatoes, now contribute most of the calories from plants to the human diet. Africa provides an example of the decrease in local crops caused by diffusion of these mega-crops. American domesticates, maize and manioc, have greatly eroded the importance of African species such as fonio (*Digitaria exilis*), finger millet (*Eleusine coracana*), and sorghum (*Sorghum bicolor*). Asian rice (*Oryza sativa*) has genetically swamped African rice (*Oryza glaberrima*) in some West African areas (Richards 1985). Mega-crops are privileged by agricultural research and marketing infrastructure, but they may also reflect deep-seated human preferences. While researching potato diversity in the Andes of Peru, I discovered that bread was one of the most prized gifts that I could bring to Andean potato farmers. Did bread's popularity come from a human preference for small grains, or was it because bread carried the taste of urban, European food enjoyed by the wealthy but rarely found on the plates of highland potato farmers? I was disturbed by how enthusiastic potato farmers were for a food that seemed distinctly inferior in my eyes to their native crop. My gifts of bread, however, are mere crumbs in a rapid and massive shift toward grain-based foods, principally rice and noodles, in the Andean region, reflecting the political and cultural ascendancy of coastal and Europeanized Peru.

Agricultural Development and Crop Evolution

A voluminous literature describes and analyzes the trajectory of agriculture from low-input, subsistence farming in preindustrial societies to high-input, commercial agriculture in industrial states. In less-developed countries, development issues dominate politics and public discourse, and because these societies tend to be agrarian, agricultural development looms large. The presumption behind the very concept of development is that all societies are capable of moving through the processes of increasing agricultural production and productivity to create greater welfare. These processes involve increasing specialization in production complemented by exchange between households, communities, regions, and nations. Spatial integration—through markets, states, common languages—replaces local self-sufficiency. Markets that allocate land, labor, and information replace nonmarket mechanisms in underdeveloped societies. Development processes in agriculture involve the progressive substitution of capital (human, technological, financial) for land and labor. Regions that have achieved high productivity have passed through stages in which animal and then mechanical labor replaced human labor, and input of nonfarm supplies such as seed, fertilizers, and pesticides made land and labor more productive. The social consequences of agricultural development include rapid population growth followed by a demographic transition to lower fertility rates, declining percentage of people in farming, the end of isolation of farming communities, and a decrease in farm wages relative to nonfarm ones.

Numerous agricultural societies on every continent experience the processes described in the preceding paragraph. The societies that first went through this process are today the wealthiest—northern Europe, North America (Canada and the United States), Japan. But regions within less-developed countries have also gone through these same processes. Examples are the El Bajio region of central Mexico, the Punjab of India and Pakistan, and central Luzon in the Philippines—all places where crop evolution can be traced over thousands of years.

Agricultural development has modified both the products and processes of crop evolution. Local crops and crop diversity have been drastically affected in some areas. The most well-known impact is genetic erosion or the loss of diversity resulting from the displacement of numerous local crops and crop varieties by fewer modern, high yielding varieties (Fowler and Mooney 1990). The virtual elimination of local maize types occurred between 1920 and 1950, as hybrid maize swept through the American Corn Belt. Similar loss has been suffered in the Punjab and the Philippines, where short stature, high yielding varieties of wheat and rice replaced local varieties. The next chapter will address the topic of genetic erosion in more detail.

Apart from affecting genetic diversity, agricultural development has also fundamentally changed the nature of crop evolution. Though Darwin's framework for describing crop evolution predated the discovery of the wild relatives and pathways of domestication and the rise of modern crop genetics, their discovery did not overturn Darwin's ideas. Nevertheless, the rise of crop genetics and industrial seed production has created a new form of crop evolution. Until the advent of crop genetics and industrial seed production, crop evolution was farmer-based and a singular process (Perales R. 1998). Farmer-based crop evolution functions on the basis of farmer control of the key processes of conscious and unconscious selection. As a consequence, farmer-based crop evolution is decentralized and relatively slow, and it produces diversity through locally rather than broadly adapted crops. Exchange among farmers accompanies decentralization so that a landrace is best understood as a "metapopulation" or a "population of populations" (see Chapter 8). Fields of landraces balance the contradictory forces of local adaptation and the flow of seeds among farmers. One force is centripetal, propelling crops toward greater diversity in fragmented landscapes. The other is centrifugal, spreading successful genotypes over large areas and thereby reducing diversity. Farmer-based crop evolution has, of course, characterized virtually all of crop evolution since the Neolithic Revolution. Gains in productivity, diffusion, and differentiation of crops were inherited by twentieth-century farmers but derived from the year-to-year management and selection of crop populations by countless generations of farmers (Evans 1993, Standhill 1976).

Table 6.8. Transformation of crop evolution

	Premodern conditions	Modern conditions
Genetic variation	Emphasis on "natural" processes mutation recombination drift Limited gene flow	Emphasis on "managed" processes Broad gene flow Transgenic transfer
Selection	Unconscious Methodical for few traits Decentralized/farmer-based	Methodical for many traits Centralized/breeder-based
Products	Many locally adapted types Heterogeneous populations Local pedigrees Slow change	Few broadly adapted types Homogeneous populations Nonlocal pedigrees Rapid turnover

Farmer-based crop evolution continues to function in some cropping systems not significantly different from the crop evolution system that existed before 1900, when Mendelian principles were rediscovered. Nevertheless, the advent of crop genetics introduced a qualitatively different type of crop evolution. One of the first and most dramatic results of the rediscovery of Mendelian genetics was its application to maize breeding in the United States, through the work of Shull at Cold Spring Harbor and Jones and East at the Connecticut Experimental Station (Wallace and Brown 1988). Within two decades, crop breeders, working for either private seed companies or public agricultural research and development organizations, had radically altered the pace and nature of crop evolution.

This novel type of crop evolution flourished in industrialized countries but exhibited tendencies that are contrary to farmer-based evolution. Table 6.8 shows the differences between farmer-based crop evolution under premodern conditions and breeder-based crop evolution. In contrast to farmer-based crop evolution, breeder-based evolution is centralized, rapid, and based on wide crosses to achieve broadly adapted plants. Unlike the diverse products of farmer-based crop evolution, breeder-based crop evolution generates populations of uniform crops that dominate large areas. Crop morphotypes are pushed toward uniformity by breeders who seek to create crops that resemble an "ideotype," such as one with a high harvest ratio (Donald 1968). Rather than metapopulations that are created and maintained by farmer seed exchange, this type of crop evolution creates a series of "islands" (farms, localities) that receive seed from a single "continent" or "mainland" — the commercial or public seed source. In other words, scientific crop breeding leads to the dominance of crop populations by a single or small number of advanced varieties, rather than to a more even distribution of numerous locally adapted varieties. Diversity among populations decreases because of centralized selection and seed production.

Under scientific crop breeding, the genetic background or pedigrees of the material found in farmers' fields becomes broader and more complex. Smale (1997) documented the increasing number of landraces found in pedigrees of improved crop varieties. While the pedigrees of landraces are unknown and obscure, they are almost certainly based on material that is less diverse and geographically more restricted. The result is that the profile of crop populations collectively grown by farmers in centers of diversity are more diverse than the populations of modern crops in industrial agriculture. But a point-by-point comparison of varieties in a traditional farm field with those in an industrial farm field may well show more diversity in the latter. Smith (1986), for instance, found that the diversity of Mexican maize far exceeded that of American Corn Belt maize, but that a race-by-race comparison showed that Corn Belt maize was more diverse than the typical Mexican race. Finally, the pace of change in crop populations increases as the life spans of varieties in farmers' fields decrease. Brennan and Byerlee (1991) observe that between 1970 and 1990 the average age of wheat cultivars in the Punjab of Pakistan decreased by nearly a year, from 11.8 to 10.9 years, as farmers adopted improved varieties. The more rapid turnover in modern crops may occur because these crops are genetically more vulnerable to pathogens, but it may also be the result of farmers changing seed to obtain higher yields. There is, of course, turnover in seed and seed ware in traditional farming systems (Zeven 1999), but this turnover is more often within a named variety rather than between varieties.

Because crop breeding has been so successful in less than one hundred years, we could assert that all crop evolution in the future will be influenced by it. This does not mean that improved varieties will completely swamp or displace local varieties. Rather, local varieties are likely to survive for several reasons. Crop breeders are unlikely to be interested or effective in every crop environment, leaving local varieties to flourish where crop improvement is not present. Nevertheless, all cropping systems are likely to be affected by the rise of crop breeding because fewer and fewer farmers live in isolation from the urban, industrial, and market-driven economies where crop breeding has succeeded. Even farmers who produce crops with local inputs and for subsistence are influenced by the price structure of crops, which in turn is increasingly determined by industrial agriculture or international trade. While crop improvement and modern varieties may not be in direct competition with local crops in some regions, the impact of more abundant crops is felt in lower wages that accrue to subsistence farming with traditional varieties.

The study of crop evolution, therefore, will necessarily relate to farmer selection under conditions that are relatively new—competition from scientifically bred cultivars in a social context influenced by national and international markets. Under these conditions, artificial selection can be understood as rational choice influenced by the interaction of three factors—production, risk, and markets. Farmers will continue to choose crop types that are best suited for their conditions. In heterogeneous conditions, they will choose different varieties for specific environments at the local level. In some situations, an improved variety will replace local material, but in others improved varieties may be chosen only for part of the farm (Brush 1995). The fragmentation of peasant farms evident in the cultivation of numerous parcels suggests that modern varieties may not be hegemonic. Farmers may lack the means to alter the conditions of their farm—through irrigation, fertilizers, or pesticides—to make modern varieties competitive.

Risk is relevant to all farmers, but it may be a strong determinant on small, subsistence farms. Local varieties may be lower yielding but more stable, thus providing some risk-lowering benefits. Other mechanisms, such as crop insurance or diversification among crops and income sources to manage risk, are costly and often unavailable. Finally, market systems, which deliver inputs or absorb the additional outputs that are linked to modern crop varieties, will not work smoothly in all environments, and in some places they may not work at all. Local tastes and preferences may not be met by improved crops or by markets, so farmers will continue to select local crops that do meet these tastes. The phenomenon of missing markets, found in Turkey's wheat market, explains much of the inertia that is observed in other traditional farming systems (de Janvry, Fafchamps, and Sadoulet 1991). Markets involve appreciable transaction costs in terms of transportation time and information—costs that some farmers cannot afford. Nevertheless, the spread of market exchange around the world makes it possible for farmers to radically reorganize their farm management to include inputs and consumption goods provided by the market rather than continuing to manage their farms based on home production. This won't happen everywhere, but it is incumbent to learn where markets will and will not function to replace local crops and diversity.

In sum, the application of Mendelian genetics to crop improvement, coupled with the rise of industrial processes and markets for agricultural inputs, has radically changed the context and nature of crop evolution. Nevertheless, agricultural development, like other human patterns, involves both linear and nonlinear processes. Linear processes include the dispersal of human ideas and technology and increasing specialization and integration. These linear processes have predictable effects on crop evolution and diversity. Nonlinear processes include cultural resistance to dispersal, the balancing of tradeoffs between competing goals (e.g., profit and food security), and the obstacles arising in habitats that are spatially structured by environmental and social heterogeneity. The result is that breeder-based crop evolution has become an important but not an inevitably hegemonic force in the future of crop ecology in centers of diversity.

7

Genetic Erosion of Crop Populations in Centers of Diversity: A Revision

Because domestication occurred in relatively limited areas and over a few generations, it created a genetic bottleneck that restricted the flow of naturally diverse germplasm from wild ancestors into crops. Nevertheless, overcoming this bottleneck and the accumulation of biological diversity in crop species following domestication are long-term and well-documented trends of crop evolution. The gradual addition of variation created rich pools of diversity during the innumerable generations of farming and the diffusion of crops away from their original homelands. Thus, crop evolution produced results within domesticated species that are similar to diversity between species arising from natural evolution in such rich areas as tropical forests.

Unfortunately, modern times have been unfavorable to diversity in both crop populations and tropical forests. Pools of biodiversity in crops and forests have eroded under the destructive forces of a large and rapidly growing human population, modern technology, and markets. In both forests and fields, these forces have lowered diversity and reversed the long-term trend of diversification. Habitat conversion of forest to field or pasture has an analog in the conversion of premodern agriculture. An increasing population needs to produce more food with finite land. Increasing communication and social mobility lead to changes in expectations and demand for higher incomes. Technological advances such as mechanization, industrially produced fertilizer, and pesticides reduce environmental heterogeneity of farming systems and perform some of the same production services as diversity in traditional agriculture. Increased productivity and commercialization, which support an enlarged and urbanized population, demand standardization of agricultural production. The application of genetics to crop improvement and the availability of large collections of crop genetic resources provide the means to produce widely adapted and high yielding crop varieties.

Modern times are characterized by worldwide pressure to replace local, traditional, and diverse crops with uniform, nonlocal, and scientifically bred varieties that result in greatly simplified crop populations. Replacement of local and diverse crop populations has occurred in many areas — the American Corn Belt, India's Punjab, and central Luzon of the Philippines are well-documented cases (Griliches 1957, Dalrymple 1986). The biological consequence is genetic erosion. Genetic erosion in crops is the loss of variability in crop populations. Genetic erosion implies that the normal addition and disappearance of crop traits and types in a population is altered so that the net change in diversity is negative. The manifest causes of genetic erosion are changes in cropping patterns and the diffusion of modern varieties from crop improvement programs. As Hawkes (1983, 109) warned: "The situation is now frighteningly clear. Genetic erosion has been taking place partly, though not entirely, as a result of the plant breeders' successful activities in breeding better varieties, which have been gradually replacing the older populations of primitive forms and land races in the regions where diversity has been the greatest." The latent cause is the condition of the modern world — population growth, technology development, markets, cultural change, and the urge to overcome poverty.

Although we lack specific data on the diversity that existed before modernization, there is no doubt that genetic erosion affected Europe and North America, where local crops were drastically reduced in brief periods. The emblematic case history of genetic erosion is the diffusion of hybrid maize across the United States between 1925 and 1950 (Griliches 1957). During this brief span, highly distinctive, local landrace populations of maize were replaced by hybrids that share a handful of parental lines developed in the latter half of the nineteenth century and that still carry the names of farmers who identified them — Reid, Lancaster, and Krug (Goodman 1995).

Agricultural changes occurring elsewhere in the world are believed to follow a similar trajectory. In particular, genetic erosion is encouraged by the development of international agricultural research centers of crops that have produced highly successful crop germplasm, which has been distributed either directly or after minor alterations at the national level (Dalrymple 1986). Short-stature wheat from CIMMYT, often referred to by farmers as "Mexican," and IR-8 rice from the International Rice Research Institute (IRRI) were precursors of generations of improved crops based on germplasm issuing from international research centers after the mid-1960s. In addition to international breeding programs, national programs were organized in many countries and supported with international financial aid for economic development. Publicly supported research and extension, which contributed to agricultural development in the United States, Europe, and Japan, was a logical model for less-developed countries facing food deficits and rural poverty. By the mid-1970s, the impacts of this so-called Green Revolution were evident in centers of crop domestication and diversity—increased production and productivity but loss of local crops. The heartland of wheat domestication and diversity in Southwest Asia is an example. By 1997, 59 percent of the Southwest Asian winter wheat area was planted with semidwarf wheat varieties (Aquino, Carrión, and Calvo 1999). This percentage is even higher in Fertile Crescent countries that have long been considered the epicenter of wheat domestication. For instance, Lebanon reported that 83 percent of its winter wheat area in 1997 was planted to these modern varieties.

As hybrid maize greatly shrank the area of traditional maize in the United States, little concern about genetic erosion was expressed. However, when modern crops made inroads into the area of traditional crops in centers of crop domestication and diversity, a worldwide effort was launched to conserve genetic resources. Returning from a field expedition for collecting barley in the Middle East in the early 1930s, the American plant explorers H. V. Harlan and M. L. Martini (1936) gave an early warning about the problem of genetic erosion in crops. Referring to barley in northern Africa and Asia they wrote: "The progenies of these fields with all their surviving variations constitute the world's priceless reservoir of germplasm. It has waited through long centuries. Unfortunately, from the breeder's standpoint, it is now being imperiled. When new barleys replace those grown by the farmers of Ethiopia or Tibet, the world will have lost something irreplaceable" (cited in J. R. Harlan 1970, 618–19). Remarkably, Harlan and Martini's warning occurred soon after the initial deployment of genetics in crop science and soon after the organization of national crop improvement programs in countries with crop diversity, such as Turkey and the Soviet Union, well before massive conversion to modern varieties.

The genetic erosion concept emerged forcefully between 1965 and 1970, a period when crop improvement had clearly demonstrated its power to transform local crop populations in both industrialized countries and less-developed regions. Of particular importance was the rapid diffusion of semidwarf wheat and rice varieties in Asia. A benchmark in defining the problem of genetic erosion for crop populations was the volume *Genetic Resources in Plants* (Frankel and Bennett 1970), resulting from the 1967 Technical Meeting of the Food and Agricultural Organization of the United Nations on plant exploration and introduction. Frankel's preface lays the groundwork for the subsequent chapters by asserting that "it is now generally recognized [that] many of the ancient genetic reservoirs are rapidly disappearing" (Frankel 1970, 2).

However, public concern for genetic erosion is similar to the situation of soil erosion in that attitudes have been fueled more by a few dramatic cases and logical suppositions than by thorough empirical research (Trimble and Crosson 2000). Phenomena such as the diffusion of hybrid maize and the 1930s dust bowl in the United States rightly alert us to problems, but extrapolation from these cases requires data and analysis across diverse environments. Unfortunately, it is often easier to rely on the emblematic case rather than build a body of cases from many places. In fact, both the record of the 1967 FAO meeting (Bennett 1967) and the Frankel and Bennett (1970) volume are distinguished by a lack of data and/or analysis of specific case studies of genetic erosion. In the subsequent volume, *Crop Genetic Resources for Today and Tomorrow* (Frankel and Hawkes 1975), three contributions partly make up for the absence of data. The first is Frankel's (1973) summary of the 1970/71 FAO *Survey of Crop Genetic Resources in Their Centres of Diversity*. This survey is described as being (of necessity) "somewhat superficial and incomplete" (Frankel 1973, x), relying mostly on the anecdotal observations of plant explorers. The second and most systematic research effort on genetic erosion reported in the 1970 survey failed to find genetic erosion in Mexican maize (Hernández X., 1973b). Nevertheless, Frankel (1973) drew the general conclusion that genetic erosion was demonstrated in the survey. The third contribution to present data on genetic erosion was Ochoa's (1975) report on potato collecting in Chile, Peru, and Bolivia. Ochoa confirms reports by others of declining numbers of primitive potatoes in South America. On the Chilean island of Chiloé, for instance, collectors found nearly the same number of primitive potato varieties (two hundred) in 1928 and 1938 but "not much more than half" that number in 1948, "even fewer" in 1958, and only thirty-five to forty in 1969 (Ochoa 1975, 167–68). Similar results are reported in northern Peru, where collection in one village yielded twenty-five native samples in 1955 and none in 1970 (Ochoa 1975, 169).

Conceptualizing genetic erosion must account for the segmented nature of crop diversity across thousands of farms and different environments, and for the negative effect of shrinking the area of local crop populations. After thirty years of crop collection and research following the FAO conferences, the concept remains more a presumption of what is likely to occur than a demonstrated fact. In re-reading the volume edited by Frankel and Bennett (1970) it is possible to see major unresolved issues that pose serious problems for the concept of genetic erosion. Efforts to estimate genetic erosion have been stymied by the lack of longitudinal (time-series) data and different taxonomic approaches. Sporadic efforts since 1970 to document genetic erosion have yielded mixed results. Hammer et al. (1996) compared collections of vegetable varieties in Albania and Italy made at different time periods and concluded that genetic erosion had occurred at similar rates (~72 percent) in both places. In contrast, ecological and cross-sectional research on major crops in a number of different centers of diversity suggest that describing genetic erosion is a more difficult task (Brush 1995). The most frequently cited evidence for genetic erosion is indirect—the diffusion of modern and high yielding varieties into areas once known for crop diversity (Hawkes 1983). Taxonomic problems of describing and analyzing crop populations and species are reviewed by Baker (1970) and Harlan (1970). Nevertheless, these issues have not led to a fuller discussion of the genetic erosion concept.

The Genetic Erosion Concept Applied to Crops

The concept of genetic erosion of primitive crop varieties is elaborated by Frankel (1970), and it relies heavily on a model of crop population structure in centers of diversity developed by others, notably Bennett (1970). Frankel's (1970) definition rests on five principles:

1. Diversity in crops exists because of adaptation by localized populations that are isolated in dissected landscapes.
2. Premodern agriculture in centers of diversity is stable in terms of production and crop composition.
3. Introduction of modern (exotic) agricultural technology, including modern varieties, leads to instability in the farming system in terms of the composition of crops and varieties.
4. Competition between local (diverse) and introduced varieties results in displacement of local varieties.
5. Displacement of local varieties reduces the genetic variability of the local crop population.

Landraces are "traditional" in the sense of being direct descendents of previous local crop populations, in contrast to "modern" crops, whose parental lines are predominantly exotic. Bennett (1970) suggests that crop diversity and adaptation are linked in two ways. First, adaptation results in "ecotype" formation and stabilization—the creation of "biotypes possessing a number of adaptive features which are genetically linked" (Bennett 1970, 170). Bennett (1970,

127) labels this "adaptive differentiation," using the same term as Harlan (1970). The differentiation of crop populations reflects the clinal and mosaic patterns of genetic variability in wild plants described by Allard (1970). Second, Bennett suggests that variability exists in cultivated populations because it confers adaptability to the population—an allusion to the diversity/stability concept.

The idea of stability of premodern crop populations derives from the supposition of adaptation. Harlan's (1975, 618) classic definition of landraces describes them as "balanced populations—variable, in equilibrium with both environment and pathogens and genetically dynamic . . . the result of millennia of natural and artificial selections." The agricultural image here is the equivalent of a plant community's "climax stage," made up of locally adapted populations that remain relatively unchanged as long as the local environment remains stable. The concept of balance repeats the definition given by Frankel and Bennett (1970, 7)—"land races are crop populations in balance with their environment and remain relatively stable over long periods of time."

In the crop model, the stability of primitive agriculture remains largely undisturbed, albeit with changes resulting from migration, gene flow, and adaptation, until the twentieth century. The watershed events were the rise of Mendelian genetics and pure line theory in modern crop biology and industrialization in Europe and North America. Earlier changes in crop populations, such as the diffusion of crops across hemispheres during the "Columbian Exchange" (Crosby 1972), paled in comparison to changes occurring at the turn of the twentieth century. The rapid transformation of European and North American agriculture presaged transformations in less-developed and genetically more diverse regions that Harlan and Martini (1936) detected. The diffusion of modern crop varieties is repeatedly identified as a critical, destabilizing element in centers of diversity, although changes such as improved soil fertility are also recognized.

Frankel's (1970) fourth principle of genetic erosion emphasizes that rivalry between indigenous and modern crops strongly favors the latter and that these two types are mutually exclusive. Exclusivity results from the fact that indigenous crops are adapted to the conditions of less-developed agriculture such as "crude land preparation . . . [and] low soil fertility" (Harlan 1975, 618). As these conditions change with improved traction and fertilizer, the existing adaptation of landraces changes from asset to liability. Modern cultivars, in contrast, were specifically adapted to higher fertility. The speed of modern cultivar diffusion indicates the determinant role of yield advantage. The replacement of landraces by modern cultivars also is associated with the transformation of agriculture from subsistence to commodity production. In the former, multipurpose crops for household consumption are needed, often supplying different products such as grain and straw. In commodity production for urban markets, grain yield becomes a singular criteria of production.

Here, integration of local farming systems into the stream of industrialized inputs, technical information, and commercial crops is assumed to reduce the number of distinct habitats in the agricultural landscape. Habitat diversity in agriculture is logically affected by the technology and social and economic context of farming systems, but the large number of physical, biological, and human factors that might be included confound the measurement of habitat diversity. The degree of intensification, measured by the frequency of cultivation in a total agricultural cycle, provides one approximation of habitat diversity within a farming system (Brush and Turner 1987). This measure is, however, less useful as we move away from shifting cultivation. Energy inputs, the source of inputs, and yield are other measures (Evans 1993), but energy flow and yield are notoriously slippery. Energy might be more usefully connected to habitat diversity by considering the balance of energy that is provided locally rather than through purchased inputs. Household production for subsistence versus commodity sales is another proxy for habitat diversity, as is the frequency of polycropping versus monocropping. Finally, measures such as farm size, number of individual parcels per farm, and the dispersion of parcels are indicative of overall habitat diversity in a farming system.

Frankel's (1970) final principle is a corollary of the first four—that displacement of indigenous crop varieties leads to loss of genetic variability—or genetic erosion. Local adaptation and exclusivity of crop type (traditional or modern) are especially important, so that reference to the presence of modern varieties in a farming system is taken as prime facie evidence of genetic erosion. Indeed, much of the evidence for genetic erosion presented in the 1970/71 FAO survey (Frankel 1973) is data on the diffusion of modern cultivars rather than on the loss of local material (e.g., Kjellqvist 1973). In other words, concern for genetic erosion benefited germplasm collection but not crop ecology or ecogeography.

The concept of genetic erosion of crop genetic resources outlined above was developed by scientists engaged primarily in plant collection for crop breeding and secondarily in research on the botany and systematics of crop species. The observation of genetic erosion was largely anecdotal and only rarely made as a result of a focused research effort. Indeed, the work of Hernández X. (1973b) on maize in Mexico is an exception that suggests problems with the concept as developed by Frankel and Bennett. The sense of urgency that imbues Frankel and Bennett (1970) explains why focused research on the topic was not an early priority. Nevertheless, the fact that large proportions of the total diversity of several crops were safeguarded in gene banks by 1985 (Plucknett et al. 1987) did not then spur crop scientists to invest in closer scrutiny of the issue of genetic erosion in centers of diversity.

One reason for the lack of research on genetic erosion is its difficult demands for data. The concept describes processes that take place over time and that require measurements of biological variability in fragmented and heterogeneous environments. Time-series data are not generally available for crops or agriculture in centers of diversity, and different or even incompatible measures of biological variability are used at different times. Data on premodern crop populations are rare because the questions posed by genetic erosion arose after these populations had been affected by technological change. Following Johannsen's experiments that he conducted at the turn of the twentieth century and that clarified the distinction between genotype and phenotype, European and American plant breeders aimed to remove or reduce diversity in crop populations, thereby improving yields (Poehlman 1995). Data on either crop diversity or the extent of modern varieties in centers of diversity were virtually nonexistent in 1970 and remain spotty and unreliable. Measures of crop diversity are costly and difficult to acquire, and they are still being developed. Baker (1970) anticipated the difficulty of measuring variability in his discussion of the need for better taxonomy of cultivated plants based on the species concept. Another problem is the complex association between area and diversity that was revealed in theoretical and empirical work in *Island Biogeography* (MacArthur and Wilson 1967). Spanning more than three decades, research on the relationship between area and diversity has generally confirmed a positive association, but a less-ambiguous association is found by looking at habitat diversity, which is a function of area (Rosenzweig 1995). However, as noted here and in Chapter 3, criteria for indicating habitat diversity in agriculture are not clearly articulated or understood. Crop scientists who championed the issue of genetic erosion tended to gloss over the complex and sometimes inconsistent nature of the area/diversity relationship.

In light of these difficulties in a formal assessment of genetic erosion, two steps can help. First, alternative methods to the direct, longitudinal observation of genetic erosion are needed. The remainder of this chapter is devoted to these methods and results obtained in research. Second, a theoretical basis to the concept must be formulated. The next chapter discusses the theoretical basis of genetic erosion.

Ecological and Cross-Sectional Assessment of Genetic Erosion

The lack of time-series data on crop populations and agricultural systems is a serious but surmountable problem in research on genetic erosion. After all, great progress has been made in understanding crop domestication and evolution without direct observation of the process. Two key methods are ecological analysis of crop populations and cross-sectional analysis of crop management among different types of farms, villages, and regions. Ecological analysis is helpful to understanding the structure and distribution of crop populations and addressing the area/diversity issue. The essence of cross-sectional analysis is to treat diversity (e.g., the planting of landraces or variability in crop populations) as a dependent variable and the supposed causes of genetic erosion (e.g., adoption of modern varieties or increased commercialization) as independent variables. In practice, cross-sectional analysis of genetic erosion relies on farm surveys, systematic crop collection, and multiple regression analysis. For instance, the effect of using modern varieties on the status of indigenous varieties or on their variability can be estimated by comparing farms that use modern varieties at different rates. Obviously, ecological and cross-sectional research can be applied only in regions where indigenous crop populations still exist. Moreover, this method of investigating genetic erosion works best in situations where both modern and indigenous crops are present. It is not appropriate in places where indigenous crop varieties have disappeared or where modern conditions are absent.

The genetic erosion model articulated by Frankel (1970) provides four hypotheses that can be tested through cross-sectional comparison:

1. Indigenous crop varieties have limited distribution.
2. Indigenous crop varieties experience slow turnover.
3. Use of purchased inputs and modern crop varieties is inversely correlated with the cultivation of indigenous crop varieties.
4. Adoption of modern crop varieties is inversely correlated with crop population diversity.

These hypotheses are similar to a large class of propositions that social scientists have examined in relation to agricultural development and the diffusion of new technology. Major themes in explaining the patterns of agricultural change include the impact of environmental heterogeneity, social constraints to adoption of technology, the role of risk, and the effect of markets and missing markets. This social science research provides a set of key variables related to genetic erosion resulting from agricultural development and technology diffusion. These variables are useful in estimating the extent of technology diffusion and the contexts that might limit it. Four classes of social and environmental variables are useful to analyzing the diffusion of new agricultural technology and its impact on local crop populations:

1. farm characteristics (e.g., farm size, number of parcels, soil heterogeneity, geographic location [e.g., altitude], and irrigation),
2. household characteristics (e.g., education, wealth, credit availability, age, and labor availability),
3. economic strategies (e.g., commercialization, risk avoidance, use of purchased inputs, and off-farm employment),
4. village characteristics (e.g., proximity to markets, presence of agricultural extension agents).

The four hypotheses of genetic erosion listed above have been researched with cross-sectional methods in several centers of crop diversity. Without exception, they are problematic.

1. INDIGENOUS CROP VARIETIES HAVE LIMITED DISTRIBUTION.

The segmentation of crop species into local, highly distinctive populations stands at the crux of the genetic erosion issue. This segmentation apparently offers a ready answer to the pivotal question, "Why so much diversity?" However, research shows that this question is not so easily answered and that a much more complex and dynamic pattern of crop population structures exists than supposed by the genetic erosion hypothesis. Crop populations of farms, villages, and regions are often dominated by a few, widely distributed varieties. Populations of "traditional" or "local" varieties show moderately rapid turnover. Exchange of varieties among farms, villages, and regions is ubiquitous and often far-flung. A logical supposition is that farmers constantly seek the most competitive and best-performing varieties and are not limited to their immediate surroundings in this search. Moreover, it is highly possible that farmers have also identified "generalist" varieties to meet heterogeneous agricultural conditions that they have limited means to control.

Two types of ecological and genetic data from a number of centers of crop diversity challenge the supposition of limited distribution of traditional crop varieties. First, evidence at the regional level indicates that a small number of landrace varieties are cosmopolitan and dominant. Second, crop research commonly concludes that most diversity is found within rather than between populations.

Generalist Potatoes

In the Peruvian Andes, potato farmers in widely dispersed agroecological zones regularly acquire seed from a limited number of localities in the

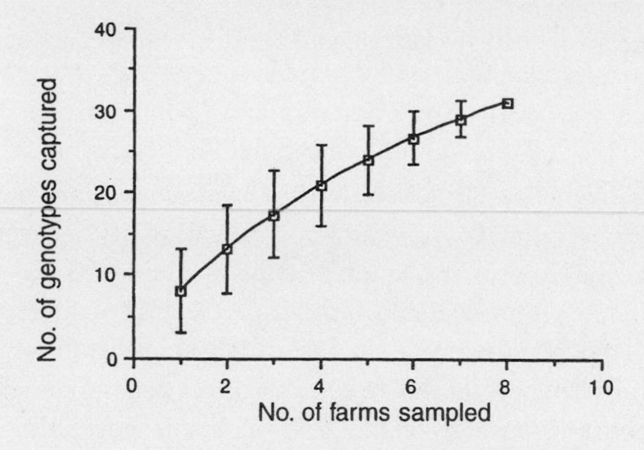

Figure 7.1 Potato genotypes found on farms sampled in Paucartambo (Brush et al. 1995) (Reprinted with permission from *Conservation Biology* © 1995 Blackwell Publishing)

region (Brush, Carney, and Huamán 1981). In central Peru, farmers on the dryer western range of the Andes traditionally acquire seed potatoes from a seed market on the wetter eastern range. Brush (1992) and Zimmerer (1991) found that individual native potato varieties are widely distributed in the regions studied. Three of the six species of cultivated potatoes studied showed adaptation to specific altitude zones, but the most common Andean subspecies, *Solanum tuberosum* subsp. *andigena,* is best described as a versatile generalist that does well across contrasting agroecological zones (Zimmerer 1991). Zimmerer's (1991) experiments with cultivars of the same species showed that they were only weakly adapted to altitude, the overriding environmental factor of the Andes. This contradicts the idea that diversity in this crop is a result of finely tuned adaptation. Villages control highly contrasting and heterogeneous landscapes with steep environmental gradients, and within villages native potatoes are frequently moved in mixed lots between fields with different soils, drainage, exposure, and altitudes. A significant number of the potato varieties found in one village are found in neighboring villages, and certain native varieties are highly cosmopolitan. The plasticity of Andean potato cultivars is suited to the common practices of moving cultivars from field to field with different microenvironments and acquiring seed from distant locations. At the genetic level, there appears to be a high rate of migration of both alleles and genotypes (Brush et al. 1995). In fact, the migration of alleles among farms is so high as to create complete mixing among farms. As a result, the number of distinct genotypes that one finds in collecting among localities plateaus at a relatively small number rather than increasing arithmetically.

Figure 7.1 shows the number of novel potato varieties found on individual farms in the Paucartambo Valley of southern Peru. By the eighth farm, all thirty genotypes (defined by isozymes) found in valley are included in the collection.

High Diversity Within Landrace Populations

A crop's diversity is ultimately derived from genetic variability among its many populations, estimated by examining the variability of specific traits and characters. Morphological or qualitative traits, such as seed and stem color or the presence of hairy glumes; quantitative traits, such as plant height or days to flowering; and biochemical traits, such as seed proteins or allozymes have been used to measure diversity in the absence of direct genetic measures. Genetic data have become increasingly available in the last decade, although they have not yet supplanted the more conventional measures of diversity for describing crop population structure. In addition to measuring diversity, the distribution of diversity among populations is relevant to identifying centers of origin, patterns of adaptation, and biological divergence between regions. A major data point in the distribution of diversity is the relative amount of diversity that is attributable to variation within populations versus the amount due to variation between populations. Theoretically, a high amount of between-population diversity indicates divergence and adaptation of local populations, while a high amount of within-population diversity suggests less divergence.

Ethiopia has been an unusually fertile area for research on crop diversity, especially in barley and tetraploid (durum) wheat. While Ethiopia is not a center of origin of either of these crops, it has been known since Vavilov's time as a region of unusual diversity and a source of valuable genetic traits. Over a dozen different diversity studies of durum wheat or barley from Ethiopia have been published since 1980 using phenotypic and isozyme analysis. Seven studies contrast within-population and between-population diversity. Table 7.1 summarizes these studies, showing that, in five out of the seven cases, within-population diversity was found to be greater than between-population diversity. Alemayehu and Parlevliet (1997, 188) observe that "within landrace variation appeared so large, that one can conclude that most plants in an Ethiopian barley landrace represent different genotypes."

High within-population diversity provides biological flexibility and resilience to the crop and allows it to resist environmental stress and variability from year to year, but the diversity is not spatially partitioned according the genetic erosion model. Biological resilience provided by diversity is especially important in premodern agriculture that lacks other means, such as irrigation, chemical fertilizers, or pesticides, to limit the effects of environmental perturbation and stress. In other words, diversity within populations might be considered to be a type of risk insurance for the farmer. However, the preponderance of within-population diversity also suggests that specific crop populations

Table 7.1. Partitioning of diversity in Ethiopian crops

	Crop		Characters analyzed		
Study	Durum wheat	Barley	Morphology	Isozyme	Within-population diversity > between-population diversity
1	•		•		•
2	•		•		
3	•			•	•
4	•		•		•
5		•		•	•
6		•	•		•
7		•		•	

(1) Pecetti and Damania 1996; (2) Tesfaye, Getachew, and Worede 1991; (3) Tesgaye, Becker, and Tesemma 1994; (4) Belay et al. 1997; (5) Bekele 1983; (6) Demissie and Bjørnstad 1996; (7) Demissie and Bjørnstad 1997.

among environments and geographic regions are not strongly distinguished from others and that populations are generalists rather than specialists in confronting the agricultural environment. The data suggest that landraces do not usually comprise narrowly distributed and geographically distinctive genotypes but rather swarms of genotypes with only weak hierarchical distribution.

High within-population diversity does not necessarily mean that the Ethiopian crops are exceptions to the well-established pattern of finding increasing diversity as larger areas and more habitats are sampled (Rosenzweig 1995). Table 7.1 is based on a limited number of observed crop traits rather than on a count of distinct genotypes that are found in the collections from different regions. Nevertheless, the Ethiopian case is a cautionary tale for estimating the gains and losses of crop traits or crop diversity either as we move across an agricultural landscape or during changes in farming practices. Reducing the number of farms that grow local or traditional crop varieties is likely to result in fewer distinct agricultural habitats and thus fewer crop genotypes, but this may not be fully commensurate with loss of variability in agronomic traits that we can measure. An analogy is the range of residents' eye colors in large and small U.S. towns. The smaller town has fewer human genotypes but not necessarily fewer eye colors.

2. INDIGENOUS CROP VARIETIES EXPERIENCE SLOW TURNOVER.

Social scientists, who began documenting farming systems in the 1970s, noted that traditional farmers in supposedly isolated and closed economies were engaged in seed networks that moved seed across wide areas and heterogeneous landscapes. By the late 1990s, this practice had become so widely reported that Zeven (1999) noted that seed replacement itself might be considered "traditional." Zeven (1999) places this practice into historical context with references to biblical times as well as medieval and pre-industrial Europe. Not only seed replacement but also long-distance movement of seed has a long historical record that is reflected among modern, albeit "traditional," farmers. Case studies of rice and maize seed management illustrate the patterns of high turnover among "local" or "traditional" crops in centers of diversity.

J. V. Dennis (1987) found that frequent variety turnover was an important part of traditional rice agriculture in Thailand. Dennis reports that farmers sow 1.7 varieties per farm, replace them, on average, every three years, and regularly acquire indigenous rice varieties from distant locations. Comparisons of rice accessions in six administrative districts in the period 1950–61 with those in 1982–83 reveal the extent of turnover. Of eighty-nine varieties in 1950–61, only fifteen were found in 1982–83, and eighty-two varieties found in these districts in 1982–83 were not present in 1950–61. The rates of variety turnover described by J. V. Dennis (1987) for traditional rice agriculture in Thailand are comparable to the rate of turnover of wheat varieties in modern agricultural systems (Brennan and Byerlee 1991).

Louette conducted intensive research on the distribution and movement of maize in the village of Cuzalapa in the west-central state of Jalisco, Mexico (Louette, Charrier, and Berthaud 1997; Louette 1999; Louette and Smale 2000). Twenty-six varieties are grown in the village, nine "major" and seventeen "minor." Six of the major varieties are "local" because they have been grown locally for at least thirty years, and three are "foreign"—recent introductions. All of the minor varieties are foreign. The six local varieties accounted for 69.4 percent of the maize area. Table 7.2 summarizes the data from Louette, Charrier, and Berthaud (1997) relating to the frequency of seed exchange. A striking finding in Cuzalapa is the relatively high rate of the use of seed lots from other communities (Louette, Charrier, and Berthaud 1997). Fifteen per cent of the seed lots of the predominant local variety were acquired from other communities, and 11 percent of all seed lots are reported to have come from other regions (Louette, Charrier, and Berthaud 1997). Seed exchange, both within and outside Cuzalapa's valley, is so important that the authors conclude that this "traditional" community is an open system. They also conclude that seed flow is high enough so that no farmer is planting seed stock bequeathed from parents.

These studies of seed exchange affirm that "traditional" farming systems are more dynamic and open than envisioned in the conventional model of genetic erosion of crop resources. Zeven (1999) describes these findings as "surpris-

Table 7.2. Origin of maize seed (percentage) in Cuzalapa, Mexico

	Own seed	Seed acquired in Cuzalpa	Seed acquired from another community
Local varieties	57.9	34.2	7.9
Foreign varieties (major)	41.2	52.6	5.3
All seed lots	52.9	35.7	11.4
Area	44.9	39.9	15.1

Source: Louette, Charrier, and Berthaud 1997

ing" and "unexplained." Nevertheless, they are so often reported as to effectively challenge the previous belief in stable, localized crop populations that underlies the genetic erosion hypothesis. These findings suggest that crops in centers of diversity are not assemblages of locally endemic or relatively static populations. Rather, they indicate that landraces are often made up of relatively plastic, "general purpose" (Baker 1965) genotypes whose adaptation is wider than local.

3. USE OF PURCHASED INPUTS AND MODERN CROP VARIETIES IS INVERSELY CORRELATED WITH THE CULTIVATION OF INDIGENOUS CROP VARIETIES.

As noted above, the use of purchased inputs is logically associated with a reduction of habitat diversity in agriculture. In developed countries, increasing integration of farms into industrial input streams and commodity flows has been associated with agricultural restructuring defined by a reduction in the number of farms and an increase in average farm size. In turn, there should be a decrease in the habitat differences that arise from farm management. The genetic erosion model is predicated on modern varieties' broad adaptability and strong fertilizer response and the lack of these advantages in indigenous varieties. However, two common aspects of farming systems in centers of diversity complicate the competitiveness of modern varieties and a simple association between the use of purchased inputs and the decline of indigenous varieties. First, the competitiveness of modern varieties is based on their yield advantage, but yield is only one of several selection criteria. Second, the weight of the fertilizer/yield response may be judged differently by farmers who operate in heterogeneous farming systems with different parcels devoted to the same crop.

Quality and stability have been identified as criteria that influence selection decisions in addition to yield (Bellon 1996). Quality characteristics include storage, cooking aspects, and demand in the market or in nonmarket exchange. Numerous ways of using and consuming staple crops exist and may involve complex selection criteria. In Peru, for instance, potatoes are evaluated not only according to production criteria but also on whether they are suited for *watia* (baking in field ovens), steaming, soups, frying, drying, or freeze-drying (see Chapter 5). *Chalo* potatoes, diverse mixtures of indigenous varieties, are valued as gift items and used as in-kind payments to attract workers at times of peak labor demand.

In Mexico, the dietary staple is maize tortillas — unleavened cakes, cooked on a griddle without oil, made with flour of maize kernels soaked in lime water to remove the hard outer layer. There is no direct link between tortilla quality and maize diversity (Perales R. 1998), but maize has many more uses in Mexican cuisine than for tortillas alone. Hernández X. (1985) observed that the diversity of maize races within and between regions can be explained partly by use. In his regional study of maize in northern Mexico, Hernández X (1985) found dents (e.g., *Tabloncillo*) used for tortillas and fermentation; flints (e.g., *Amarillo Cristalino*) used for fresh maize, sweetened flour, and animal feed; floury types (e.g., *Blanco*) used for cookies and parched maize soup (*pozole*); and popcorns (e.g., *Chapalote*) used for sweetened confections and popping. Maize is ubiquitous in Mexican cuisine, and not merely as an ingredient for tortillas. Mexico's Muséo Nacional de Culturas Populares (1982) lists 605 maize dishes, an impressive but almost certainly low estimate of the crop's versatility and the inventiveness of Mexican chefs.

A key consideration in weighing the importance of quality characteristics in selection is the ability of markets to provide for the different qualities that concern farmers in centers of diversity. Theoretically, a Mexican maize farmer who wants to have some maize for pozole soup as well as for tortillas can grow only the most profitable type and acquire the other type in the market, thus using his land and labor most efficiently. However, the farmer must be confident that the other type will be available in the market and that the cost of locating and buying it will be reasonable. In the farming villages of centers of diversity, such smoothly operating markets are frequently missing or deficient (de Janvry, Fafchamps, and Sadoulet 1991). The appropriate selection decision under such conditions is to trade yield for other quality concerns. The existence of a market for purchased inputs such as fertilizer does not ensure that a local market for specific varieties or types of varieties will exist. The response to missing markets is to maintain diverse varieties for household consumption.

The hypothesis that purchased inputs are inversely related to maintenance of indigenous crops falters on the heterogeneity and fragmentation of farming systems in centers of diversity. The critical issue is not whether purchased inputs will be used but whether they inexorably lead to the disappearance of local crops, for instance by reducing habitat diversity in farming systems. The use of purchased inputs is nearly ubiquitous in agriculture around the world, and the use of inputs such as fertilizer, pesticides, or mechanized traction can be used on both local and modern crop varieties. However, purchased inputs often benefit improved crops more than local landrace varieties (Cleveland, Soleri, and Smith 1994), so that purchased inputs are also often assumed to indicate adoption of new varieties. Whether purchased inputs flatten habitat diversity in agriculture and by how much still requires research. Moreover, the use of purchased inputs and the adoption of modern crop varieties on one part of the farm may not threaten the continued use of local crops on other parts. Bellon (1996) points out that farmers have a set of five general concerns that help explain infraspecific diversity: (1) environmental heterogeneity (soils, temperature, rainfall, etc.); (2) pests and pathogens; (3) risk management (drought, lodging, frost); (4) culture and ritual; and (5) diet. The first three of these categories are concerned with production conditions in places with steep environmental gradients and uneven socioeconomic conditions. Mountainous terrain, lack of inputs to mitigate production constraints (e.g., poor soils, water deficits), and an abundance of co-evolved pests and pathogens contribute to steep environmental gradients in centers of diversity. A result of the steep environmental gradient is the division of local farming systems into different production zones with different crops, production activities, and intensities of use (Brush 1977, Zimmerer 1996).

A common response to steep environmental gradients is to divide the farming system into numerous parcels at the farm level. Such fragmentation might follow the contours of production zones in a mountainous landscape, but it also is found where physiographic factors are less obvious, for instance across soil profiles. Moreover, inheritance practices, land tenure, and risk reduction strategies also lead to fragmentation (Cole and Wolf 1974, Goland 1993). The result is a mosaic of parcels across a village landscape in which single households cultivate a dozen or more dispersed parcels of the same crop. A consequence of fragmentation is that technology adopted for increasing yields in one zone or parcel may be less effective in others, thus dampening the impact of purchased inputs on selection.

Peru epitomizes a mountainous landscape divided into production zones. The Andes create a gradient of moisture, temperature, evapotranspiration, soils, and vegetation that is observed at different scales, from the macroscale of a trans-Andean transect to the microscale of production zones for one crop in a single valley. Starting at the scale of a village, general production zones (e.g., crops and pasture, tubers and cereals, bitter tubers and nonbitter ones) are evident, and a single zone is subdivided into specific zones for certain crops or crop varieties (Brush 1977). The potato crop is particularly affected by this fragmentation. One common division is by altitude with levels that are worked in different cycles, with different types of the tubers and different treatments (Brush 1992, Zimmerer 1996). A low- to moderate-altitude potato zone (2,500–3,000 masl) is traditionally planted in the short-cycle cropping period (August–March), with emphasis traditionally on diploids (e.g., *S. phureja*) but recently on modern varieties of *S. tuberosum* and relatively high amounts of fertilizer and pesticides. A mid-altitude potato zone (3,000–3,600 m) is planted in the long-cycle cropping period (August–June), with emphasis on *S. tuberosum* subsp. *andigena* and lower amounts of fertilizer and pesticides. A high zone (3,700–4,100 masl) is planted in the long-cycle period with emphasis on bitter species (e.g., *S. juzepczukii*), and with little fertilizer or pesticides. As Zimmerer's (1996) detailed study of one valley in southern Peru shows, these general zones can be subdivided into even smaller areas, depending on microenvironmental characteristics such as slope, drainage, and soils. Typical farm families may have as many as eight to twelve different parcels for potatoes alone, and in some areas the number is much greater (Brush, Taylor, and Bellon 1992). The number of parcels per family is partly dictated by the Andean custom of sectoral fallow managed by the community (Orlove and Godoy 1986), as well as by efforts to reduce risk (Goland 1993).

While Peru's natural heterogeneity provides an environment where farming systems are easily fragmented into production zones and numerous parcels, similar patterns exist in places with more moderate environmental gradients. In Mexico, Bellon has shown that soil differences are significant in the selection of different types of maize (Bellon and Taylor 1993). His study site in Chiapas varies less than 100 meters in elevation and has a single climate zone. Nevertheless, the farming system is subdivided into two different subsystems, *arado* and *pedregal,* that are managed differently (Bellon 1991). Six soil classes are recognized by farmers, and the communal land management system is organized so as to distribute good and poor soils evenly among the farm households. Families cultivate an average of four maize parcels, distributed between arado and pedregal and among various soil classes. As in Peru, different parcels receive different amounts of purchased inputs and are judged to be more suitable for either indigenous or modern maize varieties. Purchased inputs are used at a high rate, for instance fertilizer is used on all arado fields and virtually all (98.7 percent) pedregal fields. Indigenous varieties persist in this area because they are better adapted to the marginal soils of particular parcels and because they are more tolerant of poorly timed weeding and fertilization that characterizes poorer households (Bellon 1991). Indigenous varieties are described as *aguantadora* (tough, resistant) because of their ability to thrive on poorer soils and to compete with weeds (see Chapter 6).

In sum, field research in centers of diversity has shown that purchased inputs do influence selection decisions, but not to the degree suggested in the genetic erosion hypothesis. Selection is influenced by the heterogeneity of a farming system—natural, social, and economic. Yield is an important criterion, but only one of several that are weighed in choosing crops and varieties. The genetic erosion hypothesis fails to anticipate this heterogeneity in farming systems, selection criteria, and market conditions. This failure limits the hypothesis's ability to explain limits to the diffusion of modern varieties.

4. ADOPTION OF MODERN CROP VARIETIES IS INVERSELY CORRELATED WITH CROP POPULATION DIVERSITY.

The operative variable in this hypothesis is the decrease in area devoted to indigenous crop varieties as modern ones are adopted. One footing of the genetic erosion hypothesis is that modern varieties are in a zero-sum competition with indigenous ones. Thus far I have argued that the heterogeneity of farming systems in centers of diversity limits the diffusion of modern varieties and maintains production spaces for indigenous varieties. Nevertheless, modern varieties have diffused into centers of diversity and caused declines in area devoted to indigenous varieties. This is especially true in more optimal agronomic areas of Vavilov centers, such as irrigated paddy rice in Southeast Asia or the lower-altitude maize zones of Mexico. In other words, tests of Hypothesis 4 give contradictory results. A second footing of genetic erosion is that a direct and positive relationship exists between area and diversity. The genetic erosion hypothesis takes a simple and direct approach to the area/diversity relationship—smaller area in traditional crops reduces diversity (e.g., Hawkes 1983). As I noted in Chapter 3, habitat diversity in agriculture may be more important than area, as is suggested in ecological literature (Rosenzweig 1995), but habitat diversity in agriculture is difficult to define and measure. Indeed, modern technology may increase habitat diversity by encouraging farmers to subdivide their farms or parcels or by overcoming disease constraints that affected traditional management. As long as some areas continue to be planted in indigenous varieties, the relationship between area and diversity is complicated by the population structure of landraces and by the role of conscious selection. Because high amounts of diversity are found within populations, it is plausible that the conservation of a few populations will maintain a large proportion of a crop's diversity.

Assessing the area/diversity relationship requires judgments on (1) the measurement and significance of diversity that is found, (2) the definition of "population," and (3) the definition of area. Theoretically, the amount of diversity of a crop population increases to the limits of the population because each individual potentially is a unique genotype. Moreover, we have expanded the measurement of diversity to include morphological characterization, biochemical characteristics (e.g., isozymes), and molecular markers, and this growth offers numerous ways to increase the amount of variation that can be identified in a population. Measuring diversity depends not only on the method used but also on the unit of analysis—population, individual, genome, locus, DNA base sequence (Kresovich and McFerson 1992). Farm and village inventories of crop landraces are logical population units, but we have seen that these are often open systems—a fact that makes "population" an ambiguous concept.

While increased area will always provide new genetic variation, the question is whether adding diversity by increasing area is significant from the perspective of farmers, conservationists, or plant biologists. There are no easy or definitive answers to this question, but the practice of these groups is to limit the amount of diversity that is considered relevant. Farmers may overlook diversity within populations of field crops in applying a single folk name to the population. Conservationists recognize that some diversity is trivial or otherwise not worth capturing (Marshall and Brown 1975), and plant biologists focus primarily on species level differences and above. In other words, expanding area will add diversity, but this diversity may not add significantly to crop germplasm stocks.

The definition of area is likewise troublesome. Crop population area comprises nested units—farms, villages, microregions, nations, macroregions. All but the smallest of these are vague, and all are potentially arbitrary. Environmental, political, cultural, and economic criteria are obviously pertinent to defining crop population area, but these criteria do not necessarily coincide. In many instances, environments and cultures cross political boundaries, while in others political entities constitute distinct cultures and environments. We might focus on effective crop population area as the proper unit of analysis, defined by isolation from other populations and linked to broadly defined agricultural environments. Isolation and agricultural environment involve physical and ecological parameters, and social factors, such as cultural affiliation, may also count. This type of approach was recommended as "ecogeographic" survey and analysis (IBPGR 1985), but no crop has been thoroughly studied in this manner, and crop ecology suggests that openness rather than isolation characterizes many crop populations.

The lack of analysis of effective crop population areas leaves us without an empirical basis to evaluate the impact of declining area on diversity. Farm level studies offer insights, but when populations are parts of active seed flows, diversity measurements may be as much related to sampling procedures as to area. This sampling problem relates to conscious selection and is illustrated by case studies of cassava, sorghum, potatoes, and maize.

Research by Salick, Cellinese, and Knapp (1997) on the diversity of cassava (*Manihot esculenta* Cranz) among the Amuesha people of the Peruvian Amazon suggests that diversity in Amuesha cassava fields may be unimportant to Amuesha farmers as long as the shaman maintains the pool of diversity in his/her field that can be accessed through exchange (see Chapter 6). Likewise, potato diversity in the Peruvian Andes is not distributed by field size so much as by type of potato according to the farmers' classification (Brush 1992). Andean farmers divide their fields and potatoes into different categories, and some categories and fields are relatively uniform, while others are purposefully diverse. The overall size of a farm is not significantly correlated with diversity because of conscious selection and management of diversity in designated parcels that are a small part of the total potato area (Brush, Taylor, and Bellon 1992). Similarly, distribution of diversity of sorghum (*Sorghum bicolor*) landraces in Ethiopia reported by Teshome et al. (1999) shows that diversity of the smallest fields approaches that of the largest ones.

The role of conscious selection and crop management on diversity is also evident in maize. Researchers in Mexico (Louette, Charrier, and Berthaud 1997; Perales, Brush, and Qualset 2003) show that maize diversity (counting varieties) is most concentrated in minor varieties that are grown in kitchen gardens. In Cuzalapa (Jalisco), for instance, 14 percent of the seed lots account for 65 percent of the maize varieties in the village. Likewise, in central Mexico Perales, Brush, and Qualset (2003) report that one or two white maize varieties are dominant, although farmers commonly keep up to a dozen colored varieties in kitchen gardens. Sampling the normal field area would miss most of the diversity. There is no evidence that diversity was more evenly dispersed in earlier times.

The cassava, sorghum, potato, and maize examples discussed above show that the area/diversity relationship in crops is complicated by conscious selection and management of crop populations. In these cases, seed is handled individually in horticultural fashion rather than in bulk. Horticultural management of seed may increase the effectiveness of conscious selection in maintaining diversity in spite of decreasing area. The effect of conscious selection and management may be less noticeable in field crops in which seed is handled in bulk. Wheat and rice, for instance, may reflect a more direct relationship between area and diversity than the crops previously mentioned. However, measuring field crop diversity at the farm level is hampered by the fact that few varieties are grown on individual farms.

Compiling data on rice diversity in Southeast Asia, Bellon, Pham, and Jackson (1997) report that the average number of rice varieties per farmer is low—usually below two and never higher than three or four. Similarly, research in Turkey's Transitional Zone revealed that 85 percent of the households that planted traditional wheat varieties planted only one of them, and households kept an average of less than two wheat varieties while farming an average of seven wheat parcels (Meng 1997). In field crops, therefore, another measure besides variety number is needed to capture infravarietal diversity. In our Turkey research, for instance, qualitative characteristics of the populations of wheat were used to generate a Shannon index for each randomly sampled field. Another diversity index, using the coefficients of variation of quantitative characteristics, was also generated. However, based on data provided by A. C. Zanatta to E. Meng (1997), neither index was significantly correlated with the area of the sampled field.

Summary

The threat of genetic erosion helped spur a large international effort to collect and conserve genetic resources that are the heritage of generations of crop evolution. The concept of genetic erosion is plausible, but it was formulated without benefit of detailed case studies of crop ecology in cradle areas of crop evolution. Genetic erosion has occurred in many cropping systems, but there is now ample evidence that it is a complex and context-dependent process conditioned by local differences in environment, economy, and culture. In some cases, local conditions have limited genetic erosion. The hypotheses that are the logical underpinning of the genetic erosion concept are disputed by ecological and cross-sectional analysis. Disputing these hypotheses does not warrant a rejection of the genetic erosion concept, but it does indicate the need for greater attention to the theory and methods that we rely on to understand the interaction between crops, environment, and human behavior to describe modern crop evolution.

8

The Ecology of Crop Diversity

Resting on an analogy between crops and natural communities such as forests, the concept of genetic erosion in agriculture is deceptively simple. The concept rests on sound and firmly established connections between species' numbers, area, and habitat diversity. Like old growth forests falling beneath chain saws, traditional crops are swept aside by the tides of modernization — new technology, markets, and cultural homogenization. Crop replacement in Vavilov centers, like deforestation in the tropics, results in the loss of rare and endemic plants and perhaps the total disappearance of ancient populations and gene complexes. As noted in the previous chapter, it has been assumed that modern crop varieties, which are broadly adapted, disease resistant, and high yielding, have a natural and irresistible ability to replace traditional varieties that are locally adapted, disease prone, and lower yielding. This picture accurately reflects modernization in some farming systems, such as the American Corn Belt, irrigated rice in the Philippine lowlands, or wheat in the Indian Punjab. As we have seen, however, this picture overlooks the heterogeneity of farming systems in the Vavilov centers of crop origins and diversity. Field research in these cropping systems suggests that there is no definitive pattern of loss during modernization. Replacement of traditional crops has occurred in certain areas but not in others. In some areas, traditional crops persist within farming systems that also include modern crops. The possibility that these farming habitats and cultures may be resistant to conversion invites us to inquire into the patterns of technological change in agrarian societies. Replacement and maintenance of traditional crops is mediated by numerous factors: differences in soils and moisture availability, pests and pathogens, storage, cuisine demands, attitudes toward new technology, cultural identity, market limitations. Three general concerns can be synthesized from this list:

1. Environment—the limits of modern technology to meet production constraints on marginal land,
2. Risk—instability associated with modern technology and the ability or willingness of people to accept yield instability, and
3. Markets—whether a predictable and stable supply system is available beyond the household and village levels to provide inputs and the products and services of traditional technology such as food quality or ritual goods.

These factors have been shown to predict maintenance and loss of traditional crop varieties (Bellon and Brush 1994, Brush 1995, Meng, Taylor, and Brush 1998), but this apparent conservation may be deceptive and short-lived, and the conditions of genetic erosion may be in place but latent in centers of diversity.

As suggested in the previous chapter, genetic erosion was a plausible "folk model" among crop scientists, but it is a proposition that has not been thoroughly articulated nor explicitly tested. Arguably, this folk model is unable to deal with the complex reality of agricultural diversity and change in Vavilov centers. As concern for the loss of biological diversity of agriculture increases and investment in its conservation is augmented, a more fully articulated theory and model of genetic erosion is needed. A more explicit model of genetic erosion must encompass both the biological and social dimensions of crop ecology and population dynamics. To accomplish this, we might begin by going beyond the limited discussion of genetic erosion in crop ecology and looking to population models from ecology.

Crop Ecology

In the previous chapter we saw that the idea of genetic erosion emerged originally with organized collecting and crop breeding and gained momentum as the effects of crop breeding were observed around the world. Frankel and Bennett (1970) developed a general representation of genetic erosion of crops, but they did not explicitly describe the parameters or functional relationships of different elements of the concept such as crop environment, risk, or markets. At best, we can intuit a set of hypotheses from their work, with the possibility of misrepresentation. Previously, we saw that these hypotheses are problematic. The absence of a more explicit model is suggested by the fact that

Frankel's later and more theoretical work (Frankel and Soulé 1981; Frankel, Brown, and Burdon 1995) does not define or discuss genetic erosion in crop populations. Genetic erosion is mentioned in the later book (Frankel, Brown, and Burdon 1995), but as the erosion of heterozygosity because of genetic drift in small populations. Opportunities to build a more general model, for instance for crops with different mating systems and chromosomes above the diploid level (polyploids), have largely been avoided.

Guarino (1995) outlines a "model for quantifying the threat of genetic erosion" that lists factors to be considered and their weighting for estimating genetic erosion. Factors include a plant group's (taxon) distribution (rare, locally common, widespread, or abundant), environmental risks (drought, flooding, fires, and global warming), crop characteristics (area, availability of modern types, agricultural services, mechanization, chemical input use, and human demography), and wild species' characteristics (distribution and agricultural pressure). Each factor is given a score between 0 and 15; the higher the overall score, the greater the estimated threat that a crop or wild species will experience genetic erosion. For example, no genetic erosion is predicted for an abundant crop species or variety in an area without environmental risk and with increasing area, no modern inputs, and an increasing or static human population. This is a straightforward, indeed linear, approach to genetic erosion. Unfortunately, it is not field-tested, and thus it omits factors that field studies have shown to be highly critical to genetic erosion—fragmentation of parcels across heterogeneous field conditions, markets, and off-farm employment. Likewise, it fails to reflect nonlinear relationships and the highly correlative nature of many of the factors listed. For instance, an increasing population might trigger rapid changes in services, mechanization, and chemical inputs that could lead to rapid genetic erosion. A decreasing population in agriculture is also associated with increasing wealth, mechanization, and the use of chemical inputs (e.g., herbicides).

Another linear treatment of genetic erosion is that of Hammer et al. (1996), who compare 1941 and 1993 inventories of crop landraces in Albania, and 1950 and 1983–86 inventories in southern Italy. This comparison suggests that many landraces have been lost between collecting times. For example, in Albania, they report that five landraces of barley were collected in 1941 but only one in 1993. Hammer et al. (1996) define genetic erosion as the loss of "genetic integrity:"

$$GE = 1 - C_{t2}/C_{t1}$$

where GE is genetic erosion, C_{t2} is the number of landraces collected in the later time, and C_{t1} is the number collected initially. Using this method, Hammer et al. (1996) estimate genetic erosion in Albania between 1941 and 1993 to be 72.4 percent and that of southern Italy between 1950 and 1983–86 to be 72.8 percent.

While the genetic integrity approach of Hammer et al. (1996) has the advantage of simplicity, its usefulness for measuring genetic erosion is questionable for several reasons. There is insufficient information to judge the comparability of collections between the first and second periods. The article notes that exact duplication of collecting methods is difficult to achieve. Unfortunately, it does not tell us how similar (or different) the collections are between the two time periods. How many fields or farms were visited in each case and in each locality? While the terms "genetic integrity" and "genetic erosion" are used, the actual measurement appears to be the number of folk varieties. Folk varieties may reflect relative uniformity within a type, but they may also conceal tremendous heterogeneity within the type (Quiros et al. 1990). It is possible that folk names have changed without the loss of genetic variation—for instance when two folk names are later glossed as only one. To avoid these problems, "genetic integrity" might be based on levels of heterozygosity found at different times, comparisons of specific alleles, or morphological traits in controlled environments. Of course, it is rather easy to outline a project to measure genetic integrity over time, but financing and executing this project is beyond the reach of virtually all crop research institutions.

Qualset et al. (1997) provide an alternative and more useful view of genetic erosion. They observe that it is important to distinguish between gene *replacement* and gene *displacement*. Gene replacement occurs when one group of varieties (genotypes) of a crop is supplanted by another group—for instance when improved varieties are substituted for indigenous ones. In this case, there may be allelic erosion, although it is also likely that new alleles will be added and that the pedigree of the new varieties will be more complex and heterogeneous than the older ones (Smale 1997). It must be noted that gene replacement is a normal, indeed necessary, part of crop ecology and crop evolution associated with seed flows among farmers and farming regions. Gene displacement occurs in a region when one crop is supplanted by another, thereby eliminating the local population of the original crop. This is, potentially, a much more devastating type of change in terms of the loss of genetic variation.

Population Models

Qualset et al. (1997) note that population models, such as those from island biogeography (MacArthur and Wilson 1967) may be applicable to genetic erosion of crop species, but they do not develop this idea. Population models from ecology can be useful to our assessment of genetic erosion of crops because these models also address the issues of diversity, competition, and habitat change. Two caveats must be recognized when borrowing models from population biology for assessing crop issues. First, the population biology models are designed to analyze interspecies diversity rather than infraspecies diversity. In borrowing these models for analyzing genetic erosion, we may be forced to treat populations of the same crop species as different species. Interspecies models might help us analyze competition between different crop varieties, but they may not illuminate genetic change. Most studies of crop genetic erosion stress gene replacement because of the substitution of varieties within a single crop species. While it may be acceptable to use species models as heuristic devices, they cannot be directly borrowed for infraspecies studies because many of the parameters of species models, such as density dependence, are not relevant at the infraspecies level. Secondly, and more critically, species models are designed for situations where natural selection is the only selection force. For crops, however, we must contend with both natural and artificial selection, as Darwin (1896) informed us long ago.

Two approaches to explaining patterns of species diversity—niche theory and metapopulation analysis—are especially relevant to modeling genetic erosion. Niche theory is useful to modeling the conditions that induce genetic erosion, and metapopulation analysis can be used to describe the structure and maintenance of crop populations.

NICHE THEORY

Niche theory directs us to environmental heterogeneity as a source of diversity (Whittaker and Levin 1975). Conventionally, niche is defined by the status of the species in its community, for instance by the size of an animal and its food habits (Elton 1927). In more contemporary terms, a niche is the multidimensional space that is unique and exclusive to every species (Hutchinson 1957). The limiting factors that determine a specie's distribution and range define this space, and, from niche theory's inception, competition between species in single environments has been a major concern. The Lotka-Volterra competition model developed in the 1920s and 1930s formalized Gause's principle of competitive exclusion and was summarized by Hardin (1960, 1292) as "complete competitors cannot coexist" (see Whittaker and Levin 1975). The competitive exclusion principle rejects the possibility of two species occupying the same niche within the same geographic territory (sympatric existence) because of unequal reproduction rates between the two species (Hardin 1960). The formal model of competitive exclusion has three assumptions (Gotelli 1998):

1. Resources are in limited supply.
2. Competition coefficients and carrying capacities are constants.
3. Density dependence is linear.

Although competitive exclusion implies a zero-sum solution, equilibrium outcomes are also possible, at least in the short run.

Niche theory's stress on competition and exclusion is strongly echoed in the genetic erosion concept. Although the original papers on genetic erosion did not develop a formal theory linking crop research to niche theory, this theory seems to be a logical source of the ideas used to frame the genetic erosion concept in the late 1960s and early 1970s. The theory was developed during the same period as the establishment of crop breeding, and the niche concept was fully elaborated when concern arose over genetic erosion of crop populations. Crops and crop varieties in farming systems are easily treated as analogies to species and natural habitats. The production space allocated to a particular crop variety in a farming system resembles a niche with specific environmental limits and carrying capacity. In this case, the environmental limits are determined by not only the physical attributes of the field but also the cultural and economic concerns of the farmer (Bellon 1996). Therefore, a crop niche is created by the interplay of biophysical and social factors. Crops are fit into an agricultural niche based on their performance (yield size and yield stability) and usefulness in meeting household needs (storability, culinary quality, market value). According to the analogy of niche, the introduction of new crops or varieties into this space will trigger competition between them resembling interspecies competition. One outcome is that the existing varieties will lose this competition.

Drawing an analogy of farms or crop zones to niches is natural, but niche theory has not been as thoroughly scrutinized for its utility in agriculture as it has been in ecology. In fact, the treatment of niche theory in ecology alerts us to possible limitations for applying it to population processes in crops. Competition in agriculture is altogether different than among plants that are only under natural selection. Crops have not evolved for their competitiveness in natural selection, indeed varieties most favored by farmers may be the least competitive under natural selection. Farmers may avoid direct competition among crops and varieties because the outcome may not be favorable to farm goals. The competitive exclusion aspect of niche theory was developed for interspecies interaction where natural selection but not artificial selection operates. It is possible, indeed likely, that farms will have more than one niche for the same crop, or that farmers may create new niches for different variants of the crop. Elasticity in the number of crop niches changes the dynamic of interspecies or infraspecies competition. Competition over essential resources and space depends on resource availability among competing species, their reproduction and mortality rates, population densities, and carrying capacities (Tilman 1982). Conscious selection fundamentally alters these variables.

For crops, competition is not only for essential resources such as sunlight, water, and soil nutrients but also for space in the farming system that is allocated by farmers. In other words, competition among crops is to the farmer's favor. Niche theory is meant to address interspecies competition and not infraspecies competition. Indeed, infraspecific interaction may reduce the competitive exclusion principle to a carrying capacity model (Gotelli 1998). Nevertheless, applying the analogies of species to crop varieties (e.g., treating high yielding varieties as one "species" and traditional crop varieties as another "species"), and farm habitat to environmental niche, allows us to use niche theory and to predict genetic erosion caused by the appearance of a superior competitor in a variety's production space within a farming system.

Application of niche theory to competitive interaction between a local variety and a modern variety in the same farm habitat gives four possible solutions:

1. The local variety wins the competition and no genetic erosion occurs.
2. The modern variety wins the competition and the local variety disappears from the niche (local extinction).
3. The two varieties establish a stable equilibrium with uncertain genetic erosion results.
4. The two varieties establish an unstable equilibrium leading to the eventual exclusion of one.

The solution to competition between two species depends on the carrying capacity of each and the competition effect of each species on the other. If the isoclines that define the combination of abundance of each species interacting with the other are symmetrical (parallel), then the first solutions pertain. If the isoclines cross, then the third and fourth solutions are possible.

Recent research on the concept of niche and its relation to biological diversity has disputed the earlier emphasis on competitive exclusion. Equilibrium solutions to competition require a set of assumptions about the even distribution of resources within an environment. The inclusion of tradeoffs between competitive ability, stress tolerance, and reproductive performance in niche models solves them in favor of coexistence of different species rather than exclusion (Grace 1990). Building on the work of others, Tilman (1982) suggests that the classical niche theory of the Lotka-Volterra model should be reformulated to account for resource requirements of different species. Tilman's (1982) model predicts that the species with the lowest minimum resource requirement will be the superior competitor. For crops, this might be either use of resources, such as soil nutrients, or per unit cost of production. If use of resources is adopted as a framework for assessing competition between crop varieties, traditional varieties might be superior competitors in some farming areas. In Chiapas, Mexico, for instance, Bellon (1991) found that the traditional maize variety *Olotillo* was referred to as a "poor farmer's variety" because it competed well with weeds and did not require fertilizer to achieve expected yields. On the rain-fed hillsides of Chiapas, *Olotillo* was a superior competitor to improved maize varieties. Alternatively, if per unit cost is the determining criterion, modern varieties would be superior competitors in the valley bottoms of Chiapas, where larger farms with access to credit, mechanization, and purchased inputs are the norm. Likewise in western Turkey, Brush and Meng (1998) found that local wheat landraces were superior competitors on rain-fed hillside farms while modern wheat varieties dominated irrigated farms in the valleys. In other words, competition can be defined differently, is not limited to single resources, and is influenced by nonresource conditions such as disturbance and predators (Grace 1990). For crops, competitiveness is not defined by yield alone but by myriad characteristics relative to factors that may or may not be controlled by the farmer—sprouting vigor, precocity, drought resistance, seed viability after storage, or commercial demand (Bellon 1996, Brush and Meng 1998). In sum, crop niches are multifaceted and elastic, and Gause's predictions are largely moot since coexistence is commonplace due to the many nonselective forces and to tradeoffs.

Tilman (1994) further elaborates a modern niche theory based on a resource competition model to account for spatially structured habitats. This new version of niche theory predicts coexistence rather than competitive exclusion. In other words, spatial structure challenges the predictions of nonspatial models such as Lotka-Volterra. Spatial structure caused by habitat subdivision can occur in any environment, even though underlying physical conditions are similar (Tilman 1994). For natural communities, habitat subdivision is inevitable because of random colonization and mortality. For crops, habitat subdivision occurs with the common practices of cultivating different parcels or subdividing parcels, and because the same crop can satisfy different uses—food, fodder, commerce, construction. Coexistence of numerous species in the same general environment results from inevitable tradeoff between allocation of plant energy to roots and reproduction. Tilman (1994) observes that superior competitors for limited soil resources are poorer colonizers. Coexistence of different crop varieties in the same farming system results from the same types of inevitable tradeoff between varieties that are superior in different farm environments and for different purposes.

The spatially structured model of natural plant communities reveals a set of simplifying assumptions of the competitive exclusion models—"that resource supply rates and physical factors are spatially homogeneous, that each organism is spread uniformly throughout the environment, that resources do not fluctuate, that localized mortality does not occur, and that higher trophic levels are unimportant" (Tilman and Pacala 1993, 19). Violation of even a few of these assumptions, and accepting tradeoff in response to resource limits and reproduction, lead to models that predict the coexistence of unlimited species (Tilman and Pacala 1993). The surprising conclusion of spatially structured models is to turn conventional niche theory on its head and to allow for numerous species in single environments. The same conclusions also pertain to modeling crop competition in the environmentally, socially, and economically heterogeneous environments of the Vavilov centers of crop diversity.

Both the older and the more recent versions of niche theory provide useful insights into the diversity of crop varieties in centers of crop diversity and to competition between local and improved varieties in these centers. The classical view of niches as multidimensional spaces with competitive exclusion offers one explanation of the diversity of crop varieties. Local farming systems in centers of diversity often contain different niches in the form of microenvironments and production zones. Crop niches might be distinguished by physical and agronomic factors—soils, water availability, slope, temperature, and exposure to stresses such as wind, frost, and disease. Competitive advantage of particular varieties within the multidimensional space of each agricultural niche appears as yield potential and yield stability. The mixing of biological and nonbiotic features is problematic in delimiting niche, but the fact that different varieties have yield advantages in different microenvironments (*crossover,* see Chapter 6) is recognized and exploited by farmers, resulting in diversity. In addition to these physical and biotic dimensions, crop niches are also defined by social factors, such as labor and capital availability to the farm, market conditions, cultural obligations, and preferences. In his discussion of "farmers' concerns" Bellon (1996) alludes to these social characteristics. Concern for ritual or other cultural characteristics of the crop may, for instance, be factored into the calculation of yield, or this concern may be added as an additional niche dimension to make a new niche.

Rosenzweig (1995) reminds us that tradeoffs are implied in all adaptation, and this idea undermines confidence in classical niche theory's prediction that a superior competitor in and among niches will ultimately prevail. For crop varieties, this prediction is interpreted to mean that broadly adapted modern crop varieties will eventually eliminate traditional crop varieties. Modern varieties are configured to fill the various niches of a farming system (or when those niches are simplified by chemical inputs and commercialization). Ironically, ecological theory holds that broad adaptation is associated with poor competitive ability. Alternatively, modern niche theory described above offers a different explanation to infraspecies crop diversity, and this modern version offers a different solution to the competition between high yielding and traditional varieties. According to modern niche theory, diversity exists because of spatially structured habitats and tradeoff in resource uses and reproduction strategies of plants. Agriculture, also, is spatially structured with flexible habitats redefined by competition and with tradeoff among different crop varieties that are defined by differential ability to meet yield, resistance, and quality criteria. A visit to a peasant farm in the environments of Vavilov centers can reveal the capacity for flexibility and complexity in the arrangement of habitats within a small area. Gade (1975) depicted one such peasant farm in the Vilcanota Valley of southern Peru that illustrates this point (Figure 8.1). Similar habitat complexity has been described for virtually every Vavilov center. According to modern niche theory, the appearance of a new variety does not trigger a zero-sum competition but rather one that is likely to permit coexistence. Just as some natural habitats are dominated by single species while others are diverse, agricultural habitats are variable in their likelihood to be simple or diverse in the number of crop varieties.

The likelihood of genetic erosion follows readily from the application of classical niche theory to explain crop diversity in centers of origin. Changes wrought by the availability of modern varieties and off-farm inputs that smooth out microenvironmental differences in a farming system are analogous to the appearance of a competitor species that is not limited by previous niches and environmental changes that reconfigure niches. Modern varieties are often superior competitors to local varieties, especially in an environment that is homogenized by the application of external inputs. If a crop is planted to its maximum density or carrying capacity in a uniform environment, then the competitive advantage of modern varieties will give them an isocline above that of the local variety, resulting in the competitive exclusion of the local type.

Historical change in crops and crop diversity in developed countries and in certain underdeveloped regions appears to conform to the predictions of classical niche theory. Thus, the local and open pollinated maize populations of the American Corn Belt in 1920 succumbed to hybrid maize by 1950 (Griliches 1957), and the diverse irrigated rice populations of the lowland Philippines were replaced by International Rice Research Institute (IRRI) varieties between 1965 and 1985 (Dalrymple 1986). Given this experience, it is reasonable to apply classical niche theory to other crop regions, including centers of diversity.

Nevertheless, classical niche theory depends on assumptions that are difficult to accept in all farming systems and particularly in the complex systems of Vavilov centers of diversity. Most problematic is the assumption that competition coefficients and carrying capacities among crop species are constant.

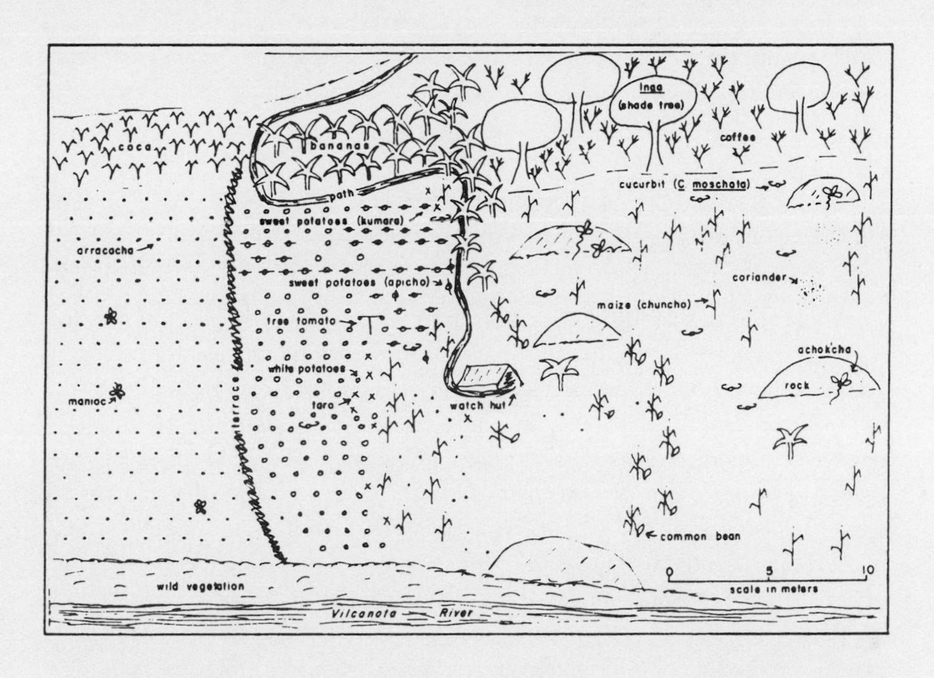

Garden Chacra at 1950 meters elevation in the Vilcanota Valley

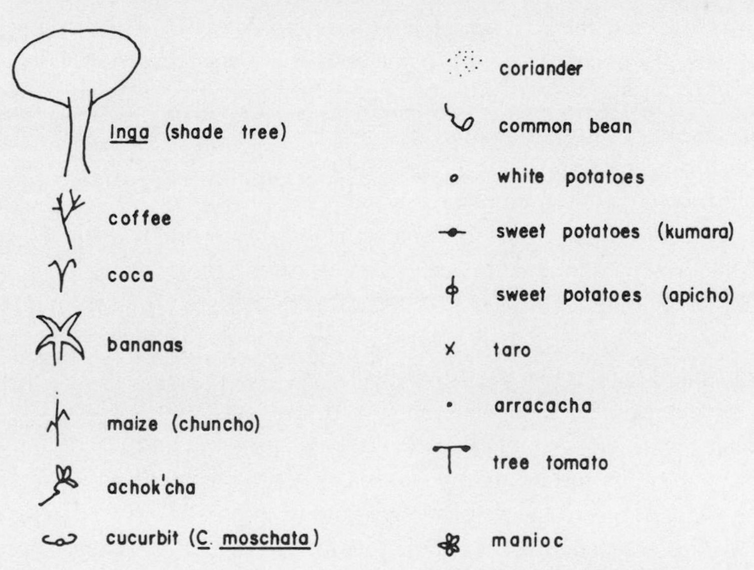

Figure 8.1 Habitat diversity in an Andean *chacra* (Gade 1975) (Reprinted with permission of D. Gade)

Other assumptions from classical niche theory (Tilman and Pacala 1993) that are implied in predictions of genetic erosion in crops include:

- resource supply rates and physical factors in the modernized farm system will be spatially homogeneous,
- resources in the modern farm system do not fluctuate,
- localized mortality of crop varieties does not occur,
- higher trophic levels (e.g., predators or human taste or markets) are unimportant.

Tilman (1982) observed that the constant competition coefficient of classical niche theory leads to models in which resource availability is also constant, and he adjusted the model to account for reduction of resources as they are used (Grace 1990). Moreover, the introduction of tradeoff in the model—between competitive growth advantage, stress tolerance, and reproduction—opens the door to competition coefficients and carrying capacities that vary in structured environments. Tilman's (1994) later and more extensive assessment of classical niche theory shows the near certainty of violating the assumptions of the Lotka-Volterra competitive exclusion model of classical niche theory. As Tilman and Pacala (1993) note, violation of these assumptions leads us to expect coexistence of different crop varieties (diversity) rather than competitive exclusion.

Field research on the ecology of landraces in centers of diversity where modern cultivars are also present provides ample evidence that the conditions of competitive exclusion are, indeed, violated. Habitats are highly structured, as evidenced by the fragmentation of fields and the practice of cultivating numerous fields in the same crop. Physical heterogeneity is not eliminated by the application of modern inputs, for instance in hill land and dry land agriculture. Resource supply rates to different fields are inevitably variable. Resources fluctuate within and between years according to the availability and timing of modern inputs, especially in locations that are removed from major transportation and input supply systems. Crop failures (localized mortality) under stress and in marginal environments are well known to farmers. Finally, human consumption is a critical element in crop selection, particularly in subsistence production. Storage, taste, cooking qualities, and use as fodder may not be more important than yield, but they are relevant to selection. In sum, the assumptions of the classical niche model are easily violated in farming systems in centers of diversity, so that we should expect coexistence rather than competitive exclusion to prevail. Fieldwork in Vavilov centers of diversity confirms this expectation.

METAPOPULATION ANALYSIS

The reduction of area and fragmentation of populations threaten species' survival and biological diversity in many different environments. Understanding the effect of area reduction and fragmentation on species survival was a principal objective of island biogeography (MacArthur and Wilson 1967). Island biogeography turns our attention to species in degraded environments. In particular, this approach was designed to understand habitat patches that remain as islands in disturbed environments. Other patches are large enough to serve as "mainlands"—areas that resemble habitats before degradation and that are able to supply species to islands as "colonists." Island biogeography may offer a useful approach to modeling crop populations in centers of diversity that are experiencing agricultural development (Qualset et al. 1997), but this approach has now been supplanted in ecology by metapopulation analysis. In an island biogeographic analogy, habitat degradation in Vavilov centers is the diffusion of modern crop varieties into systems that previously were fully planted in landraces. Villages and farms that maintain landraces are analogous to islands surrounded by a sea of modern varieties. One problem with this analogy, however, is defining what might constitute a "mainland" for a crop species.

After three decades of research on island biogeography, ecologists have developed a new approach—metapopulation analysis (Levins 1969, Hanski, and Simberloff 1997). This approach seems to be particularly useful in examining crop populations and the issue of genetic erosion. A metapopulation approach might also address important aspects of crop ecology that are problematic for niche theory, such as seed flow among regions. For instance, Zeven (1999) views seed turnover in landraces as "inexplicable," and in the previous chapter we saw that diversity assessments of crop landraces in Ethiopia revealed that most diversity is found within rather than between populations, even though these may be located in different agroclimatic zones. Metapopulation analysis predicts this phenomenon by focusing our attention on the movement of populations through recurrent recolonization (Pannell and Charlesworth 2000).

A metapopulation is defined as "a 'population' of unstable local populations, inhabiting discrete habitat patches" (Hanski 1998, 41). Each local population is extinction-prone, but migration from other local populations allows the species to reestablish itself where it has suffered extinction. Extinction, or the disappearance of a species from a habitat, is a recurrent event in many patches and is moderated by colonization to establish an equilibrium for the metapopulation. This approach has been utilized to study species in fragmented and degraded landscapes and in competition with other species. An advantage to metapopulation analysis is in avoiding the concept of mainland. From the vantage point of any one island, the mainland is merely the sum of other islands that might provide colonizing populations.

The usefulness of metapopulation analysis to understanding crop population dynamics has been noticed by crop ecologists (e.g., Zimmerer 1998), but this analysis has rarely been applied to landraces in Vavilov centers. An exception is Louette's (1999, Louette and Smale 2000) description of maize seed flows in Cuzalapa. The analogy of the crop population of a farm or village to local population is intuitively appealing. Farmers may keep seed from year to year, but loss of seed and its replacement through exchange with other farms and across different regions is a recurrent event (Zeven 1999). In other words, individual farms experience local extinction of seed, but local extinction is balanced by seed exchange (migration) among farms. The habitat of seed production and exchange may be degraded by the diffusion of modern varieties that reduce the number of sources of seed (patches) as well as their connectivity. Farmers may experience an increasingly difficult time locating seed of local varieties to replace depleted or degenerated seed. In picturing landrace populations as metapopulations, we see that their disappearance from a single farm or village may not threaten the entire landrace, but the extinction of a landrace is possible as its habitat is degraded by modernization that chokes off the supply of seed to farmers in different locations. Dispersal of landraces in traditional farming systems enables farmers to replenish seed stocks, but these dispersal rates may not be able to adjust or adapt to the modernization of agriculture.

A potential advantage of metapopulation analysis is to help bridge different approaches to genetic erosion. Hanski (1998) notes that, in large scale spatial ecology, metapopulation ecology occupies an intermediate position between theoretical ecology and landscape ecology. Theoretical ecology has demonstrated the complexity and spatial patterns of population processes that occur in idealized, homogeneous environments. At the other extreme, landscape ecology has described the complex nature of real environments but without a well-developed theoretical framework (Hanski 1998). Metapopulation analysis provides a means to link these two approaches by providing a theoretical framework to analyze population dynamics in heterogeneous environments.

In crop ecology, the original models of genetic erosion presented previously are analogous to early theoretical ecology in that they are conceptualized without environmental heterogeneity. They do not allow for the mitigation of biophysical and social factors in the process of genetic erosion. Ecological research on real farming systems, where landraces and modern varieties compete, is analogous to studying landscape ecology. This research has shown that the theoretical models of genetic erosion are difficult to apply and may be wrong, but this ecological fieldwork has not provided a convincing theoretical framework. What is now needed is a concerted effort to develop the middle ground approach suggested by metapopulation analysis.

Two premises underlie the metapopulation concept (Hanski and Simberloff 1997):

1. Populations are spatially structured into assemblages of local breeding populations.
2. Migration among local populations affects local dynamics, including reestablishment after local extinction.

Metapopulation analysis expands on earlier research on niches and competition by adding local extinction and colonization to the population dynamics of a species. Metapopulation analysis serves to describe populations in both pristine and degraded habitats. It is based on the fact that a species comprises numerous distinct breeding populations. Environmental patches experience periodic local extinction and recolonization. Whereas island biogeography is envisioned as the interaction of island and mainland populations, metapopulation analysis is based on the possibility that the entire population of a species exists on patches scattered throughout the landscape. In this approach, persistence of a regional population and possibilities of migration and recolonization are major concerns, rather than population size at a specific site. Critical variables are the fraction of population sites that are occupied, the extinction rate across patches, and the migration rate among them.

Metapopulation analysis reorients our perspective on competition and local extinction in two ways. First, multiple patches of a population spread the risk of extinction to the species. The mathematics of metapopulations are such that relatively high probabilities of extinction at the local level may be greatly reduced at the regional level. In Gotelli's (1998) model, for instance, a 70 percent probability of extinction of a local population, if shared evenly among ten different populations, is reduced to a 97 percent chance of survival of at least one population. As more patches are added, the probability of regional persistence increases rapidly (Gotelli 1998).

However, the second reorientation from metapopulation analysis implies a far less optimistic scenario. The ability of a species to survive depends on the persistence of a minimum number of patches. Although individual patches may persist, if their number is less than the "eradication threshold," the species is bound for extinction. Population size at a particular site is not necessarily a critical variable to a metapopulation's survival, but filling a minimum number of patches is required for the population to persist. This minimum is defined as the number of unoccupied patches when the metapopulation is at "a dynamical equilibrium between colonization and extinction" (Nee, May, and Hassel 1997, 125). Nee (1994) defines this minimum threshold as the unused amount of patches of a population's suitable habitat. Failure to fill the minimum number of patches establishes an "extinction debt" (Tilman et al. 1994) that must eventually be paid.

Hanski (1998) summarizes three general conclusions about metapopulation response to habitat destruction:

1. Metapopulations respond to habitat destruction in a nonlinear fashion because habitat destruction is itself nonlinear, as are its effects on migration.
2. A time lag (extinction debt) occurs between the time of habitat destruction and metapopulation decline.
3. The amount of empty habitat in a landscape before habitat destruction equals the extinction threshold, or the minimum amount of habitat required for metapopulation persistence.

These conclusions direct us to landscape heterogeneity, population patterns (e.g., replacement) over different environments, and connections among crop populations. Habitat destruction is nonlinear because habitats are heterogeneous. The survival of local populations does not mean that metapopulations will survive. Metapopulation survival may be estimated from understanding the connections between local populations and the degree to which those connections have been maintained or disrupted.

Crop ecologists have noted that metapopulation analysis is appropriate for understanding crop population processes in centers of diversity (e.g., Zimmerer 1998; Louette, Charrier, and Berthaud 1997), but the approach has not been formally adopted in any field study or theoretical exercise relating to crop dynamics. Landraces appear to conform to a metapopulation model in that they are naturally fragmented into unstable local populations that are connected through seed exchange among farmers, villages, and regions. Local extinction and reestablishment appear to be common in landrace populations. In central Mexico, for instance, farmers of maize landraces do not worry about seed loss or conservation because they believe that seed will be available from neighbors or in markets (Perales R. 1998). Further south, however, farmers in Oaxaca report that a diminished supply of seed of traditional maize varieties is a factor in the disappearance of traditional varieties (Bellon 2001).

Landraces exist as heterogeneous populations rather than as pure-line varieties (Zeven 1998), and this population heterogeneity may be due to frequent exchange and incorporation of new seed into existing stocks. Ecological research on crops has also demonstrated that environmental heterogeneity is important in population dynamics of landraces, including habitat destruction caused by the introduction of modern crop varieties. Landrace populations are stable in some environments because habitat destruction does not approach the threshold of the minimum habitat needed to maintain metapopulations. Maize in parts of Mexico (Perales R. 1998; Louette, Charrier, and Berthaud 1997) seems to conform to this description. In other cropping systems, such as native Andean potato production and Anatolian wheat production, habitat destruction of landraces—measured by the diffusion of modern crop varieties—may be at or beyond the threshold where metapopulations of local crops can survive. One research objective defined by a metapopulation analysis of landraces would be to account for the patchy nature of their undisturbed habitats in order to determine the equilibrium level of the landrace metapopulation—their extinction threshold. In other words, research on landrace survival in habitats where modern varieties are present will benefit from research on agricultural habitats where only landraces exist.

An advantage of the metapopulation approach to the issue of genetic erosion of traditional crops in Vavilov centers is that it suggests specific questions about the social context of crop ecology, thus a path of integrating biological and social ecology of crop populations. Specifically, metapopulation analysis asks us to identify the habitats of landraces, to describe the configuration of those habitats in an agricultural landscape, and to describe the dynamics of landraces in and among those habitats. The habitats of landraces are both environmental and socially defined—by agroclimatic zone as well as by markets, economic isolation, technology adoption, and culturally determined use. Likewise, the configuration of landrace habitats in an agricultural landscape might be defined as the number of farms and villages that produce landraces and exchange seed. This configuration is influenced by social factors such as investment in agricultural infrastructure, research and extension, the availability of credit and technology, and markets that supply inputs as well as outlets for agricultural produce. Finally, the dynamics of landraces in and among habitats are a function of social factors such as the demand for particular values of specific varieties and the supply of seed through trade and exchange. Demand and supply of different crop populations are influenced by their value in heterogeneous and spatially structured agricultural landscapes as well as by social conditions such as wealth, markets, and information.

While some studies of crop ecology have recognized the potential of metapopulation analysis, no crop or farming system has been fully studied with this approach. It remains as an intriguing but not fully tested way to look at crop populations and the problem of genetic erosion. The metapopulation idea has a certain bandwagon appeal (Hanski and Simberloff 1997) that should caution us. The concept is based on recolonization of lost species. Metapopulation models are designed for single species and have proven to be cumbersome for two or more species. Applying them to crop varieties inevitably involves pushing metapopulation models in directions where their usefulness has been limited. Whether lost crop varieties are replaced in the same manner has not been established by crop ecologists. Local extinction has not been easy to document, and the difficulty of documenting variety extinction is probably greater. If there is little population divergence, as suggested by the Ethiopian studies, seed movement provides scant evolutionary consequence because of the lack of gene flow. If this were the case, extinction would simply be a progressive, arithmetic loss of stands.

Summary

The assessment of genetic erosion of crop populations in centers of crop diversity is at an impasse. On one hand, a general but undertheorized model of agricultural change presents a logical argument for genetic erosion. This model has not been fully articulated or rigorously tested. It does not contend with environmental heterogeneity—either social or biophysical—in farming systems or with patterns that are evident from field research on the ecology of traditional crops. On the other hand, specific but equally undertheorized descriptions of landraces that compete with modern varieties suggest that survival of crop diversity is plausible in certain environments. One way to solve the impasse between contradicting general models and field studies is to develop a better theoretical framework for the ecology of crops in centers of diversity and genetic erosion. This framework needs to be purposively interdisciplinary, drawing on theory, methods, and data from ecology, crop science, and social science.

This chapter has suggested two ecological models that might serve to generate a more robust crop ecology. Modern niche theory and metapopulation analysis offer numerous insights and advantages to efforts to understand genetic erosion. A shared insight is that general population processes, such as genetic erosion, are affected by environmental heterogeneity. Modern niche theory and metapopulation analysis provide a middle ground between general theory and site specificity.

One limitation of both modern niche theory and metapopulation analysis for crop ecology is that neither was conceived for conscious (artificial) selection. Another limitation is that both were designed to deal with interspecific competition and processes rather than analogous processes between different populations or varieties within a species. Numerous questions surround the use of ecological models for crops. Is it appropriate, for instance, to treat types of crop varieties (e.g., landraces and modern varieties) as though they are species? This requires a constancy in defining landraces that is notably absent in crop ecology. How applicable are ecological models designed for understanding species to autogamous crops that are reproductively isolated? Should we look at modern varieties as though they are invasive species? Is agricultural modernization equivalent to habitat destruction? Although some writers have intimated this (e.g., Shiva 1997), most crop ecologists would prefer a more accommodating relationship between traditional and modern farming practices.

The application of formal population models to crops presents daunting challenges — to define key variables and specify functional relationships. The need to include both biological and social variables and functional relationships is particularly difficult to satisfy. Arguably, social science offers insights and methods that are essential to any general ecology of crop populations in centers of diversity, such as models of the diffusion and impact of modern crop varieties. However, these insights and methods were not conceived for ecological research. Obviously, there is a need to work and rework theory and methods from both ecology and social science for studying crop populations.

9

Maintaining Crop Diversity On-Farm and Off

Ex Situ *and* In Situ *Conservation*

Collecting and experimentation are companions of the human fascination with plant diversity, and in many places the sporadic, individual search for exotic plants has been followed by more purposeful and organized efforts. Gardens exist everywhere to nurture and observe the results of plant collecting. Tribal shamans, professors, and royal gardeners have all been entrusted to protect and manage exotic and everyday plants that have been collected for display, study, and use. Posey (1985) reports that the Kayapó Indians of the Brazilian Amazon maintain gardens of crop varieties and medicinal plants. Salick, Cellinese, and Knapp (1997) found that Amuesha shamans' gardens in the western Amazon are the repositories of the cassava diversity for the entire village. Gardens are reported in the records of literate societies, theocracies, royalties, empires, and industrial states.

Gardens played an important role in the growth and change of naturalist thought to modern biology with its two pillars, evolution and genetics (Findlen 1994), and they served biological science until the emergence of crop science and breeding. Originating when naturalist curiosity, religious motivation, and aesthetic impulse fueled plant collection, gardens have slipped into a supporting role in modern agriculture. Although private and public gardens helped to establish new crops and agricultural industries around the world (Brockway 1979, Plucknett et al. 1987), botanical gardens had neither the capacity nor the interest in assembling the vast amount of crop diversity required by public and private crop breeding. Collecting for infraspecific diversity of crop species gave rise to plant introduction stations where collections of thousands of accessions could be amassed for evaluation and use in breeding.

Under Vavilov's direction, the Department of Applied Botany and Plant Breeding of Russia's Department of Agriculture grew to become the All-Union Institute of Plant Industry (VIR) of the Soviet Union, and this growth was a harbinger of a worldwide scientific enterprise. Vavilov personally collected tens of thousands of accessions on his expeditions on five continents, and he directed the collection of many times that number by other VIR missions. From its headquarters in St. Petersburg/Leningrad, the VIR supervised research stations in 115 locations, where the institute's hundreds of thousands of accessions could be observed in different environments. In the United States, ad hoc plant introduction, which had been encouraged by governments since the early nineteenth century, gave way to institutionalized exploration after 1898 (Plucknett et al. 1987). The Office of Foreign Seed and Plant Introduction of the USDA hired a group of legendary plant explorers, notably Frank Meyer in China and Wilson Popenoe in the Americas (Cunningham 1984, Rosegarten 1991). Meyer's arduous expeditions in China led to some 2,500 plant introductions in the United States (Plucknett et al. 1987). His collections are credited with providing disease resistance that rescued the American spinach crop and protected apricots, plums, and peaches (Cunningham 1984). Among other crops, Popenoe introduced the Fuerte avocado, originally from Puebla, Mexico, which became the basis of the avocado industry in California and which remains the second most important variety today (Rosegarten 1991).

The initial purpose of plant introduction institutes, such as the VIR, was the acquisition of crop diversity for evaluation, direct use, and crop breeding. This purpose quickly evolved beyond evaluation to the storage and conservation of crop germplasm. Conservation was necessitated by the sheer volume of material flowing into plant introduction institutions and the time required for evaluation and breeding. Hyland (1977) notes that only a small percentage of the plant introductions to the United States under the auspices of the USDA are still living (Hyland 1977). Moreover, plant collectors began to observe the replacement of "traditional" and "local" varieties of crops by exotic ones. In South Africa in 1919, botanists noted that the disappearance of a Boer variety of oat was caused by the introduction of an Algerian oat variety, and shortly afterward Harry Harlan observed the replacement of Mediterranean cereal varieties by American cultivars (Plucknett et al. 1987). The raw material of the new crop breeding industry appeared to be vanishing before it could be used by the new science of crop breeding.

Conservation, in the form of medium-term storage, was a rationale for the organization of a network of plant introduction stations by the USDA in 1940. These storage facilities, the precursors of modern gene banks, included the first cold storage to maintain germplasm as viable seeds. In 1958, the United States established the first national gene bank involving long-term storage under controlled conditions—the U.S. National Seed Storage Laboratory (NSSL) at Fort Collins, Colorado. A leap to a worldwide network of gene banks was initiated by the United Nations' Food and Agriculture Organization (FAO) in the 1960s. The 1972 United Nations Conference on the Human Environment in Stockholm and FAO technical conferences on plant genetic resources in Rome in 1968 and 1973 were especially significant in creating an international agenda to collect crop genetic resources and to build gene banks for long-term storage (Frankel and Bennett 1970, Frankel and Hawkes 1975, Pistorius 1997). Rapid growth of this international system in a quarter century has led to the creation of over a thousand gene banks in 123 countries and with collections of over 6,000,000 accessions (FAO 1998). These gene banks include immense national facilities such as the NSSL with 268,000 accessions, the Chinese Institute of Crop Germplasm with 300,000 accessions, and a network of international gene banks such as those at the International Rice Research Institute (80,646 accessions), the International Maize and Wheat Improvement Center (CIMMYT—136,637 accessions), and the Asian Vegetable Research Institute (37,618 accessions) (FAO 1998).

NEGLECT AND ACCEPTANCE OF *IN SITU* CONSERVATION

The fail-safe assumption that guides all conservation is that biological resources are under a mounting threat, and there is no reason to believe that crops are any different. Gene banks are the best-known mechanisms to preserve the genetic diversity of crops, and their importance and value in making that diversity available to crop breeders are inestimable. Gene banks are a logical first choice for conservation because there is no other means to preserve the same amount of diversity with available scientific, institutional, and financial resources. Nevertheless, several trends suggest that gene banks will no longer continue as the only means by which to conserve crop genetic resources. The alternative means is *in situ,* or on-farm, conservation. One trend is the shift of biological and agricultural science toward ecology. Crop evolution is now viewed as more than domestication, and crop resources include many more elements than genetic diversity. Diversity is a visible portion of an enormous assemblage of elements and relationships. Crop resources are now seen to include many things that were not imagined in the 1960s—co-evolutionary relationships between different types of populations (e.g., crop, pathogen, human), indigenous knowledge, and farmer seed management. A second trend is the recognition that gene bank conservation is suitable and appropriate for some elements of the agricultural environment but not all. Wild relatives and "recalcitrant" seeds were first recognized as elements that couldn't be preserved in gene banks. This category has grown to include a host of complex relationships and processes that constitute crop evolution—co-evolutionary systems of crops, symbionts, pathogens, farmer knowledge, and farmer selection. Conservation of crop accessions in gene banks is analogous to preserving a tropical forest in botanical gardens or wildlife in zoos—adequate and necessary for some purposes but not a completely satisfactory mechanism.

However, the acceptance of conservation outside of gene banks was slow and often grudging. *In situ* conservation had been proposed early on. Zeven (1996) reminds us that *in situ* conservation in the form of farm gardens was organized in Europe between the World Wars, only to be forgotten in the ashes of World War II. Proposals for "mass reservoirs" of landraces were mentioned but quickly dismissed at the FAO technical conferences on plant genetic resources (Frankel and Bennett 1970).

Several answers might be posited to the question, "Why was *in situ* conservation delayed so long?" The most immediate and compelling answer is that on-farm conservation is incompatible with agricultural development. Because genetic diversity in crops is associated with traditional agricultural practices, it is also linked to underdevelopment, low production, and poverty. Development traditionally had been perceived as a process of technological substitution—of new crops for old—that had occurred in industrialized countries and that should rightly occur in developing areas (Cochrane 1993). Indeed, by the early twentieth century, collectors and others were already reporting the diffusion of improved varieties that replaced local ones (see Chapter 7). Scientists and policy makers who fashioned the crop conservation movement believed that world population growth, urbanization, and the desire for higher living standards would (and should) inevitably drive all agricultural systems to replace local, low yielding crops with improved crops (Harlan 1975b). By 1970, the success of the Green Revolution in Asia was interpreted as a sign of what was to happen everywhere.

Another reason for rejecting *in situ* conservation is the assumption that farmers who grow traditional crop varieties would require a direct monetary subsidy to continue this practice once improved varieties become available. Such subsidies are not only expensive but also unreliable and difficult to manage for any length of time. Finally, crop scientists who promoted conservation were not interested in conservation alone but also in obtaining genetic resources for crop improvement. As long as breeders' work is confined to experiment stations and laboratories, genetic resources that remain in farmers' fields are not directly useful for crop improvement.

By 2000, the opposition to *in situ* conservation had not only waned, but numerous national governments, international agricultural development agencies, and nongovernmental organizations were actively promoting it. Advocates of *in situ* conservation include several international agricultural research centers with large and important gene banks, including the International Rice Research Institute (IRRI), the International Maize and Wheat Improvement Center (CIMMYT), and the International Center for Agricultural Research in Dry Areas (ICARDA). Several factors reversed the negative opinion about *in situ* conservation. By the early 1980s, research in Vavilov centers indicated that the diffusion of modern crop varieties did not totally displace local varieties (Brush, Carney, and Huamán 1981). Environmental heterogeneity, lack of appropriate new varieties, local values, and missing markets resulted in partial adoption of modern technology, suggesting that local crops could survive under conditions of agricultural development. Secondly, *in situ* and *ex situ* methods were no longer perceived as exclusive alternatives but rather as complementary to each other (Shands 1991, Bretting and Duvick 1997). These methods address different aspects of genetic resources, and neither alone is sufficient to conserve the total range of genetic resources that exist. Third, a variety of methods, apart from direct financial subsidies, were proposed to promote the maintenance of crop genetic resources by farmers. Finally, the international environmental movement successfully promoted *in situ* conservation, for instance by the creation of the Convention on Biological Diversity and the Global Environmental Facility (GEF), which funds environmental conservation in developing countries. In fact, in 2001 the GEF announced a new funding program focused on biological diversity in agriculture.

Brown (1999) reviews ten postulated advantages for *in situ* conservation: (1) conservation of indigenous knowledge, (2) conservation linked with use, (3) allelic richness and genotypic diversity, (4) special adaptations, (5) localized divergence, (6) diversity to meet temporal variation, (7) continuing crop evolutionary processes, (8) avoidance of regeneration, (9) human involvement, and (10) control and benefit sharing. To understand the revised endorsement of *in situ* conservation, these advantages can be synthesized into four justifications.

First, important elements of crop genetic resources cannot be captured and stored off-site. Crop genetic resources involve more than the genetic raw material of alleles and genotypes and include related species, agroecological interrelationships, and human factors. We now recognize that ecological relationships such as gene flow and decentralized human selection and management of diverse crop resources are common components of crop evolutionary systems that generate crop genetic resources. A critical difference between *ex situ* and *in situ* conservation is that the former is static and designed to avoid loss or degeneration by maintaining the genetic material in the state in which it was collected. In contrast, *in situ* conservation is dynamic and meant to maintain a living and ever changing system, thus allowing for both loss and addition of elements of the agroecosystem (Bretting and Duvick 1997, Wilkes 1991). The dynamic aspect of *in situ* conservation is one of its most difficult attributes for planning and evaluation. The success of *in situ* conservation cannot be judged only by the number of alleles or genotypes preserved. It might also be measured by the number of farmers within a target area who maintain local crop populations and manage those populations according to local criteria and practices. Alternatively, the success of *in situ* conservation might be measured by the use of local germplasm in breeding programs that result in new crops but that do not replace the crop population of a region.

A second reason for promoting *in situ* conservation is that gene bank collections fail to capture diversity or resources that are generated after the collection has occurred. Recurring collection is rare and limited by many obstacles both within gene banks and outside them. Gene bank conservators believe that they have almost complete inventories of the diversity of leading crops (Plucknett et al. 1987), but continuing crop evolution, different sampling procedures, and poor documentation may indicate otherwise (Frankel, Brown, and Burdon 1995). Indeed, comparative studies of *ex situ* and *in situ* maintenance show a steady divergence of crop populations from the same place (e.g., Soleri and Smith 1995; Tin, Berg, and Børnstad 2001). Estimates of the percentage of crop diversity collected in gene banks are derived from a consensus among scientists rather than from a thorough analysis of genes in the bank and genes in farmers' fields. New resources become available because of a variety of mechanisms—mutation; recombination; gene flow between wild, weedy, and cultivated populations; somatic variation; and exchange from outside of the collection region.

Third, all forms of conservation are vulnerable, and gene banks are subject to numerous risks—genetic drift within collections, loss of seed viability, equipment failure, security problems, and economic instability. Gene banks depend on volatile public and political support. Even large and prestigious institutions can suffer sudden reversals of fortune, endangering their collections. The fate of the collections of the VIR in the political and financial collapse of the former Soviet Union has been publicized (Raman, Zimnoch-Guzowska, and Zoteyeva 2000), and many smaller, provincial collections are vulnerable without notice. Gene banks are usually inadequately funded (NRC 1993), so that storage and regeneration facilities are limited, evaluation is partial, and equipment is obsolete or inadequately backed up. While the purpose of *in situ* conservation is not to preserve alleles and/or genotypes per se, regions where successful on-farm maintenance of genetic diversity occurs provide potential stores for re-collection of genetic resources.

Finally, service and political reasons bolster *in situ* conservation. Improved crops have increased food supplies and have benefited farmers in most countries, but many farmers have been bypassed by this technology. In a de facto sense, the limits of crop improvement in reaching all farmers affect on-farm conservation. However, few scientists or policy makers concerned with either genetic resources or poverty would accept this as a strategy for conservation. That is, conservation cannot be achieved at the expense of the poor. *In situ* conservation has been proffered as an ally of agricultural development for areas bypassed by conventional technology improvement schemes and as a way to achieve development without Green Revolution technology (Altieri and Merrick 1987; Cleveland, Soleri, and Smith 1994). Because it is dynamic (Bretting and Duvick 1997), *in situ* conservation can theoretically generate far more diversity and involve farmers in improving local crops through "participatory crop improvement." Underlying all strategies for on-farm conservation is the belief that they must improve food availability and income.

Politically, there is now a fairly strong international mandate for *in situ* conservation. This mandate includes the Convention of Biological Diversity and the Global Environmental Facility mentioned above. The GEF has funded projects dealing with the *in situ* conservation of crop genetic resources, including wild crop relatives, in Turkey, Ethiopia, Peru, Lebanon and Jordan, and the Fertile Crescent of the Near East, and the organization of a regular funding program for biological diversity in agriculture will increase funding and projects. Additionally, the FAO's International Undertaking for Plant Genetic Resources for Food and Agriculture has provided a forum for supporting both gene bank and *in situ* conservation.

SCIENTIFIC FRAMEWORK FOR *IN SITU* CONSERVATION

While the forces of genetic erosion—population growth, technology development, and trade—appear to be tectonic and irresistible, so too does the fractal nature of agricultural environments seem to invite opportunities to conserve traditional crops. There is ample evidence that *in situ* conservation is indeed practiced in a de facto sense by farmers who use modern technology, including improved crop varieties, and who are engaged in commercial production. Nevertheless, the ecology of crop diversity is poorly understood. For instance, as suggested in the previous chapter, it is unclear whether we should approach diversity from a modified niche theory or on a metapopulation basis. Just as the future of natural evolution is unpredictable, so too is the future of crop evolution inscrutable, but this scientific dilemma should not deter us from attempting conservation. Two scientific challenges immediately confront the planning and implementation of *in situ* conservation of crop diversity: (1) the scope of conservation, and (2) how to promote on-farm maintenance of diversity.

In situ conservation is not a sectoral policy of alternative agricultural development but rather a targeted program working in specific areas. Determining the scope of *in situ* conservation requires data and analysis from crop genetics, population biology, and social science. Several criteria delineate its scope: the crop species to be conserved, the location and distribution of conservation target areas, the physical size of these areas, and the number of farms and communities that are included. Besides daunting technical issues in determining the biological and social scope of *in situ* conservation, financial, institutional, and political factors are also likely to have weight. Although a number of *in situ* projects have been initiated in the last decade, these have been exploratory and focused on developing scientific capacity rather than on full-scale implementation of on-farm conservation programs. The next step will be to move beyond the exploratory stage (Maxted et al. 2002). An early task will be to identify specific crop populations and farming communities to be involved in conservation.

Because the measures of crop evolution or potential for evolution are not developed, the choice of areas for on-farm conservation will logically rely on genetic diversity data and analysis. The choice of crop populations for *in situ* conservation might utilize the same theoretical basis as that for collecting for gene banks. Ideally, collections are based on the premise that some populations will be more appropriate than others. Marshall and Brown (1975) identified two critical population parameters for this choice: (1) the extent of genetic divergence among populations, and (2) the level of level of genetic variation within a population. Divergence among populations is a measure of the frequency (rare to common) and distribution (local to widespread) of distinct genetic components, from alleles to crop populations. Using Marshall and Brown's system, four different distributions exist: rare and local, rare and widespread, common and local, and common and widespread. The critical targets for collection are populations with alleles that are locally common, and this same criteria might be applied to the choice of crop populations for *in situ* conservation. Thus, a crop population with a high diversity of genotypes that are common in the population but not found elsewhere would be selected for on-farm conservation. Surveying national collections for distribution and frequency of distinct crop genotypes is a direct method of determining the number and location of populations to be considered for *in situ* conservation, but many gene banks have not characterized their collections to provide this type of baseline data. Another criteria for selecting crops might be the threat of genetic erosion, although we have seen previously that this is difficult to estimate (see chapters 7 and 8).

The choice of farming communities is equally important for planning on-farm conservation. Social scientists have demonstrated that a number of criteria are associated with the maintenance of crop diversity: cultural autonomy in terms of local language, an orientation toward subsistence production, and the absence of markets (e.g., Brush, Taylor, and Bellon 1992; Zimmerer 1996). Arguably, the best choices of farming communities for *in situ* conservation are those that have a high probability of choosing to grow local varieties and distinctive crop resources (Aguirre G., Bellon, and Smale 2000). The cost of implementing on-farm measures in these communities will logically be lower than the cost in places where farmers are likely to choose modern varieties over local types. One study that investigates the probability of selecting local crops rather than improved ones was conducted in western Turkey on selection of different types of wheat. Meng, Taylor, and Brush (1998) could identify both villages and households with a 95 percent probability of selecting local wheat varieties. Factors such as farm size, educational level, and market access were especially significant. Likewise, this study was able to identify regions and villages where greater diversity in wheat populations, measured by qualitative and quantitative characteristics, was found. This Turkish study thus provided the criteria for choosing both crop populations and farming communities for on-farm conservation.

Institutional Framework

Because of its novelty and history of neglect, *in situ* conservation lacks the institutional framework that has been developed for gene banks. *Ex situ* conservation occurs at both national and international levels, and since the creation of the Consultative Group on International Agricultural Research (CGIAR) network and the International Board for Plant Genetic Resources (IBPGR), international support and leadership have been important to national programs. This follows from the fact that germplasm resources are an international public good and can be maintained in an artificial environment. *In situ* conservation depends on farmer participation, and, therefore, must rely principally on national agencies for planning and implementation (Maxted et al. 2002). National agricultural research programs (NARs) often maintain regional germplasm collections and data banks on agriculture, and these will be crucial for deciding where to implement *in situ* programs. NAR scientists are usually the ones most familiar with the genetic resources and the needs of farmers.

However, an obstacle to involving national agencies has been the assumption that on-site conservation is antithetical to the primary development goals of agricultural ministries. Financially strapped ministries in developing countries are likely to see *in situ* conservation as a luxury that benefits other countries. Farmers who are the most likely candidates for *in situ* conservation usually live and work in marginal areas where national agricultural research programs are limited. These farmers are often members of minority ethnic and language groups and commonly are not represented in the lists of the personnel of agricultural ministries or research institutions. The physical and social discomforts of working with these farmers have impeded many development programs (Chambers 1983), and we should expect the same impediments for on-farm conservation.

The international mandate embodied in the Convention on Biological Diversity, the Global Environmental Facility, UN agencies, and international agricultural research centers is a first step in developing an institutional framework for on-site conservation. The GEF role is especially relevant because of the funding it makes available. A second step has likewise been taken—exploratory research on crop ecology and conservation, and strengthening scientific capacity in Vavilov center countries for *in situ* conservation (Jarvis and Hodgkin 1999).

A third step is to engage participation of public and private (nongovernmental) agencies and organizations (Maxted et al. 2002). Nongovernmental organizations (NGOs) are also essential to the success of *in situ* conservation because of its decentralized and participatory nature. Local organizations might include marketing cooperatives for traditional varieties or cultural heritage groups. These groups are advantaged by having much closer ties and better access to farmers who produce traditional varieties. Nevertheless, NGOs may lack the necessary institutional capacities to support *in situ* conservation: longevity, articulation with gene banks, and interdisciplinary skills across crop science and social science.

In Situ *Strategies: Supply and Demand*

In situ conservation is defined as the maintenance of genetic resources in natural settings. For crop resources, this means the continued cultivation of crop genetic resources in the farming systems where they have evolved, primarily in Vavilov centers of crop origin and primary diversity. While *in situ* conservation occurs on a de facto basis, the use of the term here applies to public and private programs to promote on-farm conservation. The success of these programs depends on a broad political and scientific consensus that also must anticipate the need for new agricultural technology including modern crop varieties. *In situ* conservation must complement rather than oppose other strategies to increase food production and income. Thus, *in situ* conservation is here conceived not as a general strategy for developing an agricultural sector, but rather as a targeted program to conserve key elements of the crop evolutionary system in areas that are significant for generating crop genetic resources. An assumption is that the logical starting places for these programs are Vavilov centers.

The logic of on-farm maintenance of local varieties is that this practice carries much of the baggage of crop evolution—farmer knowledge, selection, seed flow. Natural selection continues under all circumstances, but decentralized, methodical, and unconscious selection by farmers requires that they maintain and produce seed. Broadly, two different approaches are possible to encourage local seed production: (1) to increase the value of and demand for local crops, and (2) to increase the supply of seed of these crops.

The demand-side approach has received most of the attention and support because of the assumption that local crops have lost value in the presence of increased population, markets, and modern crop varieties. The essence of this approach is that marginal improvement in the value of local crop types will induce farmers to maintain them rather than switch to improved varieties. Three general strategies have been developed to increase the value of local crops: (1) diversity fairs, (2) market development, and (3) participatory crop improvement (PCI) (Almekinders and Elings 2001). The third strategy, PCI, illustrates the opportunities for and challenges to on-farm maintenance, and this strategy will be discussed in detail.

Diversity fairs were the first organized program in the contemporary era to promote local crops in Vavilov centers (Gonzales 1999). A natural pride in showing diversity and an interest in observing it are sufficient to stimulate enthusiasm for the diversity fair. Promoters have also found that public gifts, such as materials for school construction or repair, add to the effectiveness of the diversity fair. Marketing of products for environmental, humanitarian, and other social causes has proven to be a successful method to support public causes. "Green marketing" in particular has been developed (Wasik 1996; Peattie 1995). This type of marketing identifies the product with particular qualities of consumer interest, such as organic or dioxin-free, or with beneficial characteristics, such as biodegradable (Wasik 1996). Regulated labeling guarantees the products of this type of marketing in some places and for certain qualities. The growth of both national and international trade associated with environmental or social causes has been sustained and robust for a decade or more, showing that consumers are willing to support these causes.

Developing the local and national market for landraces may be accomplished in several ways. Identifying special "niche markets" where landraces are in demand, and gathering information on the marketing channels that bring landrace produce to market, both can suggest bottlenecks and constraints to the market, for instance the lack of adequate storage and transportation facilities or information, or inadequate supplies of landraces for market. Supplies may be inadequate because of the lack of credit or other inputs. Market constraints might be overcome directly as part of *in situ* conservation projects, for instance through promotional campaigns to consumers, helping to increase production of landraces for market, and labeling or appellation. An example of a successful green marketing program for promoting *in situ* conservation has been established in the United States to maintain and utilize ancestral maize by Cherokee farmers in North Carolina (Brown and Robinson 1992). Maize landraces of the Cherokee had been all but eliminated by the diffusion of hybrid maize and contamination with commercial varieties. Maize scientists located remnant populations on a few farms and in collections. Through controlled genetic crosses, Brown and Robinson (1992) were able to reestablish pure Indian flour maize, which is now grown and marketed by the Cherokee Boys Club in North Carolina.

Appellation relies on legal enforcement, market demand, and a willingness to pay additional costs for the guarantee of the label. The appellation system is well developed for high-quality food products in Europe, such as wine, cheeses, and meats (Bérard and Marchenay 1996). To date, the appellation system is based on geographic location or manufacture to ensure quality and authenticity. In at least one instance, however, a certification system has been developed to guarantee the origin of plants on the Hawaiian Islands (Meilleur 1996). An appellation or certification might be attached to genetic resources for the purpose of financing *in situ* conservation. There is currently no national or international market for crop genetic resources, and the crop breeding industry is not likely to generate a market (see Chapter 10). The transaction costs of establishing and maintaining legal mechanisms to increase the value of local crops are probably high and above the potential market price of crop genetic resources. Consequently, marketing and labeling landraces for conservation is apt to succeed only by association with other proven consumer preferences such as high-quality or organic products.

PARTICIPATORY CROP IMPROVEMENT FOR CONSERVATION

Perhaps the most important strategy for increasing the value of local crops is to use them as the basis for crop improvement programs, especially with the participation of farmers who will benefit from the results. Formalized cooperation between farmers and plant breeders in such activities as identifying crop improvement needs and priorities, selecting varieties or populations, and evaluating varieties may obviate the need to choose improved crops and varieties that cause genetic erosion (Almekinders and Elings 2001, Eyzaguirre and Iwanaga 1996). Improving local varieties and populations through procedures such as mass selection or interpopulation evaluation and selection may be especially suitable for marginal environments where conventional breeding has had limited success. Collaboration between crop breeders and farmers thus provides not only a context for *in situ* conservation but also an instrument to reach new groups of farmers. Participatory crop improvement for conservation is important because it involves existing agricultural research and development institutions, such as commodity improvement programs, that are linked to gene banks. Because crop improvement is proven and effective in many environments, it has constituents among farmers, scientists, and policy makers, making it a potentially important ally for conservation. Indeed, the premise of *in situ* conservation—that development and conservation can be compatible—makes collaboration between conservators, farmers, and crop breeders a logical choice for on-farm maintenance.

Collaborative efforts for improving local crops exist mainly in pilot projects and proposals, and expanding and institutionalizing these efforts rest on their ability to improve crops in new areas. They must also prove themselves adequate for *in situ* conservation of crop genetic resources. Collaboration between crop breeders and farming communities for crop conservation depends on identifying modes of cooperation that are novel to each side. Crop breeding is well established in most of the countries in which Vavilov centers have been established; indeed, it is thought to be a force behind genetic erosion. Unfortunately, conventional crop breeding has several characteristics that make it incompatible with the collaborative farmer/breeder programs envisioned for *in situ* conservation. Conventional crop breeding utilizes genetic resources from both local and foreign sources to create broadly adapted varieties that respond well to inputs such as fertilizer and irrigation and that return a higher yield than local varieties. In many developing countries, the use of foreign germplasm available through the international agricultural research system is more important than the use of local germplasm (Gollin 1998). Moreover, this applied science relies on controlled observation on agricultural experiment stations followed by on-farm testing and diffusion. In contrast, farmers' methods of crop selection, evaluation, and seed exchange rely on informal seed flows, uncontrolled observation, and postharvest selection. These methods have succeeded because of the large number of repetitions under different and often stressful conditions. The issues facing proposals for collaboration between the two are whether a third alternative will be acceptable to both crop breeders and farmers, and whether this alternative will conserve crop diversity.

Two distinct types of collaboration for conservation are distinguished in participatory crop improvement (PCI) by the timing of the association between crop breeders and farmers (Almekinders and Elings 2001). Participatory varietal selection (PVS) involves farmers later in the crop improvement process, after crop breeders have identified and preselected nonsegregating populations or genotypes that might be promoted for adoption (Whitcombe et al. 1996). Participatory plant breeding (PPB) involves farmers early in the crop improvement process, working with crop populations typically managed by peasant farmers (i.e., genetically variable, segregating material). Actually, most experience with PCI is with varietal selection (Almekinders and Elings 2001, Eyzaguirre and Iwanaga 1996), but with time more cases of breeding are available (e.g., Smith, Castillo G., and Gómez 2001).

Participatory varietal selection was developed as a means to produce novel crop varieties that will be more acceptable to farmers than conventional crop breeding done without the collaboration of farmers (Sperling, Loevinsohn, and Ntabomvura 1993; Whitcombe et al.1996). The essence of PVS is to involve farmers at the stage when crop breeders have identified improved crop materials through selection or breeding and are deciding which to promote or to develop further. In Rwanda, for instance, Sperling, Loevinsohn, and Ntabomvura (1993) describe the advantage of involving women in the selection of different varieties of beans that had been identified by crop breeders. Indeed, the women selected varieties that subsequently proved to be superior to the ones that breeders had selected. In India and Nepal, Whitcombe, Joshi, and Sthapit (Whitcombe et al. 1996; Joshi and Whitcombe 1996; Sthapit, Joshi, and Whitcombe 1996) describe the use of PVS to identify varieties of rice and chickpea that were acceptable to farmers in areas where conventional crop breeding had limited success. According to Joshi and Whitcombe (1996, 461), "[T]he methods of PVS employed were designed to identify and overcome constraints that caused farmers to continue to grow landraces." In the Indian case (Joshi and Whitcombe 1996), the PVS materials were improved varieties that had been bred outside of the three states where the PVS experiment was conducted (Gujarat, Madhya Pradesh, and Rajasthan). The successful rice cultivar (Kalinga III) came from the Central Rice Research Institute in Orissa, and the successful chickpea came from the international breeding program at the International Crops Research Institute for the Semi Arid Tropics (ICRISAT). In the Nepalese case (Sthapit, Joshi, and Whitcombe 1996), the material used was a pure line rice cultivar that had been isolated from a landrace originally introduced from India.

There is solid but limited evidence that PVS is an effective method of identifying improved varieties that are acceptable to farmers in places where conventional breeding has had limited success. Farmers are always on the lookout for new and better varieties, and PVS is a way to harness their knowledge of local conditions and criteria and apply it to a wider pool of improved varieties than is usually available. Participatory varietal selection also fits relatively easily into the institutional framework of crop improvement, where comparison between alternative improved varieties on experiment stations is routine. The novel aspect of PVS is that farmers are present and active in the comparison of cultivars that are close to release. Participatory varietal selection closely matches established crop breeding programs because it resembles multilocation variety evaluation methods and replacement of local varieties.

PPB, on the other hand, is a more radical departure from the current practices of both farmers and crop breeders. This method seeks to involve farmers in crop improvement from its early stages by employing two different strategies. One strategy is for breeders to work with farmer material (i.e., mixed and segregating crop populations) rather than with nonsegregating populations developed by breeders. This was the approach of the Milpa Project in the Valley of Mexico to improve local, open-pollinated maize populations (Smith, Castillo G., and Gómez 2001). Here, methods such as mass selection and interpopulation evaluation and selection were employed to create a new population or variety that is superior to the typical seed grown by farmers. Another strategy of PPB is to train farmers to improve their selection and management of seed, for instance by adding preharvest selection during the growth to the normal postharvest selection practiced by farmers. Participatory plant breeding employs techniques that were important in earlier phases of crop breeding in industrial countries but have been eclipsed by methods such as controlled crosses among pure lines (Poehlman 1995). The potential of PPB to improve farmers' seed with these methods is indicative of the limits inherent in farmer methodology, such as postharvest selection. To date, very few programs employing a PPB approach have been implemented or evaluated. While the Milpa Project has demonstrated that PPB can achieve significant improvement in the yields of local crop populations, it remains an experiment rather than an institutionalized program to change Mexican crop improvement programs or farmer practices.

The common elements of PVS and PPB include collaboration between farmers and crop breeders, and the objective of identifying improved crops. Both require understanding of the farmers' selection criteria and an appreciation of how farmers acquire and manage seed. The differences between the two methods, besides the timing of farmer involvement and the type of crop material used, are the relative emphasis on crop improvement versus conservation of local material and the institutional adjustment required. The discussion and research surrounding both methods have focused on methodology for involving farmers and for gaining an understanding of the ways that farmers select and manage seed. In Mexico, for instance, research on maize selection has shown that farmers vary in their criteria but maintain phenotypically homogeneous populations through the exchange of seed and cross-fertilization among different seed lots (Soleri, Smith, and Cleveland 2000; Louette and Smale 2000). Selecting only for ear and kernel characteristics at postharvest time, Mexican maize farmers achieve a high degree of morphological uniformity in their outcrossing crop, a skill that has been recognized for many years (Anderson 1946, Bellon and Brush 1994). Researchers interested in collaboration with farmers for improvement and conservation have pointed out that the farmers' selection criteria and practices are not well connected to improving the maize plants in farmers' fields to achieve higher yields (Soleri, Smith, and Cleveland 2000; Louette and Smale 2000).

Less attention has been paid to meeting conservation goals and to institutionalizing the approach among farmers or crop breeders. *In situ* conservation is justified not by its ability to conserve crop diversity per se but because of its potential to maintain crop evolutionary processes that gene banks are not meant to preserve. Above all, *in situ* conservation is aimed at maintaining farmer selection and exchange of local crop varieties. The requirement that farmers' crop varieties be competitive provides the rationale for collaboration with crop breeders, but the criteria of using local crop resources strongly recommends against PVS as it has been practiced (e.g., Whitcombe et al. 1996). The conservation goal cannot be met if the goal of the breeders is to disconnect farmers from their local crop populations in favor of improved varieties that have been developed but not successfully promoted by crop breeders. Moreover, the PVS strategy to delay farmer involvement until the later stages of selecting improved types effectively removes them from the most critical stages of crop improvement, giving farmers a menial rather than a collaborative role. Thus, from a conservation perspective, PPB is the proper choice for collaborative programs between farmers and crop breeders. Unfortunately, it is much more difficult to imagine the institutionalization of PPB than PVS.

Two constraints confront institutionalization of PPB that do not confront PVS. First, it requires a reorientation of the standard practices of both farmers and breeders. Second, the end product of PPB is ambiguous. PVS avoids these obstacles because it is essentially a modification of established breeding practices that involves farmers in limited and well-defined ways, allowing breeders to follow their normal practices. The demands of collaboration are minimal on both parties. Second, PVS has an unambiguous goal—to identify superior crop varieties that will be adopted by farmers and that will replace existing varieties—business as usual for crop breeders. In contrast, PPB implies significant modification in the normal practices of both crop breeders and farmers. It requires that breeders work with crop populations from farmers' fields rather than with "elite" breeding lines that are nonsegregating and often well known and characterized by other breeders. And depending on where the crop populations are evaluated and by whom, PPB implies that crop breeders will exercise less control over experiments than is customary in conventional practice. Starting with farmers' material rather than breeding lines adds several years to the breeding program and may isolate the breeder in terms of the farm sector that will benefit. For farmers, PPB involves an uncustomary and perhaps unwelcome effort to produce seed. Rather than selecting seed from the harvest pile or purchasing it from a neighbor, market, or seed supply company, PPB may involve farmers in the complex and laborious process of identifying seed through systematic observation during the crop's growth stages. Moreover, the goal of PPB is unlikely to be simply a better crop variety but rather a new way to approach crop breeding—that is, a process that will be managed by farmers. Unlike its rival, PPB faces daunting uncertainties about the ability of either agricultural research institutions or farmers to implement and sustain collaboration and new ways of producing seed.

In sum, PPB is the more appropriate method for conservation but the more difficult choice for institutionalizing collaboration between crop breeders and farmers. In seeking ways to reconcile the conflict between PVS, which is institutionally more convenient but inappropriate for conservation goals, and PPB, we might examine other participatory methods in agricultural development. A logical choice for comparison is farming systems research (FSR). Experience gained in FSR provides many social and institutional lessons that are directly applicable to collaboration between farmers and crop breeders for conservation purposes.

Lessons from Farming Systems Research

Farming systems research was a response to uneven diffusion of agricultural technology and reported negative social and environmental impacts of the Green Revolution (Dewalt 1985). In particular, FSR was proposed as a way to confront charges that conventional agricultural research is either inappropriate or damaging to small-scale farmers (Griffin 1974, Chambers 1983). This movement sought to bridge the gap between commodity-specific, experiment station–based research and the complexities of small-scale, subsistence-based production in unfavorable environments. Seven major components constituted the FSR approach (adopted from Simmonds 1985, 32):

1. creation of multidisciplinary teams within the research program,
2. identification of the target system or "recommendation domain,"
3. socioeconomic research on technology use and adoption,
4. identification of economically sensible innovations,
5. experiment station and on-farm research,
6. recommendations to national agricultural extension services, and
7. monitoring and evaluation.

The aim of FSR was to develop technologies for resource-poor farmers by linking research and farmers in a decentralized approach oriented to specific regions and farm sectors. The major departures from conventional research envisioned by FSR fell into two areas: (1) priority setting based on interdisciplinary, farm-level research, and (2) on-farm research and evaluation. The premise behind FSR was that marginal socioeconomic and environmental conditions required that technology be developed within the context of the farming system rather than in a commodity- or factor-specific context. The FSR strategy was directed at agricultural scientists in conventional research settings—universities, government institutes, and international centers. The aim of FSR was to provide input to experiment stations for determining research objectives. This input was to be generated from multidisciplinary field research.

Farming systems research existed as an organized research movement for approximately ten years (1978–1988), was promoted by major international donors and research organizations, and was the basis of numerous projects in dozens of countries. FSR had important impacts in some research programs and generated useful technology (Merrill-Sands et al. 1991). FSR has been institutionalized in some agricultural research systems, for instance in the creation of social science units that are integrated with crop research. Tripp et al. (1990) identify a number of positive results of the FSR movement. FSR drove home the idea that farmers are heterogeneous and that special approaches are necessary for small-scale farmers in marginal environments, an important client group. Knowledge of traditional practices such as shifting cultivation and agroforestry increased because of FSR research. Particular contributions of FSR include the rapid rural appraisal (*sondeo*) technique, improvement of on-farm testing, enhanced roles for social scientists in agricultural research, and a number of locally and regionally specific technologies that have been generated by FSR and adopted by farmers (Tripp et al. 1990, 390–92). Although the heyday of FSR has passed, it helped to redirect international agricultural research and remains an important, if less visible, research strategy (Byerlee and Husain 1993). In summary, FSR had more success in orienting research methodology than in technology development.

In spite of its successes, reviews of the FSR experience indicate a pattern of problems. Tripp (1991) points out that FSR's inherent nature isolated it from conventional agricultural research and led to its institutional marginalization. Key characteristics of FSR which isolated it were its emphasis on holism and location specificity, and its focus on resource-poor farmers and on-farm testing. Proponents made exorbitant claims for FSR, and slowness in fulfilling these promises compounded its novel and anomalous nature (Farrington 1988). Following nearly two decades of FSR, several of the institutional limits which were to be addressed by the FSR movement are still present. Agricultural research largely remains disciplinary, commodity specific, and based on experiment stations. More importantly, the key problem that triggered FSR remains critical in most parts of the world: agricultural research is least effective at reaching farmers whose land holdings are small and of uncertain tenure, who farm in difficult or marginal environments, and who practice subsistence agriculture.

Farming systems research provides many valuable lessons to participatory research for *in situ* conservation of crop genetic resources. Participatory research shares many of the goals, methods, and ideals of FSR and is perceived as a replacement for FSR in agricultural development research (Ashby and Sperling 1995). Both FSR and PCI seek to generate improved crop varieties which are more adapted to the conditions faced by small-scale, subsistence-oriented farmers in less-developed countries, especially in marginal agricultural environments. Participatory plant breeding goes beyond FSR in attempting to conserve the genetic resources of local crops which might otherwise be threatened by agricultural modernization or socioeconomic change. The chief methodological innovation of PCI is to involve farmers in the early stages of selection and testing of new crop material. After thoroughly reviewing FSR programs in nine countries, Merrill-Sands et al. (1991) elaborate a number of important lessons about carrying out on-farm and participatory research. These lessons concern (1) institutional issues of incorporating on-farm research and interdisciplinary research and (2) farm level issues about involving farmers directly in the research process.

INSTITUTIONAL LESSONS

Proponents of FSR found themselves in an uphill struggle to maintain the integrity of on-farm research, whose most important clients are small-scale farmers in marginal environments. Agricultural research institutes did not readily institutionalize one of the primary goals of FSR—to internalize the feedback from on-farm research to priority setting, planning, and programming of research on experiment stations (Merrill-Sands et al. 1989). Three institutional issues have been identified as problematic in integrating on-farm and experiment station research (Merrill-Sands et al. 1989):

1. conflict between on-farm and on-station researchers,
2. lack of effective exchange of specialized knowledge and information, and
3. specialization in on-farm research.

Conflicts between on-farm and on-station researchers arise because of different research objectives, methods, and criteria for evaluation. On-station research is oriented toward technology with wide adaptability and testing under carefully controlled conditions. On-farm research is oriented toward technology with more narrow adaptability, and toward testing under conditions where control is difficult to obtain. In competition for limited budgets, on-farm research appears to be more expensive in terms of reaching beneficiaries. Specialized and commodity-based researchers have full research agendas that may not directly benefit from investing in location-specific, on-farm research. Long-term research programs of on-station scientists are not easily interrupted or redirected by on-farm research needs, and useful information seems to flow one way: toward on-farm research but not in return.

Besides incorporating on-farm research into the program of agricultural institutes, another goal of FSR was to build effective interdisciplinary collaboration, especially the inclusion of social scientists. The experience has been that interdisciplinary research is easily thwarted; disciplinary scientists tend to seek autonomy, and social science presence is often fleeting (Merrill-Sands et al. 1989). The complexity and cost of social research is easily underestimated, for instance in defining "recommendation domains." This is one reason for the popularity of "rapid rural appraisal." Baker (1991) notes that the strongest advocates for participatory research are foreign technical assistants, an ephemeral constituency. The rhetoric of reversing normal agricultural science (e.g., Chambers and Jiggins 1987) did not endear the participatory research agenda to institute scientists.

The FSR experience gave no definitive answer on whether on-farm research should be the responsibility of specialized teams to work on-farm or the responsibility of individual scientists who combine on-farm and on-station research. Specialization creates barriers to integrating on-farm and experiment station research, and it strains the staffing situation of most agricultural research systems (Merrill-Sands et al. 1991). Neither senior nor junior scientists find the prospect of assignment to on-farm research sites attractive. On the other hand, problems resulting from the incorporation of on-farm methods into the programs of individual scientists include the lack of interdisciplinary communication and a pattern of neglecting on-farm activities in tight research schedules. Merrill-Sands et al. (1991) conclude that incorporation of on-farm activities into the program of an individual scientist is appropriate when the mandate is narrow and clearly focused, for instance on a single commodity.

FARM LEVEL LESSONS

On-farm research was one of the significant and innovative ideas of FSR. Indeed, some proponents declared farmer participation to be the harbinger of a new paradigm in agricultural research (Chambers and Jiggins 1987). However, in spite of these proclamations, FSR did not sweep away the older research establishment based on experiment station research and following hypothesis testing from theoretical deduction. In Baker's (1991) terms, reorientation of normal agricultural science, rather than reversal, appears to characterize the role of farmer participation at the farm level. Reviews of the FSR (Merrill-Sands et al. 1989, Baker 1991) describe three problem areas in participatory research:

1. sustaining farmer involvement,
2. selecting farmer collaborators, and
3. synthesizing and using information from farmers.

A common experience among FSR projects was that enthusiasm of both scientists and farmers for participation is greatest during the diagnostic stage but that it flags during the trial stages (Merrill-Sands et al. 1989). Many factors discourage sustained farmer involvement in on-farm research. Farmers may be too busy to participate in meetings, site visits, and other activities (Baker 1991). In fact, participatory research seems to rest, in part, on a very old bias about the amount of time and the value of time available to peasant farmers—the myth of the idle peasant (Brush 1977). Peasant farmers are interested in new technology, and they often carry out experimentation, but this does not mean that they have the time or other assets to add a new type of activity to their schedules. Limits on farmer time may conflict with the interest of agricultural researchers in multiple trials. Peasant farmers are, in fact, seldom idle and may find the demands of on-farm research taxing and inconvenient. Inconvenience may be exacerbated by the slowness of research and by small, incremental gains in the agricultural research process. Farmers may ask whether their scarce available time can better be spent on other activities rather than in doing on-farm research with visiting scientists.

The complexity of small-scale farming communities in marginal areas was one of the pitfalls of FSR, responsible for the profusion of increasingly complex and impenetrable models searching for elusive "recommendation domains." Inter- and infrahousehold differences in respect to technology choice are critical to sustaining farmer interest in participation and to achieving confidence in the usefulness of technology resulting from participatory research. Yet, Merrill-Sands et al. (1991) note that FSR has a weak record in systematic methods to choose farmer collaborators. Selection was often ad hoc and biased in favor of males, the wealthy, and politically active farmers. These biases undermined the usefulness of the information and contradicted the general goals of FSR to reach farmers who are bypassed by normal agricultural research. Unfortunately, rigorous attention to farmer selection extends the diagnostic stage and limits the time for on-farm technology development in the life of most projects.

The synthesis and use of information obtained from on-farm research is the final farm-level problem. The mandate of most agricultural research programs using public funds is to attain the greatest social benefit. This benefit is defined differently according to constituency, and small-scale farmers in marginal lands represent one of several constituencies for national and international research programs. Using farmers' assessments and difficulty in controlling the on-farm research environment made ordinary extrapolation to broader client groups all but impossible (Baker 1991). Indeed, difficulty in institutionalizing participatory research may be traced to this basic problem of quantification and replication, a problem that is not only close to the core of the sociology and ideology or science but that is also connected to the efficient use of public funds for research.

LESSONS FOR *IN SITU* CONSERVATION

The experience of FSR provides three lessons for participatory plant breeding as a method to achieve conservation of genetic resources in local crop populations.

1. Institutionalization of collaboration between crop breeders and farmers for *in situ* conservation will be most likely to succeed if the goal of conservation is clearly articulated.

The FSR experience suggests that institutional obstacles are inevitable in implementing a program that runs counter to the normal science of the core disciplines, and acceptance of a novel program is facilitated by adopting an incremental approach. Normal science in crop breeding relies on carefully controlled experimental conditions that are best achieved at research facilities rather than in the chaos of farmers' fields. Moreover, crop improvement and conservation are conventionally perceived as being incompatible, so breeding for conservation may be easily perceived as an institutional anomaly. Rather than dismissing these constraints, PCI will succeed only if they are understood and addressed incrementally. One reason for the early acceptance of PVS was that it avoided the stigma of incompatibility by making crop improvement rather than conservation the objective of participation. The anomalies of conservation and on-farm research may not be dispelled within breeding programs because they have other objectives and clienteles. On-farm and breeding activities may well be novel and disruptive within conservation units, but less so than they would be in existing breeding programs. The anomalies may be lessened if PCI is placed within a conservation unit rather than adding conservation responsibility to a breeding unit. In sum, the aim of PCI for *in situ* conservation is to find an institutional niche which is most compatible with the existing framework of crop improvement programs.

Special attention is required to define conservation goals in a participatory framework. While production objectives (yield, stability, quality) are familiar to plant breeders and farmers, conservation of genetic diversity is novel and potentially confusing. A solution is to clearly define a set of conservation products that should result from participatory plant breeding. These products should go beyond crop improvement per se and include such elements as the use of local crop materials and the strengthening of farmer selection and seed management. Difficulties with defining conservation products should not be swept aside as scientists and farmers agree on production goals. There may well be resistance because it requires long-term goals, trade-offs, and novel types of data and analysis.

2. Rigorous procedures for selecting farmer collaborators are essential.

A primary problem for FSR was adequate sampling in order to generalize from the relatively few on-farm trials that could be carried out. On-farm research must be planned so that it can be extrapolated in the same way as experiment station research. The same cautions pertain to participatory plant breeding for conservation. An additional complexity here is extrapolating for conservation as well as production objectives. Typical ad hoc selection biases include choosing wealthy, politically active, and educated men with larger farms. These indeed may be the best types of farmers to carry out *in situ* conservation, but such a choice needs to be justified. Sampling for conservation introduces a complex and troublesome set of issues in population biology, which few crop research programs can adequately address. Nevertheless, ignoring sampling is worse than doing it inadequately.

3. Systematic monitoring of farmers' reactions to conservation and participatory plant breeding is mandatory.

The task of monitoring farmers is not usual to plant breeders, but it is necessary to PCI for *in situ* conservation for two reasons. First, this is an experimental procedure, and it will require constant modification and adjustment. Second, a major goal of PCI is to initiate self-sustaining practices by farmers. In order to judge the sustainability of PCI we must have an accurate idea of farmers' reactions, reservations, and preferences. Farmer-based plant breeding may simply not be worth the farmers' time or may falter because of migration, lack of transfer across generations, or one of numerous and unforeseen problems. Success in participatory breeding will be important, but learning from failure can be equally valuable.

Summary

Landrace maintenance occurs in a de facto fashion (Brush 1995), so that we may ask whether *in situ* conservation might best be achieved by benign neglect of farming systems where genetic resources are diverse and intact. Benign neglect, however, ignores the presence of significant threats to genetic resources, especially population growth, markets, and infrastructure development, including new technology. Although landraces are still cultivated in many areas, these threats may overwhelm them. Intervention is, therefore, justified to save landraces. Nevertheless, such interventions are experimental and incur the risk of making genetic erosion worse than it otherwise would have been.

A number of different strategies are proposed to promote *in situ* conservation. One strategy is to increase the supply of traditional crop material to farmers who rely on seed exchange but face increasing difficulties in obtaining landraces. A second and more elaborate strategy is to increase the value of local crops so that farmers will maintain these rather than switch entirely to improved varieties. Seed fairs, green marketing, and participatory plant breeding are all designed to increase the value of local crops.

Participatory plant breeding has been identified as a method to achieve two goals which are often assumed to be contradictory—improving crop production and maintaining genetic diversity (Eyzaguirre and Iwanaga 1996). By definition, on-farm conservation of genetic resources requires farmer participation, but farmer selection of improved varieties developed on an experiment station does not promote conservation of local varieties. The collaboration between farmers and plant breeders must, therefore, be in the PPB mode. Participatory crop improvement has been shown to be successful in meeting production goals (Sperling, Loevinsohn, and Ntabomvura 1993), but it has never been evaluated for achieving conservation goals. The obstacles faced by FSR will also confront PCI, and they will probably be more severe because of the addition of *in situ* conservation as a goal. Successful pilot programs in PCI will depend on overcoming institutional and farm-level obstacles which the FSR movement encountered. Luckily, the rough road ahead has already been partly illuminated.

Participatory crop improvement has received attention because it provides a role for crop breeders who are already connected to genetic resource conservation through gene banks. The challenges facing PPB are representative of obstacles that any strategy for *in situ* conservation will face. These strategies rely on an institutional framework that is often burdened by lack of resources and demands from clienteles who are not interested in conservation or marginal farming sectors. Creating new institutions for *in situ* conservation may avoid these problems in the short run, but these difficulties are likely to emerge under any institutional framework.

10

Rights over Genetic Resources and the Demise of the Biological Commons

The synthesis between Darwinian evolution and Mendelian genetics (Huxley 1942) transformed biological science and quickly impacted applied botany, particularly the nascent area of crop breeding. Understanding inheritance opened new approaches to crop improvement based on controlled crosses of pureline varieties (Poehlman 1995), and fusion with Darwinism fundamentally altered the perception of crop diversity and evolution. The synthesis pointed the way to a more systematic collection and use of genetic resources. After 1900, crop improvement programs took advantage of not only the understanding of the principles of inheritance but also the recognition that crop variation across environments and between populations is itself a resource. Understanding the principles of inheritance allowed crop scientists to focus on single traits rather than on the plant as a whole. This ability converted exotic plants into sources of useful traits that could be brought into varieties adapted to different environments. Recognition of variation as a source of new and useful traits thus gave plant collecting and introduction new dimensions and meanings.

The ensuing changes, from plant collecting to the capture of genetic resources, and from plant selection to plant breeding, laid cornerstones of new attitudes toward plant resources and traditional farmers. One new perception was a vastly enhanced appreciation of the value of plant resources. Another was that crop scientists would control crop evolution in ways hitherto impossible. The utilization of crop germplasm in more effective and centralized crop improvement programs changed the ancient practice of seed saving, at least for some crops, and this change affected crop evolution itself. The pace of population change became more rapid, artificial selection became more centralized and more narrowly directed, and the previous noise in crop evolution from millions of individual farmer decisions theoretically was reduced to decisions of a few breeders. However, crop science's new power to transform crops also proved to have a Damocles outcome because it unleashed the specter of genetic erosion.

A consequence of anticipated demand and decreasing supply of diversity was to increase the implicit value of plant resources. This value was implicit because there was no market for genetic resources that might set their price. Increased value is clearly shown in the organization of more ambitious and thorough crop collection and conservation efforts, epitomized by Vavilov and the creation of the All-Union Institute of Plant Industry (VIR) of the Soviet Union. Another early indication of increasing value was the lament of Harlan and Martini (1936) that genetic resources were disappearing. Eventually, crop conservationists began to see farmers as partners in an expanded effort to conserve genetic resources and as stewards of crop evolutionary processes that could not be captured in gene banks. Concern for the loss of landraces and the need for conservation was mentioned at the International Agricultural and Forestry Congress in Vienna in 1890 and addressed at a special conference on the topic organized in 1927 by the International Agricultural Institute in Rome (Zeven 1996). The Rome conference recommended that landraces be maintained at school gardens or similar locations in their native region (Zeven 1996). By the end of the twentieth century, conservationists had ambitions for effecting on-farm conservation (Bretting and Duvick 1997, Brush 1999) in Vavilov centers.

The rise of crop breeding after 1900 also changed perceptions about "ownership" of living matter. Before the changes wrought by the Darwinian/Mendelian synthesis, crop improvement was a perpetual, albeit low level, activity practiced by farmers everywhere and was rewarded only by higher yields, prestige, and perhaps local seed sales. By the second quarter of the twentieth century, crop breeding and seed production were joined in a new commercial sector (e.g., Fitzgerald 1990). Inevitably, some nations accepted crop breeding as one of the inventive arts, akin to chemistry or mechanics, and thereby suitable for privileges of intellectual property accorded to other creative enterprises. The United States created the world's first legal framework for ownership of plant material with passage of the 1930 Plant Patent Act. Whether plants should be private property, and what type of plants might be owned, have emerged as issues that deeply affect research and conservation of crop genetic resources and fracture communities of scientists, farmers, and consumers according to geography and ideological orientations.

Common Heritage

Although traditional and subsistence-oriented agriculture persist in many places and are practiced by a significant portion of the world's farmers, the metamorphosis of agriculture that occurred in industrialized countries after the turn of the nineteenth century is seen as an adumbration of transformations everywhere. The reflex to predict genetic erosion in the wake of the Green Revolution is one forecast among many that assume an industrial and capitalist transformation of agriculture everywhere. These forecasts are braced by a human population that is surging toward ten billion, the success of manufacturing relatively low-cost industrial inputs and making them available in most farming systems, the revolution in human communication, and integration into more closely knit economic systems at both the national and global levels. An additional metamorphosis in the organization of agriculture under the ascendancy of nation-states and markets is the waning of an ethos that sees genetic resources as a common heritage.

Common heritage was the *ex ante* governance of biological resources until the last quarter of the twentieth century. Embargoes on the export of plant resources have been reported or imagined. One widely repeated case of fabricated plant theft was the removal of rubber from Brazil (Dean 1987). The story was concocted by Henry Wickham, the collector who sought to aggrandize his exploit, and uncritically reiterated by scholars intent on discrediting plant collecting (Brockway 1979, Kloppenburg 1988b). As Dean's (1987) history shows, theft was not part of the story of rubber. Nevertheless, real prohibitions on plant collecting and export have recurrently appeared. In 1787, when he was Washington's Minister to France, Thomas Jefferson violated an Italian prohibition on exporting seed of its high-quality rice by smuggling two bags of seed to the United States (Root and de Rochemont 1976). The attempt to maintain monopoly position in producing industrial plant resources is illustrated by nineteenth-century Peruvian and Bolivian embargoes of *Chinchona* seedlings that could be used to break their supremacy in the production of quinine (Musgrave and Musgrave 2000). Recently, Ethiopia has prohibited export of coffee plants (*Coffea arabica*) (Fowler and Mooney 1990), and India banned the export of the seeds of black pepper (*Piper nigrum*). But banning the collection and export of plant resources is rare and tends to emphasize wild species that provide industrial material rather than crops. This lends credence to the argument that common heritage was the norm for food crops.

Common heritage refers to the treatment of genetic resources as belonging to the public domain and not owned or otherwise monopolized by a single group or interest. Common heritage is similar to common property regimes that anthropologists and other social scientists have described for nonmarket economies (McCay and Acheson 1987; Ostrom, Gardner, and Walker 1994). Neither common heritage nor common property implies a lack of rules (*res nullius*) governing the use and management of common assets, a fact that has been often misunderstood (Hardin 1968, Kloppenburg and Kleinman 1987). Rather, they imply community management (*res communis*) that involves regulated access to common resources and reciprocity among users. In the case of genetic resources, open access was balanced by generalized reciprocity among farmers and plant breeders across economic sectors and national borders. Plant collectors gathered material that is freely exchanged within farming communities, and collectors continued this free exchange with crop breeders everywhere. Cleveland and Murray (1997) assert that exchange cannot be construed as evidence for the lack of indigenous notions of ownership of biological resources, and they note several instances of indigenous peoples' versions of intellectual property rights. However, the rarity of farmers' efforts to monopolize seed and knowledge sharply contrasts with the ubiquity of open exchange.

Collectors have historically worked under the ethos of science in the public interest. The privatization of seed is limited to a few areas and crops, and public sector breeding and other crop research continues even where seed companies exist. Public sector research and the free availability of genetic resources from gene banks denote a clear reciprocity to farmers and countries that provide genetic resources. Technical assistance for agricultural development and international crop improvement programs, such as those at IRRI or CIMMYT, represent concrete reciprocity in connection to open access for genetic resources. The wide diffusion of modern crop varieties from international breeding programs is one indication of the extent of reciprocity under common heritage (Byerlee 1996). Another is the importance of international collections in the crop breeding programs of Vavilov center countries. As an illustration of reciprocity between plant collectors and farmers, Evenson and Gollin (1997) document the importance of the rice collection at IRRI to countries in the center of rice domestication and diversity. Vietnam obtained 98.1 percent of the rice lines used in its breeding program from IRRI, and other countries had comparable percentages of using IRRI material: Bangladesh—

81.6, Burma—74.4, India—65, Philippines—81.8. In contrast, U.S. rice breeding obtained only 13.6 percent of its breeding material through IRRI.

Conceptualizing genetic resources as common heritage began formally when the phrase appeared in the 1970s in the discourse around international responsibility for the global environment (Cunningham 1981). Common heritage was a response to concern over environmental degradation in territories not controlled by any one state, such as the open seas or Antarctica (Joyner 1986). Because resources such as the atmosphere and oceans provided benefits to all humans, protecting them was presented as a common obligation. In negotiating international agreements such as the Law of the Sea, five elements of common heritage emerged (Joyner 1986):

1. areas defined as common heritage would not be subject to appropriation by private or public interests,
2. all people would share in the management of common territory,
3. economic benefits from the exploitation of common territory would be shared internationally,
4. common territory would be used for peaceful purposes only, and
5. scientific research in common territory would be freely and openly accessible.

PLANT RESOURCES AS COMMON HERITAGE

Common heritage for plant resources implies open access and nonexclusion to seeds and plants from farmers' fields, with due recognition of the importance to farmers of seed, food, and undisturbed fields. Prior to intellectual property regimes, the genetic sequences embedded in seed were owned collectively and thus represented a common pool resource. This logic derived from the natural and amorphous processes of crop evolution: mutation, natural selection, exchange, and decentralized selection. Because no person or group controlled crop evolution, it is inappropriate for anyone to claim authorship or ownership. Historically, the sociology of agricultural science supported the logic of common heritage in three ways. First, there was an ethos of public service, which survives despite the current rush to patent. Second, progress in crop improvement is made predominantly by incremental contributions rather that by heroic breakthroughs. Third, access to basic theory and other key elements in science customarily was free.

The essence of common heritage for plant genetic resources is very similar to the "communism of science" observed by Merton (1973). This social ethos is that while ideas and theories may be credited to a particular scholar or scientist, they cannot be individually appropriated because of their collective and incremental nature. In a similar vein, plant resources spring from collective sources of knowledge and nature. Plant genetic resources are "products of nature" (Bozicevic 1987) that belong to the commons because of the impossibility of determining the contributions of artificial selection versus natural selection, or untangling the history of collective invention. This regime survives among farmers and in the practice of gene banks to freely distribute genetic material to bona fide plant breeders anywhere.

Common heritage is logical within farming communities where seed is often exchanged or shared and provenance is ambiguous. It becomes problematic across social and political boundaries where differences in culture, wealth, and management of genetic material become visible. Extended to nations, this principle means that plant resources should be freely and openly exchanged, but Mooney (1979) and Kloppenburg and Kleinman (1987) have shown that international exchange can also be viewed through a lens that emphasizes differences in wealth and intellectual property. Unfortunately, international reciprocity involving information exchange, technical assistance, and breeding material is indirect and rather easily overlooked.

Defining common heritage is similar to belated and sometimes desperate efforts to demarcate the public domain after the expansion of private property (Litman 1990). The concept of common heritage was absent from the early modern texts about crop resources such as the two volumes from FAO conferences (Frankel and Bennett 1970, Frankel and Hawkes 1975). Rather, the term was borrowed from the international discourse on development and the environment that culminated in the 1972 U.N. Conference on Human Environment in Stockholm.

In contemporary parlance, common heritage genetic resources are an international public good (Kaul, Grunberg, and Stern 1999; Kanbur, Sandler, and Morrison 1999). In the conditions that prevailed before 1990, crop genetic resources satisfied the two key criteria that delineate public goods: nonrivalry and nonexcludability. Rivalry is absent because crop resources, either as seeds or specific genetic traits, can be used simultaneously without affecting their availability. Nonexcludability arises from the history of open diffusion, replication, hybridization, and collection of crop populations which makes it difficult for any one person, group, or nation to exclude access to the genetic codes embedded in seeds. It is also evident that the period of common heritage management provided an international benefit of immeasurable proportions. The availability of crop resources outside of their original hearths provided food sources that altered human history. The "Columbian Exchange" (Crosby 1972) not only benefited Europeans but it also made new staples, such as maize, beans, sweet potatoes, and potatoes, available to Africa and Asia. More recently, the collection of genetic resources under common heritage led directly to increased food availability around the world through breeding high yielding varieties whose pedigrees include germplasm from numerous countries (Smale 1997, Brush 1996).

Intellectual Property over Plants

The demise of common heritage that was seeded in the creation of intellectual property over plants became fully manifest in 1992, when the Convention on Biological Diversity defined genetic resources as belonging to sovereign states. Numerous factors pressed for closure of the genetic commons of agriculture. Closing the gap between wealthy and poor nations by leveraging access to genetic resources provided a rationale for closure. Likewise, promoting conservation by giving farmers ownership of genetic resources justified closing the genetic commons. Proponents of indigenous rights and neoliberal economics urged closure to solve inequity and genetic erosion via the market (Sedjo 1988, Posey 1990, Vogel 1994). But the germ for this vision is the creation of intellectual property rights in industrial countries for the purpose of recognizing and promoting crop breeding.

NATURAL RIGHT OR NATIONAL CONVENIENCE?

An enduring question concerning intellectual property is whether this right derives from a natural source as opposed to a utilitarian one. Defending natural rights or promoting social welfare are each compelling reasons for closing the commons. The natural right argument derives from the Enlightenment's view that ownership of the products of personal labor is a natural right (Kramer 1997). The social utility argument in favor of intellectual property, which is explicit in the U.S. Constitution, views it as a device to achieve economic benefits for society. Article I, section 8 of the U.S. Constitution authorizes Congress "to promote the progress of science and useful arts by securing for limited times to authors and inventors the exclusive right to their respective writings and inventions." The rationale for granting monopoly rights to inventors and authors, famously put by Joan Robinson (1956, 87), is that "by slowing down the diffusion of technical progress it (the patent system) ensures that there will be more progress to diffuse."

The utility argument has risen to prominence since the beginning of the twentieth century. This ascendancy rests on two pillars of scholarship that delineate property as a social institution. One is Hohfeld's (1913) landmark analysis of property as a system of rights and obligations among property holders and others. The second is Coase's (1960) analysis of social cost and the role of property rights in the negotiation of socially optimal solutions to conflicts. The fact that intellectual property is conditional bolsters the social utility interpretation. The right is finite and of limited duration, and claimants must satisfy seemingly arbitrary application criteria, such as being the first to file. Nevertheless, the natural rights argument remains persuasive, especially among groups that have not conventionally been awarded intellectual property for inventions and creative endeavors (Posey 1990, Greaves 1994). Included here are farmers and indigenous people whose inventive activities tend to be informal, unrecorded, and collective. The coincidence between the rise of industrial capitalism and the use of intellectual property rights bolsters the claims of social utility, and these claims have been widely and loudly repeated by private interests who invest heavily in research and development. Whether it is intellectual property or some other constellation of factors that have stimulated the inventiveness of advanced industrial countries is a matter of some dispute (Levin et al. 1987), but the arguments favoring broader intellectual property and stricter enforcement currently prevail.

PLANT PATENTING AND VARIETY PROTECTION

There is no doubt that an increased flow of new crop varieties, which are higher yielding, more disease resistant, or drought tolerant, are socially useful. This utility is the justification for extending intellectual property to plant breeders (Mastenbroek 1988). Plant patenting joined a growing family of state-protected monopolies over knowledge. The first recognition of "plant breeders' rights" was the 1930 enactment of the limited right to "plant patents" in the United States. This right was intended to protect the work of breeders of ornamental plants, such as roses and camellias, and was limited to asexually propagated and nonfood plants. Although the 1930 Plant Patent Act intended to cover only a small and narrowly defined group of plants, it opened the door to sweeping changes in the ownership of life.

Since 1930, the right to claim intellectual property over the products of plant breeding has been inexorably expanded both in terms of which "products" are eligible for ownership and the amount of protection that is afforded by the state. In 1970, Congress passed the Plant Variety Protection Act, which created a system of "Plant Variety Certificates." These certificates were limited as intellectual property because of two exemptions in certification. The first exemption allowed farmers to save and replant seed of certified varieties and to sell limited quantities of seed they had harvested. The second allowed plant breeders to use certified varieties to create new types of plants, without needing to license the certified variety from its originator. These conditions appear to limit the economic value of plant certificates (Lesser 1994b).

By the turn of the twentieth century, mild forms of protecting the creative work of plant breeders, such as plant patenting in the United States under the 1930 Act and plant variety protection under the 1970 Act, had given way to more comprehensive and restrictive ownership rights over life forms, including food plants. In 1994, the United States amended the 1970 plant variety protection provisions to limit the farmers' exemption. After 1994, the harvest of certified seed could only be used to replant the farmer's field (Kimpel 1999). In its 1980 *Diamond v. Chakrabarty* decision, the U.S. Supreme Court allowed industrial-type patents, known as "utility patents," on life forms, and the U.S. Patent Office Board of Appeals extended this right to crop plants in its *Ex parte Hibberd* decision in 1985. Since then, utility patents have been granted for novel varieties of many species, for specific traits, and for plant breeding processes (Baenziger, Kleese, and Barnes 1993). The U.S. Supreme Court reaffirmed the legitimacy of utility patents for crops in *JEM AG Supply v. Pioneer Hi-Bred International* No. 99–1996 (December 10, 2001). Plant variety protection and utility patents vary according to their criteria and the right to exclude others from using the protected plant (Table 10.1). Thus, in the United States, utility patents require the applicant to demonstrate utility, novelty, and nonobviousness, while plant certificates require the plant to be stable, uniform, novel, and distinct (Baenziger, Kleese, and Barnes 1993). The farmers' and breeders' exemptions of plant variety protection imply a far less exclusive right of "ownership" for holders of plant certificates as opposed to utility patents. The more restrictive aspects of a utility patent imply greater economic benefit for the holder.

Moreover, the practice of recognizing plant breeders' rights has been widely adopted outside of the United States. The international Union for the Protection of Plant Varieties (UPOV), organized in 1961, promoted a more restrictive approach to plant variety protection than did the 1970 U.S. law. The 1994 adjustments in the United States brought its practices in line with UPOV. Perhaps most importantly, plant breeders' rights are part of the package of policies required for membership in the World Trade Organization (WTO). Plant breeders' rights are included in the Trade Related Aspects of Intellectual Property Rights (TRIPS) agreement of the WTO requiring countries seeking membership and trade benefits under the WTO to have a national system of plant variety protection or plant patenting. While the TRIPS agreement allows countries to fashion their own (sui generis) approach to plant breeders' rights, the need to conform to international standards creates very heavy pressure to adopt a system that resembles the UPOV approach. Countries looking for ways to protect plant breeders' rights can choose among several distinct tools that have been developed in other nations, but utility patents and plant variety protection are the most important and common methods. Progress has been slow in extending plant variety protection beyond industrial countries because of the lack of national intellectual property regimes, their cost, and wide-

Table 10.1. U.S. Breeders' Rights—selected alternatives

	Plant patents	Plant variety protection certificates	Plant utility patents
Eligible materials	Asexually propagated cultivars	All sexually reproduced or tuber-propagated cultivars	Seeds, plants, plant parts, genes, or physical traits for inbreds, hybrids, and cultivars
Criteria for protection	Novelty, distinctiveness	Novelty, distinctiveness, uniformity, and stability	Novelty, utility, nonobviousness
Protection	Right to exclude others from asexually reproducing, selling, using, or importing products from protected plants	Right to exclude others from propagating or selling, importing or exporting the protected cultivar and its harvested material	Right to exclude others from making, using, or selling protected material.
Examples of protected plants	Roses, grapevines, strawberries, peaches	Wheat varieties, inbred lines used to produce hybrid seed	Hybrid maize line Bean cultivar Bacteria-resistant gene Herbicide resistance Process for microplant propagation
Exemptions	Essentially derived cultivars not protected	Essentially derived cultivars are protected Farmers' exemption (farmers may plant harvested seed) Breeders exemption (breeders may use protected material to breed new cultivars)	Essentially derived cultivars are protected No exemptions

Source: Kimpel 1999, Jondle 1989

spread reticence to allow ownership of plants and other life forms. Like other forms of intellectual property, the existence of plant breeders' rights correlates with a country's wealth (Lesser 1998). Table 10.2 shows the diffusion of plant breeders' rights. While poverty slows the adoption of these rights, it is not an insurmountable obstacle.

THE CASE AGAINST PATENTING PLANTS

A central issue dividing supporters and opponents of intellectual property turns on the nature of the commons and the right of the state to restrict

Table 10.2. International adoption of plant breeders' rights (2000)

Country income level (totals)	Number of countries with plant breeders' rights	Percentage of countries with plant breeders' rights
Low income (*n* = 56)	28	50
Lower-middle income (*n* = 45)	16	36
Upper-middle income (*n* = 32)	19	59
High income (*n* = 40)	29	73
Total number of countries = 173	Countries with PBR = 92	% of countries with PBR = 53

Source: World Development Indicators database, World Bank [July 2001] at: http://www.worldbank.org/data/databytopic/class.htm; Classification, members, and observers of World Trade Organization [July 2001] at: http://www.wto.org/english/thewto_e/whatis_e/tif_e/org6_e.htm

access by granting special concessions to individual interests. Unlike tangible property, ownership of ideas and information cannot be easily defended by mere possession. The fact that ideas or information can be duplicated virtually without cost means that they are naturally part of the commons once they are shared. This same consequence applies equally to the products of ideas—e.g., poems, songs, software, and seeds. Once an artist has released an artwork to the public without copyright, it belongs to no one. This has been the case for the vast majority of written expression (Rose 1993). The concession to limit free access to ideas is the essence of intellectual property, such as a copyright that restricts the freedom to reproduce artistic expression without paying the author for the privilege.

To the many critics of intellectual property for plants, utility patents are an egregious violation against the commons of living beings and genetic resources (Mooney 1979). A serious flaw in the patenting process is the limited ability of patent application examiners to determine whether the "invention" meets the criteria of novelty and nonobviousness. Determination of prior existence in the public domain is seldom straightforward, and the patent system relies on challenges rather than definitive evidence of novelty. A problem is that challenges are costly and require tracking of patents that are issued. An example of the potential problems with utility patents for plants is United States Patent 5,894,079, applied for in 1996 and issued to Larry M. Proctor on April 13, 1999. The patent is for a field bean cultivar, named "Enola," whose distinctive characteristic is yellow-colored beans. According to the patent application, Proctor found yellow beans in a packet of mixed beans purchased in 1994 in northern Mexico. By successive selection during two years, Proctor claims to have "invented" a nonsegregating yellow bean. Besides awarding ownership of this variety, the patent also awards ownership of "a method of producing a field bean plant comprising crossing a first parent field bean plant with a second parent field bean plant, wherein the first field bean plant is the field bean plant of claim 2 (the Enola variety)."

Yellow beans have been known and described in the literature of New World agriculture since early Spanish colonial times. For instance, yellow beans are mentioned in the sixteenth-century Florentine Codex of Sahagún (1979, 8th Book, 13th Chapter). Yellow beans were exported from Mexico to the United States for four years prior to Proctor's claimed invention, and genetic tests indicate that the patented variety is indistinguishable from other yellow beans in Mexico and the United States (Bassett et al. 2002). Moreover, the method of producing a pureline variety by selecting among crop progenies has been standard practice among farmers and crop breeders for well over one hundred years. The patent was immediately challenged by the International Center for Tropical Agriculture (CIAT) in Cali, Colombia, which maintains the world collection of beans, including numerous yellow types. This case shows the ease by which the patent system can be abused, and the cost and uncertainty of challenges in such cases underscore the awkward nature of the system.

Closing the Genetic Commons

The idea of common heritage appeared belatedly in the discourse on conserving crop genetic resources and soon foundered on the shoals of national and private interests. Ultimately, common heritage was expendable because it was associated with genetic erosion and exploitation. As long as genetic resources were common property, incentive to conserve them was insufficient, and farmers lacked the means to balance equities in their exchange with plant breeders. Theoretically, legal initiatives favoring plant breeders might also be applied to farmers who are stewards of important resources and practitioners of successful crop selection. This idea resonated with the wider public concerns about environmental degradation and indigenous rights. A perceived obstacle to finding legal and financial solutions to environmental problems such as the loss of biodiversity was ambiguity over ownership. As long as biological resources were common heritage goods, nations richly endowed with genetic diversity had no incentive to conserve them. Indeed, the lack of ownership was seen as an incentive to overexploit biological resources (Sedjo 1988). Moreover, the existence of plant patenting or variety protection in wealthy countries and its lack elsewhere presented the specter of exploitation of poor farmers who followed common heritage practices (Fowler and Mooney 1990). How could common heritage be legitimate for landraces when appropriation as private intellectual property existed for breeders' varieties?

When applied to rivalry goods, such as the territory of Antarctica, prohibition against private property is clearly intended in common heritage, as seen in the Joyner (1986) definition above. However, when applied to nonrivalry goods, such as genes or language, the private property prohibition of common heritage is muddled. A long and richly argued debate over copyright (Rose 1993, Kent and Lancour 1972) and other forms of intellectual property (Machlup and Penrose 1950) shows that compatibility is possible between public domain and private property in goods. Predictably, a fundamental disagreement about the public domain and intellectual property emerged between industrial nations with intellectual property regimes and nations without these regimes. Once plants and their component parts became acceptable material for intellectual property, the term *common heritage* acquired a dual meaning and use.

Legal discourse in industrial countries with a long record of intellectual property accepted the coexistence of the public domain and intellectual property by acceding to the difference of "products of nature" and "products from nature" (Bozicevic 1987). This distinction rests on the conception that genetic resources are public goods in the public domain that can be manipulated into novel and unique forms by skilled persons. Industrial countries with intellectual property for plants take common heritage to mean the conventional open movement between nations of resources embodied in crop landraces and wild relatives. According to this view, crop genetic resources are "products of nature," (Bozicevic 1987) and because landraces are shared among farmers they are thus already in the public domain. Manipulating these resources in crop breeding creates "products from nature" that can be claimed as intellectual property.

The contrary view to reject intellectual property (e.g., Alford 1995) holds that the very existence of intellectual property in plants is a fundamental breach of common heritage. Here, common heritage means open access to all genetic resources, including breeding material that had been swept into intellectual property in industrial countries (e.g., Mooney 1979, Shiva 1997). According to the anti-intellectual property view, the distinction between "products of nature" and "products from nature" is unacceptable, and intellectual property in plants is seen as theft (Shiva 1997). This second view holds that wealth and political influence determine which resources are public and which are private. This view considers property in life to be an arbitrary privilege allowing an individual monopoly over complex entities that have evolved naturally and collectively. Moreover, intellectual property over crop plants is seen to threaten the well-being of some of the poorest members of global society because it limits access to seeds. This view harks back to pre-1930, when there was no system of intellectual property to take plant resources out of the common pool. This position soon leads to the argument that common heritage and intellectual property are incompatible.

The intellectual and political divide between these views overshadows discussions of the future of crop resources. Despite the recognition of the importance of international public goods, the subtleties that characterize public goods and the public domain were lost in the political rhetoric of the international debate over plant resources. The contest between these two views takes on political significance because of the increased value of genetic resources, the need to secure a firm financial base to support conservation, and the need to provide for continued access and movement of genetic resources between countries. The charge of "biopiracy," which emerged in the early 1990s (Odek 1994, Shiva 1997), epitomized the demise of the open collection of plants and the impoverishment of discourse by sloganeering. Few people bothered to ponder the nature of the *ex ante* system of common heritage (Brush 1996). Rather, discourse on biological resources shifted rapidly to control and ownership by different actors: nation-states, indigenous peoples, seed companies, international organizations, or research institutions.

The 1992 U.N. Conference on Environment and Development (UNCED) finalized the Convention on Biological Diversity (CBD), which addressed the negative consequences perceived of common heritage: resource degradation and exploitation. The clearest and perhaps most important provision of the CBD was to define biological resources as belonging to sovereign states, thereby effectively terminating common heritage among nations and the relatively unrestricted flow of genetic resources. The action of the CBD is, however, merely a fragment in an international discourse about crop genetic resources that includes political polemic (e.g., Shiva 1997), scholarship (e.g., Kloppenburg 1988a), and negotiation (International Institute for Sustainable Development [IISD] 2001). The longest running discourse has been conducted by the FAO, which has overseen negotiation about the International Undertaking on Plant Genetic Resources for Food and Agriculture since 1983 (Esquinas-Alcázar 1998). Negotiations were recently completed at the FAO to create a multilateral system for sharing agricultural germplasm as part of the International Undertaking on Plant Genetic Resources for Food and Agriculture (IISD 2001 and below). Despite supporting the principle of sharing, several countries chose not to include their resources in the common pool, among them soybean, tomatoes, groundnuts, and tropical forages (Charles 2001, IISD 2001).

Unfortunately, conceptualizing crop genetic resources as common heritage leaned toward rules derived for rivalry goods rather than toward the rules regulating public goods in the public domain. Common heritage for plant resources conceivably could involve open access but not the absence of property over any and all biological material. Thus, if property in plant materials were conceived in the same terms as the more widely accepted copyright, a consensus on common heritage and property might have been achieved. By analogy, authorship or musical composition create ownership via copyright, but they do not threaten the common ownership of the original resources of language and music. If plant resources were seen as analogous to language, appropriation of novel plant materials resulting from individual invention would not threaten the original common resource. However, the motifs of tangible property and theft dominated this discourse and thwarted consensus about protection of the public domain while promoting social utility.

Logically, the failure to reach a consensus about property affects the accessibility of genetic resources, and the lack of consensus about the legitimacy of intellectual property in plants has undermined the collection and transfer of plant genetic resources in two ways. First, sovereign ownership, perceptions of increased value, and fear of theft prompted many nations to elaborate plant collecting rules that inevitably encumbered research. In Peru, for instance, two offices within the Ministry of Agriculture engage in the permitting process for plant collecting. The National Institute for Agricultural Research (INIA) and the National Institute for Natural Resources (INRENA) both give permits. The former requires collectors to sign a material transfer agreement stipulating that they will not seek intellectual property and will neither commercialize nor transfer the material to others without INIA's permission. INRENA's permit, on the other hand, restricts collection in several different ways. Only authorized material can be collected. No information from local communities is to be solicited. No transfer of material (or parts, such as chromosomes) is allowed. All of the collected material is to be delivered to a national scientific entity.

Secondly, plant collecting and exchange were negatively affected by the tendency of nations to see germplasm as a tool to gain economic and political advantage (Fowler and Mooney 1990). Ethiopia closed shipment of its wild coffee germplasm in an effort to gain advantage in the lucrative coffee trade. Without these resources other coffee producing countries, such as Colombia and Costa Rica — with significant coffee production but a narrow biological base in the crop — might lose market share because of difficulties in confronting diseases. The United States has gone through periodic episodes of limiting access to its large collections on political grounds. China has been restrictive with germplasm relating to soybean. Difficulty in germplasm exchange is also evidenced by gene banks seeking to obtain material from other banks. The U.S. Department of Agriculture, for instance, reports a noticeable increase in denials and pending decisions, despite its strong record of providing material to other countries (Charles 2001).

ANTI-COMMONS IN CROP GENETIC RESOURCES

The breakdown of common heritage, which is indicated by the CBD and the hindrance of plant collecting and germplasm exchange to avoid claims of biopiracy, has the potential negative effect of creating an "anti-commons" (Heller 1998, Heller and Eisenberg 1998). Hardin's parable of the tragedy of the commons (1968) disparaged the lack of private property as leading to the overuse and decline of natural resources. This prompted a generation of research on the relation between environmental degradation and social resource management (e.g., McCay and Acheson 1987; Ostrom, Gardner, and Walker 1994). Hardin's parable resonated with the rise of neo-liberalism's embrace of private property and markets as a means to solve social problems (Coase 1960). This philosophy stands behind the impulse to seek equity and conservation through the market. As Coase (1960) pointed out, establishing property rights is a necessary step in creating a mechanism to negotiate solutions to problems of unaccounted costs, such as genetic erosion.

Ironically, using sovereignty and market-based benefit sharing to address the perceived problems of genetic resources have arguably lead to an anti-commons that is as bad or worse than the tragedy of the commons. If overuse and degradation because of the lack of private property defines the tragedy of the commons, underuse and the loss of the benefits of use define the tragedy of the anti-commons (Heller 1998, Heller and Eisenberg 1998). This situation arises when "multiple owners each have the right to exclude others from a scarce resource and no one has an effective privilege of use" (Heller and Eisenberg 1998, 698). The original case that generated the idea of the anti-commons was the difficulty in creating a viable commercial real estate market in Russia after the collapse of Soviet communism. This left a confusing plethora of rights and situations where different people held partial rights to the same asset (Heller 1998). No one has ownership of the full "bundle of rights" that constitute property (Hohfeld 1913), but rather different persons own different parts of the bundle, and with these they can block others from using or benefiting from a particular piece of real estate such as a Moscow storefront.

Heller and Eisenberg (1998) take the anti-commons concept beyond real property in underdeveloped markets and extend it to the case of biomedical research in the United States. Two mechanisms are cited as contributing to the anti-commons in U.S. biomedical research: (1) the existence of too many "concurrent fragments of intellectual property rights in potential future products," and (2) the multiplication of licenses by "upstream patent owners" on future discoveries (Heller and Eisenberg 1998, 699). In this case, intellectual property is allowed on incremental achievements in the inventive process. An inventor must license components on the possibility of future utility. Examples in biomedical research are patented gene fragments that later become useful in diagnostic tests. This case contrasts sharply with the case of collective invention in British metallurgy described by Allen (1983), when incremental discovery was not seen as worthy of patent but contributed to significant technological progress. In the case of the Moscow storefront, the anti-commons grows out of a set of inadequately developed property and market relationships, while in U.S. biomedical research it stems from too many people claiming property rights. The anti-commons problem identified by Heller and Eisenberg (1998) seems to corroborate Merges and Nelson's (1994) apprehension about possible negative effects of an overly broad scope in patent decisions.

The current difficulty experienced in the collection and exchange of crop genetic resources suggests the possibility of an anti-commons that is similar to post-Communist Russia (Heller 1998). The demise of common heritage has left nations with numerous claimants to the right to restrict access to genetic resources. The domain of crop resources is crowded with real and potential claimants to ownership rights. Private parties claim the materials of crop breeding—elite breeding lines, gene fragments, genes, and genotypes—as trade secrets, patented plants, certified varieties, and utility patents. The CBD recognizes national sovereignty over biological resources. In some regions, such as in the Andes, international treaty blocs have agreements on the management and exchange of crop material. Germplasm, often in the form of duplicate sets, is stored in local, national, and international gene banks that may have different charters governing access. Many countries recognize, either legally or in a de facto sense, the right of a community to control entry and the activities of plant collectors on village lands. At all of these levels—farming community, breeder, national government, local gene bank, national gene bank, regional treaty block, international gene bank—the potential exists to exclude the right to collect genetic material or to access material already collected.

The existence in Peru of two different agencies within the same ministry that give permits for collection and exchange is an example of the confusing situation that confronts a plant collector or breeder seeking access to collections. The admonition by INRENA not to solicit information from local persons in Peru is modified by the condition that, if information is exchanged, written permission from the village is required. This may not stop plant collection or information gathering but it can later choke off the use of material or information. Likewise, a plant breeder who has obtained plant variety protection under Peruvian law faces ambiguity in providing that variety to others. Does the variety belong to the plant breeder or to the state? INRA's material transfer agreement is ambiguous in this case, but it is clear in stipulating that INRA must agree to exchange. Moreover, Peru is bound by regulations of the Andean Pact concerning genetic material. The seemingly innocuous idea of requiring material acquisition and transfer agreements to indicate prior informed consent to collection and exchange (Barton and Siebeck 1994) may lead to an anti-commons situation by providing different players the opportunity to exclude others in the use of genetic resources. Exclusion may happen accidentally because an official is afraid of liability for the consequences of signing transfer permits. The slowdown in the exchange of crop germplasm (Charles 2001) seems to confirm the problem of an international anti-commons in which no one benefits from the resource.

Ironically, a slowdown in crop germplasm exchange is likely to hurt poor countries in Vavilov centers more than wealthy industrial countries without indigenous crop resources. Industrial countries have established effective crop collections that are used not only by their national breeding programs but also by programs elsewhere. Gollin (1998) studied the flow of rice landraces for crop breeding among countries within and outside the crop's center of diversity in Asia. His study follows the work of Evenson and Gollin (1997) on the importance of IRRI rice collections to poor countries in the Asian Vavilov center. Gollin (1998) reveals that the poorest countries are net borrowers from other countries, including the United States. His findings are summarized in Table 10.3.

Opening Access

Recognizing genes as economic resources and intellectual property effectively foreclosed a return to the *ex ante* practices of common heritage. However, in the face of an anti-commons that freezes the flow of crop genetic resources, efforts have been made to maintain open access to resources. Within a decade of the 1970s' FAO conferences on crop diversity and conservation, political issues emerged that were absent and seemingly unanticipated in the original conferences. Long-established oppositions—North/South, First World/Third World, developed/underdeveloped—soon framed discussions about crop resources. By the mid-1980s, genetic resources were construed as oppositions in terms of biological endowments, poverty, and property (Mooney 1983, Kloppenburg and Kleinman 1987). Simple but misleading dichotomies were used to construe a genetic divide between "gene rich" vs. "gene poor," or "gene producing" vs. "gene using," countries. For instance, the United States is a supposedly

Table 10.3. Summary of international flows on rice landrace ancestors, selected countries

Country	Total landrace progenitors in all released varieties	Own landraces	Borrowed landraces	Own landraces used in other countries	Net lending (borrowing), as share of total landraces
Bangladesh	233	4	229	10	(0.940)
Brazil	460	80	380	43	(0.733)
Burma	442	31	411	9	(0.910)
China	888	157	731	2052	1.488
India	3917	1559	2358	1749	(0.155)
Indonesia	463	43	420	420	0.000
Nepal	142	2	140	0	(0.986)
Nigeria	195	15	180	0	(0.923)
Pakistan	195	0	195	10	(0.949)
Philippines	518	34	484	229	(0.357)
Sri Lanka	386	64	322	57	(0.687)
Taiwan	20	3	17	669	32.600
Thailand	154	27	127	220	0.604
United States	325	219	106	2420	7.120
Vietnam	517	20	497	89	(0.789)

Notes: In the last column, all numbers are given as shares of landraces used in domestic varieties; figures in parentheses are negative numbers. Numbers may exceed 1 if a country is a large net lender of landraces. Positive numbers indicate that a country is a net lender; negative numbers indicate that a country is a net borrower.

Source: Gollin 1998

"gene poor" nation, but because of its extensive and accessible collections it provides crop germplasm to "gene rich" nations.

The encounter between "users" and "suppliers" of genetic resources sharpened with the knowledge that genetic diversity is threatened and in demand for crop improvement. Vavilov center countries sought political leverage with their genetic endowments, and the large stockpiles of genetic resources in northern and international gene banks became a liability to realizing a payoff. Indeed, of the 6,159,248 accessions inventoried among all gene banks in 1996, slightly more than half (3,447,469) were held by gene banks in Europe, North America, Japan, and international agricultural research centers of the Consultative Group on International Agricultural Research (CGIAR) (FAO 1998). Moreover, crop scientists believed that, by the mid-1980s, a large portion of the total diversity of the world's major crop species had been captured and stored in gene banks of major industrial countries and international agricultural research centers (Plucknett et al. 1987). Examples of these are the National Seed Storage Laboratory of the United States, which maintains 268,000 accessions of different crops in its Fort Collins, Colorado, facility; and the world rice collection of 86,000 accessions at the International Rice Research Institute in Los Baños, Philippines. The success in preserving crop diversity *ex situ* might seem to solve the conservation problem, but it potentially devalued the remaining resources in Vavilov center countries. However, reliance on gene banks as the sole method of conserving crop resources was undercut by three factors: (1) lack of secure funding, (2) need for international germplasm exchange, and (3) ecological perception of crop resources as environmental assets that require *in situ* conservation. These factors ensured an enlivened debate over the ownership and access to genetic resources with platforms for the status quo of free exchange, the abolition of intellectual property for plants, taxation on collection and use of genetic resources, Farmers' Rights, and the creation of new forms of intellectual property for indigenous people.

Equity is now included with conservation and access as a major topic in the international dialogue about crop resources. Two distinct solutions to resolving the inequity implied by the association of genetic diversity and human poverty have emerged out of the international discourse. The first is bioprospecting, which offers a market-based and bilateral solution. The second is a nonmarket and multilateral approach with provisions for Farmers' Rights.

BIOPROSPECTING

Among its many recommendations for fulfilling the Convention on Biological Diversity, *Global Biodiversity Strategy* (WRI, WCU, and UNEP 1992) urges that conservationists "base the collection of genetic resources on contractual or other agreements ensuring equitable returns" (World Resources Institute, World Conservation Union, and United Nations Environment Programme 1992, 94). Intimating that noncontractual approaches are inequitable, the *Global Biodiversity Strategy* accedes to the idea of plant collection as a potential act of "biopiracy" (Shiva 1997, Odek 1994). Contracts between collectors and "owners" (states and/or communities) are proposed as a direct way to end piracy, by formalizing the relationship between broadly defined "sources" and "users" of genetic resources.

An early implementation of this recommendation was the agreement between the American pharmaceutical company Merck and InBio, a quasi-public nongovernmental organization in Costa Rica (Reid et al. 1993b). The agreement between Merck and InBio specifies that InBio would provide Merck with extracts from natural preserves in Costa Rica to identify commercial natural compounds or genetic traits. Merck would provide direct payments to InBio ($1,135,000) and an undisclosed share in future royalties (Reid et al. 1993a). The agreement between Merck and InBio spawned similar arrangements by private pharmaceutical companies (e.g., Shaman Pharmaceutical, King et al. 1996) and public institutions (e.g., the U.S. National Institutes of Health, Mays et al. 1996). Agreements attempt to balance equities by providing payment for access to biological resources and are referred to as "bioprospecting" (Mays et al. 1996). Although bioprospecting has been used only for pharmaceutical compounds from wild plants, it is also plausible for agricultural germplasm (Sedjo 1988, Vogel 1994).

Bioprospecting solves the imbalance between providers and users of genetic resources by establishing a mechanism to transfer some of the public value of the resources into private benefit through contracts between local people and plant collectors. Theoretically, this solution is applicable across vastly different ecological and cultural conditions, assuming that states and markets exist to play supporting roles. The contract involves offering direct benefits (money, training, community development assistance) and long-term benefits (a share in royalties) in exchange for the right to collect both knowledge and genetic resources. These agreements seemingly satisfy the spirit and letter of the CBD (Gollin and Laird 1996; Moran, King, and Carlson 2001). Contracts acknowledge control by local people over genetic resources by requiring consent to obtain and use the resources. Equally important, bioprospecting contracts increase the private value of biological resources and knowledge to indigenous people through both material and nonmaterial compensation.

Bioprospecting is one of the more fully developed proposals to transform common heritage into a market stream. Gollin and Laird (1996) observe that the global-local link is viable only with direct involvement and support of national legislation. While bioprospectors envision a contract between a provider of genetic resources (e.g., a local community) and a private and/or public institution, it carries implications far beyond the immediate context. One implication is for national legislation to provide for enforcement of the contract through such mechanisms as litigation, arbitration, or mediation. Another is to make intellectual property available to bioprospectors.

The Predicaments of Bioprospecting

Bioprospecting contracts and their attendant intellectual property represent a version of recolonization, in which elites and privileged groups affix their own seals to cultural goods that have been collectively generated and shared (Coombe 1996). A biocontract makes the co-signers partners in appropriating genetic resources but does not fundamentally change the terms of the *ex ante* relationship between peasants and private firms. The "market" in this context is a monopsony with one "buyer" (a pharmaceutical or seed company), who controls vital downstream information, and numerous "sellers" (indigenous people or farmers) who lack information.

Three predicaments beset bioprospecting. First, it unfairly privileges certain communities. Second, it jeopardizes seed flow among peasant farmers. Third, it is unlikely to result in conservation.

Albeit novel, bioprospecting is a form of international development "assistance" involving a cast of participants that is similar to previous development programs: state agencies, international donors, corporations, and bilateral aid missions. Like previous efforts for international development, bioprospecting will be scrutinized by critics who recognize that the lofty goals of international assistance often camouflage conflict, hierarchy, and exploitation. In some cases, negative results of development assistance are objects of discourse and protest by the poor, but in others development silences and depoliticizes the poor. Scott's (1985) exegesis of the differing histories of the Green Revolution in a Malaysian village is justifiably regarded as a classic. He finds at least two Green Revolutions in local memory. "Winners," those with land and capital, see the Green Revolution as progress, while tenant farmers, "losers," regard it as a mechanism to enrich the landlords and further marginalize the poor by reorganizing labor and eliminating social obligations. Likewise, Dove (1996) writes of peasant understanding that "gifts" such as development assistance are likely to facilitate resource extraction and lead to increased marginalization, inequality, and impoverishment.

Other observers of the "development industry" have described its tendency to distort local politics in favor of elites. Ferguson shows that the apparatus for eliminating poverty through development programs in Lesotho is also "a machine for reinforcing and expanding the exercise of bureaucratic state power, which incidentally takes 'poverty' as its point of entry — launching an intervention that may have no effect on the poverty but does have other concrete effects" (Ferguson 1990, 255). Ferguson argues that by construing Lesotho as a country that is not yet part of the modern world, the "development industry," such as the World Bank or bilateral aid agencies, colludes with the state so as to depoliticize and silence the poor and thus to consolidate state power and further enrich the elite. Similarly basing his critique in Foucaultian theory, Escobar (1994) examines the discourse of "sustainable development" with particular attention to Colombian agriculture and locates it within the hegemony of national elites and international powers. Following Ferguson's (1990) path, Escobar (1994) describes the way that the "development" discourse obscures a history of hierarchy, domination, and exploitation and thereby suppresses other voices and solutions to poverty.

Bioprospecting has the look and feel of being another tool of the "development industry." Like the discourses about poverty and hunger that framed agricultural development in Lesotho, Colombia, and Malaysia, the idea of bioprospecting (e.g., Reid et al. 1993a) emerged in a context with different sectors and voices: nation-states and their political elites, transnational corporations, biologists and social scientists, various types of farmers, and indigenous people. Similarly, the discourse about the potential and plight of genetic resources construes an image of these resources that advantages certain sectors, particularly the state and transnational corporations, while silencing other sectors, principally the poor and marginalized sectors. This silencing is accomplished by appropriating the poverty of peasant farmers and indigenous people as the interest of the state in nationalizing biological resources, but this appropriation obscures the ways that biological resources are managed by the small farm sector.

Bioprospecting contracts for crop genetic resources would change not only the relationship between farmers and collectors but also among farmers and farming communities. This approach involves removal from the public domain and monopolization of common assets as intellectual property. This arrogation arises in the contracting phase of bioprospecting when certain individuals, groups, or communities are franchised while others are left out. Evidence of this problem is found in the difficulties encountered by several International Cooperative Biodiversity Groups (ICBG) funded by the National Institutes of Health, the National Cancer Institute, The U.S. Agency for International Development, and the National Science Foundation (Mays et al. 1996). Established to promote bioprospecting pharmaceutical compounds, these consortia involve actors from both the public and private economic spheres and follow the model of the Merck/InBio agreement. Many of the ICBGs have run afoul of local and regional politics and been labeled "biopirates" because the contracts involved biological resources and indigenous knowledge that were a common trust rather than the domain of a single group (Nigh 2002). In one instance, the Maya ICBG in Chiapas, Mexico, was terminated because of local disputes of participation, benefit flows, and patenting (Nigh 2002).

The difficulties experienced by the ICBG program suggest that contracting for agricultural germplasm will be problematic. If resources collected under contract lead to a patent, communities who are not parties to the contract are effectively precluded from commercializing their knowledge or resources. While the community may not directly claim intellectual property, the exclusionary effect of the contract is still felt because the seed company will seek exclusion. The objection asks why one community is favored over others in a benefit stream derived from collectively created and managed resources.

Genetic diversity in agriculture provides numerous benefits: agronomic performance in different environments; opportunity for hybrid vigor; insurance against pests, pathogens, and the vagaries of climate; quality food for home consumption; gifts; and rituals. Nevertheless, it is difficult for each farmer to maintain a maximum of diversity, so the average number of varieties per household is often low (e.g., Bellon, Pham, and Jackson 1997). Because diversity is so important, all farming cultures practice the exchange of seeds among farmers and alleviate the need for a single farmer to maintain the maximum amount of diversity. Moreover, the chance of losing seed is a constant threat, and farmers everywhere need to acquire new seed periodically. A result is that seeds flow frequently and freely within and among villages (Zeven 1999).

A small percentage of seed exchange within and between villages can create large-scale population impacts over wide regions. Louette, Charrier, and Berthaud (1997) report that slightly more than half (53 percent) of the farmers in their Mexican study site used their own seed, while 36 percent obtained seed within the community and 11 percent obtained seed from the outside. Seed of the most widely grown varieties is likely to have been obtained from farmers outside of the Cuzalapa (Louette, Charrier, and Berthaud 1997). Brush, Bellon, and Schmidt (1988) found that 82 percent of maize farmers in their study area in Chiapas, Mexico, changed seed of *Tuxpeño* regularly, and 38.5 percent changed landrace seed regularly. A similar situation exists in the Andes. Brush, Carney, and Huamán (1981) and Zimmerer (1996) found that some villages in highland Peru are renowned as sources of seed potatoes, and Valdivia et al. (1998) describe a peasant-based international flow of seed tubers of oca (*Oxalis tuberosa*) between Peru and Bolivia.

Peasant seed systems distribute good cultivars over a very wide area. Names of varieties are ephemeral in this exchange, especially if a cultivar is cosmopolitan over a large geographic region. Within a short time, farmers lose track of where a particular type of seed originated. Seed exchange within and between villages means that the crop varieties on single farms and villages are part of metapopulations rather than single, autonomous populations of landraces. These metapopulations are interconnected by migration of seeds and genes, accounting for the high percentage of overall diversity found in a single population (Chapter 8). The existence of metapopulations means not only that numerous villages are necessary to maintain a viable population of landraces but also that collecting in one place may capture most of the diversity of a region. However, collecting in one village succeeds only by dipping into a circulating stream of seeds and genes, and this makes the benefit-sharing contract unfair.

The inequity of contracting with a single farmer or community might be avoided by including partners who represent the broad collectivity sharing seed. Larger geographic units may tend to have more biological diversity and thus be more attractive to prospectors; larger units would reduce competition and price cutting, along with the number of conflicts about the origin of biological resources. For instance, a federation of Mayan-speaking communities of Chiapas could negotiate prospecting contracts. Alternatively, the state government of Chiapas might decide that no biological prospecting could be done except under contract with the state government, a decision that would require the agreement and support of the federal government of Mexico.

Stepping up from the community level to a regional level in contracting for genetic resources might address the equity problem of community-based contracts, but this step raises other problems. The first is determining where and how to draw boundaries around metapopulations. While many races of maize in Mexico are bounded regionally, their boundaries are fuzzy and permeable. The wide diffusion of *Tuxpeño* throughout lowland Mexico shows how a race can spread beyond its original domain. Larger biocontracting units would dissipate the financial rewards to individual farmers, thus detracting from the incentive to conserve. They presuppose political and financial cooperation among villages and perhaps even among ethnic groups, which contradicts a long history of intercommunity hostility (P. A. Dennis 1987). Finally, larger units in biocontracting would involve high overhead costs to governments and bureaucracies, which would dilute the financial benefits to farmers.

We have seen previously that farmers regularly depend on the flow of seed from local and nonlocal sources. Bioprospecting potentially jeopardizes peasant-based seed systems because of the chance that individual farmers or communities will perceive a benefit to monopolizing their seed. Restrictive policies of nations such as Ethiopia and China show that monopolies over genetic resources are attempted even in the absence of contracts (Fowler and Mooney 1990). The experience of ICBG programs show similar sensibilities among communities that otherwise appear to be cohesive and homogeneous. As argued above, the shift to define ownership of genetic resources leads toward an anti-commons that has restricted the flow of resources among nations. A similar situation among peasant villages could have very negative consequences to farmers and consumers who have little margin for error in their household budgets.

The conservation predicament of bioprospecting derives from its reliance on a steady demand from users of genetic resources. While contracting appears to be a direct and simple solution for promoting *in situ* conservation of landraces, it entails questionable assumptions about the demands for genetic resources for crop improvement, about the costs of administering a system, and about the conservation results of contracts. Proponents of the contracting solution have focused their attention on the supply side of genetic resources, neglecting the demand side. However, three factors will dampen the willingness of seed companies to invest in bioprospecting contracts. First, breeders who use crop genetic resources have a strong preference for well-known, "elite" germplasm and an aversion to unknown germplasm found in landraces. Second, large stores of crop genetic resources are held in public gene banks. Third, biotechnology enables alternative sources of germplasm and thus undercuts the demand for crop-specific germplasm. These demand-side problems are well illustrated in the case of maize.

A robust seed market for maize exists, but breeding for this market relies on elite breeding lines from public and private breeding programs rather than from landraces. Goodman (1985, 501) reports that "foreign germplasm accounts for less than 01% of the U.S. maize germplasm base," and he argues that this level is unlikely to increase. Mexico shares this small fraction with numerous other countries. Many large breeding companies, such as Pioneer Hi-Bred International, claim to be self-sufficient in maize germplasm. Maize breeders who are interested in acquiring exotic germplasm from landraces look to public gene banks in the United States, in Mexico, and at CIMMYT for free germplasm. The FAO (1998) review of the global status of crop germplasm resources reports that the United States has 14,091 maize accessions, Mexico 10,028, and CIMMYT 13,070. Plucknett et al. (1987) estimate that public collections contain 95 percent of all maize landraces. In addition to being free, germplasm from gene banks may be characterized for agronomic traits, which makes it more useful than material collected on farms. In 2001, negotiations at the FAO were successfully completed for the International Treaty on Plant Genetic Resources for Food and Agriculture (ITPGRFA) (FAO 2001). Among its other provisions, the treaty establishes a common pool approach for the genetic resources of 36 crop and 29 forage plant genera, and maize is included in the list of crops covered.

Markets for germplasm from maize landraces are further weakened by the potential of biotechnology to provide useful genes from nonmaize sources, thus bypassing farmers altogether. The incorporation of pest resistance from soil-borne *Bacillus thuringiensis* into maize is an example of commercializing genes from nonmaize sources. While transgenic crops are limited in traits and species, biotechnology is sufficiently well established to dissuade companies from investing in bioprospecting contracts to provide germplasm.

Transaction costs, or the costs of administering contracts, pose another obstacle to schemes to finance conservation through farmer contracts with seed companies (Visser et al. 2000). Seed companies that claim an exclusive discovery based on germplasm obtained from one group of farmers may have to defend their invention against similar discoveries based on germplasm obtained from different farmers. This would require more knowledge about landraces than is now available, and enforcement would be complex and costly if other farmers and public programs contest uniqueness claims. Costs incurred in creating and administering a legal infrastructure, monitoring compliance, and resolving conflicts are likely to absorb much of the profit which might have ended in the pockets of farmers.

Finally, the limited duration of contracts does not bode well for conservation. Contracts are likely to be for single collecting events rather than for continued access (Barrett and Lybbert 2000). Conservation requires a long-term commitment that will continue after the contract lapses. Farmers may be long-term beneficiaries of the development of commercial products as co-inventors or royalty partners, but this does not require them to conserve the original plants which were collected under the contract. Furthermore, a seed company is likely to be more interested in using a resource than in conservation. From a seed company perspective, the need for *in situ* maintenance expires once the resource is incorporated in the working collection. Delays of a decade or more in identifying and deploying a useful trait are likely to weaken incentives to invest in germplasm conservation, especially in a competitive and unstable seed market.

To summarize, bioprospecting is a well-intended effort to balance equities between North and South, but it is an unattractive means to this end. Contracting is apt to be a narrow and arbitrary sharing of benefits and may well aggravate conflicts in regions rich in genetic resources. Equally distasteful is the possibility that the lure of financial benefits from bioprospecting will interfere with the exchange of seeds within and among communities. Contracting disguises the North/South division while internalizing the political tensions over benefits to poor farming communities. Lastly, problems of weak and unstable demand are likely to discourage a conservation response by farmers.

THE MULTILATERAL APPROACH WITH PROVISION FOR FARMERS' RIGHTS

Bioprospecting provides a bilateral approach to conserve crop genetic resources and balance equities between the providers and users of those resources. An alternative solution is to broaden the multilateral system that was created during common heritage to conserve crop resources. In particular, multilateral support for *in situ* conservation policies aimed at increasing the food and income potential of traditional crops within the small farm sector can avoid the equity and efficacy predicaments of bioprospecting.

Creating a multilateral approach for dealing with genetic resources struggled with fundamental and long-standing disagreements over whether intellectual property on life is appropriate and, if so, over what materials and under what conditions. Although intellectual property dates to the time of the Enlightenment and the emergence of modern capitalism, it was narrowly defined and generally did not encompass materials that were judged vital for human existence such as pharmaceuticals and food plants. Indeed, many developing countries continue to debate the issue of whether to allow intellectual property for these human essentials.

One objection is that plant patenting, plant variety protection, and utility patents recognize the accomplishments and contributions of plant breeders but not those of farmers. Farmers, especially in centers of great crop diversity such as Vavilov centers, engage in creative management of crop evolution, including purposeful hybridization and selection that look similar to the work of plant breeders. Indeed, it has become fashionable to discuss "farmer/breeders" as a version of "bare foot doctors" (Cleveland, Soleri, and Smith 1994; Smale et al. 1998). Farmers have clearly been immensely successful in identifying, selecting, and diffusing crop varieties of exceptional value, but this success has been collective and incremental. However, the conventions of intellectual property do not recognize collective invention (Allen 1983). These conventions stand on the Enlightenment's foundations of recognizing individual rights, epitomized by the image of honoring the heroic, lone inventor.

The most egregious scenario in the contradiction between farmers' management and plant breeders' rights is the possibility that a farmer's variety will be picked up by a plant breeder and designated as intellectual property without alteration or with only minimal changes. The Enola bean is one case that seems to confirm this scenario, and numerous challenges to plant patents charge that they appropriate existing knowledge and plant varieties (Shiva 1997). The patent system relies on legal challenges to reveal that "inventions" are already known and available in the public domain. Unfortunately, these challenges require information and costly legal assistance that is unavailable to farmers in less-developed countries and is perhaps even beyond the means of agricultural research institutions in the United States and elsewhere. The logical alternative to abolish or severely limit intellectual property over plants is contrary to the historic trend in expanding and strengthening this type of property and to the political force behind international trade.

A multilateral approach to conservation and equity has been fashioned in an intermittent and somewhat uncoordinated process involving the FAO, a handful of nations with crop diversity, the CGIAR, the Global Environmental Facility (GEF), and various nongovernmental organizations (Fowler and Mooney 1990). The process of creating a multilateral approach was initiated in the FAO in 1983 with the creation of the Commission of Plant Genetic Resources, later changed to the Commission of Plant Genetic Resources for Food and Agriculture. The conference that established the commission also initiated an International Undertaking on Plant Genetic Resources with a resolution affirming that "plant genetic resources are a heritage of mankind and consequently should be available without restriction" (FAO 1987). In 1989, the FAO Commission on Plant Genetic Resources reinforced the concept that plant genes are a heritage of mankind with the concept of Farmers' Rights defined as "rights arising from the past, present and future contributions of farmers in conserving, improving, and making available plant genetic resources, particularly those in centres of origin/diversity. These rights are vested in the International Community, as trustee for present and future generations of farmers, for the purpose of ensuring full benefits to farmers, and supporting the continuation of their contributions" (FAO 1998, 278).

The ostensible purposes of the International Undertaking were to conserve crop germplasm and ensure its exploration and availability, but the resolution on Farmers' Rights also recognized equity interests of farmers by asserting the intention to "allow farmers, their communities, and countries in all regions, to participate fully in the benefits derived, at present and in the future, from the improved use of plant genetic resources, through plant breeding and other scientific methods" (FAO 1998, 278).

Mooney (1996, 40) asserts that Farmers' Rights "originally [were] an NGO proposal . . . to counter the movement for Plant Breeders' Rights." Thus, Farmers' Rights might be seen as a ploy to return to the pre-1930 status, before the U.S. Plant Patent Act ushered into the world intellectual property over plants. Associating Farmers' Rights with the assertion that plant genetic resources are a heritage of humankind might be construed to mean that all genetic resources were common property, including plant varieties and genetic sequences owned as intellectual property in industrialized countries. While Farmers' Rights do not refer to exclusionary or commercial outcomes, they could imply that farmers should have free access to breeders' materials just as breeders have had free access to farmers' materials. If the common genetic heritage included both landraces and elite breeding lines of seed companies, it would effectively vitiate plant breeders' rights.

Predictably, the tendency in Farmers' Rights discourse to conflate open access of common heritage with terminology borrowed from intellectual property has proven to be problematic. The concept intends full access to genetic resources, but the use of terminology that is similar to long-established restrictions in plant breeders' rights has caused confusion and considerable delay in the wide acceptance of the concept of Farmers' Rights. If farmers have the right to all plant genetic resources, it would be pointless for breeders to claim monopoly rights over their creations. While conservation and anti-intellectual property agendas seemed to be logical partners to some supporters of the International Undertaking (e.g., Mooney 1983), they turned out to be seriously mismatched, and the thinly veiled attack on plant breeders' rights was predictably opposed by industrial countries. The ensuing conflict between defining resources in farmers' fields as nonexcludable "products of nature" and plant breeders' varieties as excludable "products from nature" delayed the final acceptance of the International Undertaking for a decade and a half.

In 2001, the International Treaty on Plant Genetic Resources for Food and Agriculture (ITPGRFA) was completed and has now been signed by seventy-nine countries, including the United States (FAO 2003). The ITPGRFA takes a multilateral approach that reaffirms common heritage for the crop genera that are included in the list of crops covered by the pact. States retain sovereign rights over their genetic resources, including the right to designate genetic material and whole plants as intellectual property. The core provisions of the ITPGRFA (Articles 10–12) place the resources of thirty-six genera of crops and twenty-nine genera of forages in the public domain and guarantee access to these resources for breeding and research. Germplasm from the multilateral system will be available with a Material Transfer Agreement (MTA) that may include provisions for benefit sharing in the event of commercialization. The Treaty stipulates that "[r]ecipients shall not claim any intellectual property or other rights that limit the facilitated access to plant genetic resources for food and agriculture on their genetic parts or components, in the form received from the Multilateral System" (ITPGRFA Article 12.3 d). The phrase "in the form received" may be interpreted as allowing intellectual property once significant, inventive manipulation has occurred (CIPR 2002). The FAO serves as the proprietor of the international crop collections that are held in trust by the CGIAR, and the CGIAR system has repeatedly confirmed its adherence to open access to these collections.

Article 13 of the ITPGRFA lays out a procedure for benefit sharing by stipulating that commercialization of a new plant variety will trigger a financial contribution to the multilateral system. Again, the approach is multilateral rather than contractual between the genetic resource provider and the person who commercialized a product using that resource. The level, form, and conditions of payment (for instance, whether small farmers are exempt) is not resolved in the treaty and will be subject to further negotiations within the Governing Body of the International Undertaking. The benefit-sharing mechanism of the ITPGRFA faces serious logistical difficulty because of the long lag time between access to genetic resources and commercialization. Moreover, identifying the contribution of a specific resource within the complex pedigree of an improved crop variety poses a major obstacle to negotiating benefit sharing. Nevertheless, the treaty provides a mechanism for negotiating these obstacles while access to crop resources remains open.

This treaty grew out of nearly two decades of negotiation at the FAO concerning an international system for managing crop genetic resources. While the CBD sovereignty clause invited the rise of bilateral agreements, four factors pushed treaty negotiation toward a multilateral framework. First, replacing the open system with one defined by bilateral contracts would entail steep transaction costs that might exceed the value of the resources. Second, the process of creating a new access regime based on bilateral contracts threatened to interrupt germplasm exchange because of an anti-commons (Heller and Eisenberg 1998) resulting from the claims of different parties to control over access (Correa 2000). Third, increasing evidence suggested heavy dependence by poor countries on outside germplasm resources (Fowler et al. 2001), contradicting the earlier conclusion (Kloppenburg and Kleinman 1987) that industrial countries were more dependent on germplasm from developing countries. Fourth, accessions from large and valuable collections of the CGIAR network and industrial countries, such as the National Seed Storage Laboratory of the United States, remained openly available to crop breeders.

Negotiations related to the International Undertaking at the FAO (Fowler and Mooney 1990) bogged down because of uncertainty over whether a new international order for crop genetic resources reconfirmed or undermined plant breeders' understanding of common heritage. The ITPGRFA overcame the conflict by shifting emphasis toward open access to crop resources and away from the issue of compensation. Avoiding the long-term disputes about patenting life forms and gene sequences also aided the agreement on the status of international collections. Finally, by separating the issue of gene bank access from Farmers' Rights and accepting the co-existence of Breeders' Rights and common-pool rights, the ITPGRFA gained acceptance from more than one hundred countries and avoided any specific national opposition.

Farmers' Rights and International Protection of Traditional Agricultural Knowledge

The FAO Commission's International Undertaking on Plant Genetic Resources provided a forum to discuss equity interests of farmers in developing nations and gave rise to the Farmers' Rights movement. FAO Commission Resolution 8/83, which established the International Undertaking on Plant Genetic Resources in 1983, stressed the common heritage principle that plant genetic resources should be available without restriction and provided a sweeping definition of genetic resources as incorporating not only wild and weedy crop relatives and farmers' varieties, but also newly developed "varieties" and "special genetic stocks (including elite and current breeders' lines and mutants)" (FAO 1987). Non-governmental organizations that presented the idea of Farmers' Rights to the FAO Commission in 1985 were antagonistic to Breeders' Rights (Mooney 1996) and perhaps believed that international acceptance of Farmers' Rights would undermine individual rights (Fowler 1994).

The gambit to undermine Breeders' Rights through a binding international resolution endorsing unrestricted access to all genetic material failed because of opposition from states that provide for Breeders' Rights and the availability of large stocks of genetic resources in open collections that are linked to international agricultural development. FAO Resolution 5/89 resolved that the two types of rights were not incompatible and defined Farmers' Rights as: "rights arising from the past, present and future contributions of farmers in conserving, improving, and making available plant genetic resources, particularly those in centres of origin/diversity. These rights are vested in the International Community, as trustee for present and future generations of farmers, for the purpose of ensuring full benefits to farmers, and supporting the continuation of their contributions" (FAO 1998, 278). Farmers' Rights differed from Breeders' Rights in that they were to be vested in the "International Community" rather than with individuals. However, by not specifying what genetic materials were covered or who could claim ownership, the FAO definition created a problematic category. Farmers' Rights have remained an elusive goal. Their early association with the anti–Breeders' Rights agenda, and their ambiguities regarding materials and holders of the rights, thwarted its acceptance as an international principle or program. Following the ITPGRFA negotiation, the fate of Farmers' Rights will be determined at the national level.

The nature of the rights conferred by Farmers' Rights hinges on the economic benefit provided in the past, but no estimate of value or widely accepted method to estimate value of crop genetic resources is available. Estimating the historic contribution of farmers' varieties ideally requires one to separate the economic contribution of germplasm from other factors such as the development of physical infrastructure and human capital. Likewise, estimating the cost of Farmers' Rights is hampered by the lack of a program for how the stream of benefits to farmers might be used to achieve conservation goals.

Bioprospecting contracts potentially offer a mechanism to provide equity and stimulate conservation by increasing the value of biological resources, but this mechanism is likely to be ineffective for addressing equity and conservation issues relating to crop germplasm. Because collecting genetic resources tends to be "single shot" (Barrett and Lybbert 2000), fees for these collections are unlikely to have a long-term conservation effect. Contracts are likely to favor single communities arbitrarily or regions that have no special claim to crop germplasm, and Barrett and Lybbert (2000) argue that bioprospecting windfalls may be exclusionary or even regressive. The reaction of groups that were excluded from bioprospecting agreements confirms that exclusion is a liability (Nigh 2002). If conceived as a market situation between community "sellers" and seed company "buyers," Farmers' Rights exist in a monopsony environment in which a multitude of farmers with genetic resources face an extremely limited set of potential "buyers" for their resource. Mendelsohn (2000) observes that this situation leads to market failure and argues that a monopoly acting on behalf of farmers is necessary.

Possible titleholders of Farmers' Rights include farming communities and states (Correa 2000). Intercommunity exchange and seed flows make claims by one community for rights to a specific landrace or other crop resource open to challenge from other communities. The same may be true at the international level, where informal seed movement also exists (e.g., Valdivia et al. 1998). Transaction costs to settle such disputes may be higher than the value of the right, and arbitrary allocation presents ethical problems of favoring one community over others. Because of possible international disputes or price competition, some regions, such as the Andean nations, have initiated a consortium approach to providing biological resources (ten Kate and Laird 1999), but other communities are unlikely to adopt a similar approach because of the costs associated with the number of possible participants and other factors.

The content of Farmers' Rights is equally ambiguous. Characterization of gene bank collections is limited, and much of the material is stored without adequate documentation to identify source farmers (Peeters and Williams 1984). Defining knowledge rather than genetic resources as the subject matter of Farmers' Rights is equally problematic because farmers' knowledge is local, widely shared, changeable, and orally transmitted. Lastly, Farmers' Rights do not specify whether they cover wild relatives of crops, which have provided valuable traits to crop improvement but which are not always known or used by farmers. The final criterion that distinguishes Farmers' Rights from intellectual property is their duration (Correa 2000). The monopoly right of a grant of the intellectual property is made to be temporary as a way to balance the goal of increased invention over the goal of open competition. The unlimited duration of Farmers' Rights foregoes this balance, a policy of dubious merit if other communities or nations have valuable genetic resources or prove to be more effective conservationists.

The ITPGRFA moves away from a binding international resolution to create Farmers' Rights and assigns the realization of Farmers' Rights to national governments. The treaty inveighs on its Contracting Parties to provide for these rights in three ways: (a) protection of traditional knowledge; (b) equitable participation in sharing benefits; and (c) participation in making decisions related to the conservation and use of plant genetic resources for food and agriculture (FAO 2003). As in the *ex ante,* common heritage period, farmers are not granted the right to exclude others from using, or benefiting from, crop resources. Negotiating Farmers' Rights at the national level faces obstacles that were not critical in the international arena, such as political weakness of the traditional farming sector, urban and consumer demand for low-cost commodities, and the need to promote agricultural development. Although the CBD does not distinguish crop genes as a special category of biological resource, negotiations for Farmers' Rights will have to acknowledge the regime established by the ITPGRFA. Research on crop populations in traditional farming provides three lessons that will weigh on Farmers' Rights negotiations. First, crop genetic resources are collective inventions and metapopulations rather than assets that are privately derived and managed. Second, developing nations have benefited from adopting new technology, including new crop varieties, but landraces still exist in specific agricultural niches. Third, demand for crop genetic resources from outside sources is greatest in developing countries.

Experience gained in research and negotiation about possible mechanisms to protect farmers' knowledge offers four guidelines for crafting national Farmers' Rights programs. First, the goals of Farmers' Rights are to balance Breeders' Rights and encourage farmers to continue as stewards and providers of crop genetic resources. Second, Farmers' Rights are held collectively rather than by individual farmers or communities. Third, Farmers' Rights are not exclusive or meant to limit access to genetic resources. Finally, mechanisms are needed for sharing benefits received by the international community from genetic material from farmers' fields or international collections.

FARMERS' RIGHTS AT THE NATIONAL LEVEL

India's Act No. 123, 1999, for The Protection of Plant Varieties and Farmers' Rights, recognizes (Article 16d) Farmers' Rights in four ways (India 1999). First, farmers' roles as keepers of genetic resources and sustainers of crop evolution are to be recognized and rewarded through a National Gene Fund that will be financed by annual fees levied on breeders of registered varieties in proportion to the value of these varieties. Benefit sharing to communities that provided germplasm used in a registered variety will be determined according to the extent and nature of the use of genetic material in the registered variety (Article 26[5]). Second, India's Act No. 123 establishes the farmers' exemption that was present in early plant variety protection regimes (Baenziger et al. 1993), allowing farmers are entitled to "save, use, sow, resow, exchange, share or sell his farm produce including seed of a variety protected under this Act in the same manner as he was entitled before the coming into force of this Act" (India 1999, Article 39iv). Third, breeders are required to disclose in their application for registration information regarding tribal or rural families' use of genetic material used in the breeding program. Failure to disclose this information is grounds for rejecting an application for variety registration. Fourth, any interested party may file a claim on behalf of a village or local community stating its contribution to the evolution of a registered variety. If this claim is substantiated, the breeder is required to compensate the National Gene Fund.

The Organization of African Unity's African Model Legislation for the Protection of the Rights of Local Communities, Farmers and Breeders, and for the Regulation of Access to Biological Resources (OAU 2000) establishes Farmers' Rights in four ways. First, farmers can certify their varieties as intellectual property without meeting the criteria of distinction, uniformity, and stability that breeders must meet. This certificate provides farmers with "the exclusive rights to multiply, cultivate, use or sell the variety, or to license its use" (OAU 2000, Article 25). Second, farmers are given the right to "obtain an equitable share of benefits arising from the use of plant and animal genetic resources" (OAU 2000, Article 26). The African Model Law (Article 66) establishes a Community Gene Fund to accomplish benefit sharing and to be financed by royalties fixed to registered breeders' varieties. Third, farmers are guaranteed an exemption to Breeders' Rights restrictions, to "collectively save, use, multiply and process farm-saved seed of protected varieties" (OAU 2000, Article 26 [1e]). Fourth, farmers' varieties are to be certified as being derived from "the sustainable use of a biological resource" (OAU 2000, Article 27). This certificate does not imply financial reward.

The ITPGRFA, Indian Act 123, and the African Model Legislation accept the co-existence of Breeders' Rights along with Farmers' Rights and intend to accomplish benefit sharing through a centralized funding mechanism linked to Breeders' Rights. This same benefit sharing mechanism is present in the Genetic Resources Recognition Fund (GRRF) of the University of California, which imposes a licensing fee on the commercialization of patented plant material involving germplasm from Developing Countries (ten Kate and Laird 1999). This mechanism is a generic tool for reciprocity rather than one to reward specific farmers or communities. The African Model Legislation goes beyond all other documents in signifying individual communities as the beneficiaries, and Indian Act 123 combines both the generic and specific uses of compensation through the centralized gene fund. Farmers' Rights are also provided in farmers' exemptions to restrictions embedded in Breeders' Rights. Contradicting the view that Farmers' Rights are not a form of intellectual property (CIPR 2002), the Model African Law goes beyond the ITPGRFA and Indian Act 123 in granting exclusive rights to farmers over their varieties.

Two factors indicate that taxing certified crop varieties will offer meager resources to finance Farmers' Rights. First, plant variety certificates in industrialized countries have either relatively low or negligible value. Lesser (1994b) determined that the price premium associated with soybean certified seed was only 2.3 percent in New York State and concluded that this form of protection is too weak to be an incentive to breeders. Second, modern breeding programs are increasingly dependent on the use of "elite" breeding lines that are several generations removed from farmers' varieties and that show increasingly complex pedigrees involving crop genetic resources from many sources (Smale et al. 2002). Although India is a net exporter of landraces as breeding material, foreign landraces are as important to India's rice program as national landraces (Gollin 1998). Because African agriculture is heavily dependent on crops originating in other regions, dependence on international germplasm is high. For instance, in Nigeria's rice breeding program, 180 out of 195 landrace progenitors used in breeding were borrowed from other countries (Gollin 1998). Estimating the contribution of a single landrace or collection to the value of a modern variety has not been accomplished and is likely to become more difficult as pedigrees become more complex.

Summary

The increasing value of crop resources that resulted from the rise of formal crop breeding, the diffusion of modern varieties, and the availability of plant breeders' rights in the twentieth century spurred a movement to clarify ownership of genetic resources and broaden the pool of people who benefit from their availability and use. This movement resulted in the Convention on Biological Diversity and the International Undertaking on Plant Genetic Resources for Food and Agriculture. The movement was based partly on the notion, which was prevalent around the time (1992) of the UNCED meeting in Rio, that the *ex ante* system of common heritage was exploitative of farmers in Vavilov centers because of its lack of explicit reciprocity.

While the charge of biopiracy is still commonly voiced, we now more fully recognize the benefits of open access to crop resources in an international public domain. One benefit is public breeding programs, such as those at the

CGIAR centers, aimed at farm sectors that are poorly served by other agricultural research. These research systems depend on an open and readily accessible flow of crop germplasm. Indeed, some of the most important flows of germplasm are from collections in industrialized countries and international programs to developing countries, including "gene rich" countries. Moreover, we now recognize the dangers of an access system that depends on market negotiations between purported "owners" of genetic resources and "users." This system is likely to abuse the rights of people who have long been involved in the common pool of genetic resources but find themselves arbitrarily excluded in contracting. Another consequence that is discussed above is the creation of a genetic anti-commons for crop resources. A market-based solution to social cost may have philosophical merits in economic theory (Coase 1960), but history suggests that societies often opt for nonmarket solutions to provide many vital social services, such as public safety, education, lighthouse navigation, and agricultural research. Providing access to crop genetic resources and providing means to conserve them can, likewise, be managed by nonmarket means. The significance of agreements and programs such as ITPGRFA and the GEF program for financing *in situ* conservation of agricultural biodiversity is to reassert the value of the nonmarket approach that was the basis of common heritage. Farmers' Rights remain as a moral but largely rhetorical recognition of the contribution of farmers to the world's stock of genetic resources, and they provide only a limited mechanism to share benefits from using crop genetic resources or to promote their conservation.

11

Locating Crop Diversity in the Contemporary World

The Neolithic legacy of crop diversity in Vavilov centers is a dynamic system of protean human and biological components rather than a static inventory of genes or varieties. The evolution of both crops and cultures makes it all but impossible for the contemporary societies in Vavilov centers to be the same cultures or have the same crop inventories as those in Neolithic times. Likewise, it is improbable that today's farmers or their crops are direct, lineal descendants of those times. Nevertheless, successive farming cultures in Vavilov centers have sustained the crop evolutionary system that began with domestication. The biological components of this system include diverse crop populations, pests, pathogens, and symbionts that have co-evolved with wild progenitors capable of gene flow with their domesticated kin. The human components include local and decentralized selection, knowledge systems, seed flow, and exchange. Crop diversity is woven into the fabric of the cultures of Vavilov centers in ritual, cuisine, and memory (Nazarrea-Sandoval 1995), although culture is never a sealed vessel or a static list of traits or knowledge. Crop diversity survives in many places and may actually be increasing in modern agricultural systems. Indeed, maintaining diversity is as much an economic imperative as it is a biological one for the viability of agriculture. Diversity survives because of rational choices of farmers and because of the role that crop diversity plays in the warp of culture.

Diversity as an Intentional Choice

Associating crop diversity with "traditional" agriculture can obscure the fact that maintaining diversity is an active and purposeful part of farm management and not an unthinking or reflexive act by farmers who lack alternatives. Farmer selection, seed flow, and the ability to effectively distinguish among different crop types can as easily push crop populations away from diversity as toward it. Conscious selection throughout crop evolution provided constant opportunities to select against diversity and to greatly reduce diversity in favor of broadly adapted, high yielding types. However, crop evolution guided by farmer selection not only overcame the bottleneck of domestication but also led to greatly augmented diversity. Part of this augmentation owes to the diffusion of crops out of their original hearths, but the great assemblages of diversity identified by Vavilov and other crop scientists in centers of domestication testify to a source of diversification within cropping systems as well as between them.

Crop diversity has been part of cultural adaptation to heterogeneous and often fickle environments, especially in mountainous terrain where the boundaries between production zones are unstable. Selection for specific agronomic environments, such as soil types or altitude zones, has been described in several studies of low-input agriculture (e.g., Bellon and Taylor 1993, Richards 1986, Brush 1992). A positive interaction effect from crop diversity under farmers' field conditions and management is theoretically logical and supported by limited field evidence (Zhu et al. 2000, Wolfe 1985). The role of crop diversity in farmers' risk management strategies has been proposed as a primary explanation for the persistence of diversity (Clawson 1985; Cleveland, Soleri, and Smith 1994), although risk remains a theoretically thorny concept (Goodman 1975). Risk aversion conforms to the "moral economy" model (Scott 1976), which explains apparently uneconomic behavior as an adaptation to the difficult conditions of peasant life. Thus, farmers with limited access to specialized inputs value specific crop types for their ability to tolerate drought or resist pests and may discount the relatively low yield of these resistant crops.

Ecological studies of landrace management suggest that planting as uniform stands of single types rather than as mixtures is frequent. Diversity is advantageous to farmers in several ways, but many farm households cannot maintain high levels of diversity and rely on exchange or markets to provide diversity. "Crossover" in performance of different varieties in distinct social and agronomic environments has been widely observed (see Chapter 6), and diversity at the household level is explained by farmers' understanding of this effect.

Because landraces are variable populations of different genotypes, they may inherently afford an interaction effect, but farmers do not seem to search for this effect in their selection or planting practices.

Culture and Crop Diversity

Marginal environments and poverty, however, are only partially able to account for the diversity of crops that are maintained in Vavilov centers. Cultural identity almost certainly plays a role, although this factor is often submerged by a research methodology designed to study individual selection and make cross-sectional comparisons of variety choice. In these analyses of rational choice, culture becomes a residual factor used to explain the diversity that has not been explicated by individual decision making. Farmers may not perceive or assert that they maintain diversity because they are Tzeltal or Quechua but because it is an advantage in their farming environment. Culture's link to crop diversity is observed in gift giving, ritual, prestige, and cuisine. Most ethnographers who have studied crops have emphasized material and ecological issues rather than the crop as a cultural object. Use has usually been studied in terms of the choice between local and improved varieties rather than diversity within landraces. Thus, mixed lots of chalo potatoes are culinary superiors to improved types, and they are preferred as gifts and for attracting laborers who are paid in kind (see Chapter 5). Theoretically, however, a single local variety might suffice to meet ritual and culinary purposes. Indeed, the success of prized landraces, such as the widely diffused *huayro* potato from the central Peruvian highlands, and "foreign" maize in Cuzalapa, Mexico (Louette 1999), suggests that the contrast between local and nonlocal categories is fluid and ephemeral. Indeed, improved varieties can lose their identification over time as they are managed along with local types. A case in point is the *Rocamex* maize variety of Chiapas, which many farmers think of as a local variety, but in fact is an advanced generation of an improved, open-pollinated variety released by a joint program between Mexico's Ministry of Agriculture and the Rockefeller Foundation to boost Mexican agriculture in the 1950s and 1960s (Bellon and Brush 1994).

While culture is a difficult factor to connect to crop diversity, suggestive case studies allude to a role for culture. Anthropologists frequently link food and cuisine to culture (e.g., Weismantel 1988), but relatively few ethnographies have examined keystone crops in a cultural tradition. Exceptions include Ohnuki-Tierney's (1993) study of rice in Japan and Sandstrom's (1991) study of maize rituals in Náhuatl communities in highland Mexico that require different varieties of colored maize—red, black, yellow, and white. Another connection of crop diversity to culture is social identity that is linked to specific crop varieties. Sutlive (1978) mentions that Iban lineages in Sarawak identify with specific rice cultivars. In highland Vietnam, Phan Trieu (2001) found an association between specific varieties of upland rice and matrilineal kinship. In the Mnong communities of Daklak Province, farmers inherit rice according to their matrilineal affiliation. Thus, mothers, sisters, and daughters share a pool of varieties, while their brothers and sons share a different pool. Phan Trieu (2001) found that the matriline was significant in determining the similarity and differences of the rice varieties grown among ninety-two households studied in Krong No District. Since sons and brothers are required to marry outside of their natal matriline, they acquire a different set of rice varieties when they establish a new household.

Cultural diversity has been correlated with biological diversity (Maffi 2001), including crop diversity, but it has not been well studied as a functional variable that contributes to diversity. One difficulty in identifying a role for culture in crop diversity is the need to separate culture and environment as functional determinants. Anthropologists observed long ago that distinct cultural groups often occupy different environmental areas across a wider landscape (e.g., Kroeber 1939, Barth 1969), a fact that complicates the question why specific crop varieties are associated with cultural groups. Apart from distinctive traits that may be maintained because of cultural factors (e.g., lineage identity, cuisine, ritual use), culture organizes seed exchange and knowledge about crops and, thereby, should also influence diversity. Crops are metapopulations that are shared among farmers, and seed exchange is likely to favor kinsmen and neighbors of the same ethnic group. Seeds are often part of the dowry that new households receive, and this dowry is often passed unilineally as illustrated in the case of highland Vietnam discussed above.

Finally, information about crops, such as yield, susceptibility, and quality, is costly to obtain and evaluate. Anthropological theory (Boyd and Richerson 1985) suggests that culture can be understood as a mechanism that organizes the flow of information that is essential for survival. Cultural transmission is biased and can maintain cultural difference, for instance when individuals conform to local practices, because this is less costly than individual experimentation and learning (Boyd and Richerson 1985, 1995; Soltis, Boyd, and Richerson 1995). Atran et al. (1999) report that distinct folk ecological models affecting land use among the Lowland Maya of Guatemala's Department of El Petén are significantly influenced by ethnic group affiliation. Three cultural groups (native Itzaj Maya, Spanish-speaking immigrant Ladinos, and immigrant Q'eqchi Maya) live in a similar environment and exploit it in similar ways but have different views on human impact and how to maintain the health of the forest ecosystem. Distinctiveness in cognitive models of the environment was maintained by cultural segregation. Social networks of the different groups living in the region did not include any persons outside of an individual's ethnic group. This shows that socially acquired information is strongly bound to the ethnic group. Under these restrictions, it is logical that farmers will reduce the costs of obtaining essential information about a crop variety's performance by seeking it locally where it is provided by people who are trusted and where it can be observed in familiar circumstances. Thus, the cost of information should bias farmers toward local varieties and thereby help to segregate crop populations according to cultural group.

Poverty and Underdevelopment

Both practical and cultural reasons explain the persistence of crop diversity in particular farming systems. Nevertheless, we must acknowledge a wholly different interpretation of the persistence of crop diversity that connects it to poverty and isolation from more prosperous and developed sectors. According to the poverty explanation, crop diversity is a vestige of underdeveloped agriculture that locks farmers into a vicious cycle of low production, underutilization of resources, and poverty. Frankel, Brown, and Burdon (1995) succinctly summarize this concept when they argue that "it is the farmer who foots the bill" for maintaining crop diversity. The correlation of crop diversity with poverty seems inescapable. The most diverse farming systems are also ones with high infant mortality, illiteracy, limited life expectancy, subsistence orientation, and low income. They are isolated from more prosperous and agriculturally developed sectors by geographic barriers, poor infrastructure, the absence of markets, and local cultural attachment. The poverty explanation links to the classic and deeply held view of modernization theory that societies can be splayed along a historic continuum from underdeveloped to developed (Arndt 1987). When one contemporary society shares traits with a historic stage of another, it is thought to show evidence of being slower to develop.

The syllogism of modernization theory holds that early stages of agricultural development (e.g., in Europe or North America) were characterized by greater crop diversity, and because contemporary societies show this same characteristic, they must therefore be at an earlier stage of potential development. Of course, these premodern stages are also stigmatized by the extreme poverty and debased conditions of the Hobbesian vision of savage society. Modernization theory holds that the obstacles to selection of more productive seeds, such as poor soils or lack of improved seeds, can be overcome in a series of steps that begin with a move away from general subsistence production and toward specialized production and exchange. This shift is fully accomplished by subsuming the allocation of productive inputs (e.g., land, labor, and seed) into markets and producing commodities for the market rather than food for the household. This shift is accompanied and strengthened by development of infrastructure (e.g., roads, schools, agricultural research facilities) that links local production to a system that eventually becomes global. Modernization is propelled by the need to feed more people through more efficient use of agricultural resources and the universal desire to create greater wealth. The loss of diversity is a necessary cost in this process. This logic is entrenched in a philosophy of achieving social justice and alleviating poverty, goals that have extended and fortified modernization theory from its critics (Comaroff and Comaroff 1993).

Modernization theory provides a powerful narrative about human history and social differences, and it forms a subconscious stratum beneath much agricultural research focused on agriculture in poor countries, including Vavilov centers. This stratum is the girding beneath the idea of genetic erosion. Accordingly, diversity is borne as a cost by farmers who are extremely poor. The inevitability of the loss of diversity is verified by the convictions that poor societies will (and should) choose the technology which affords the greatest efficiency and wealth, and that higher yielding crops and inputs will be available everywhere. Accordingly, all farmers will succumb to the momentum and tug of social and technological change that propels agriculture into a stream of globalized, industrial inputs and outputs, and away from a mosaic of local and self-sufficient patches. This momentum in agricultural change is fueled by two predominant factors that affect human affairs in all aspects and places: increase of the human population, and integration of that population into a more singular world system. At the time of the first crop domestication, some ten thousand years ago, our numbers had not yet reached five hundred million. World population now approaches seven billion and is predicted to go to ten or eleven billion before the end of the twenty-first century (Harris 1996, Cohen 1995). The rapid and sustained acceleration in human population growth is a lasting reminder of the amazing achievement of crop domestication.

Integration of dispersed and isolated human populations, like population growth, is a distant outcome of the Neolithic Revolution. Crop domestication and sedentary life that came with agriculture triggered the process of absorbing local cultures and societies into larger regional polities. Globalization—measured by the relative importance of foreign trade in national economies, movement of people, or the diffusion of crops—is not particular to the end of the twentieth century (Geyer and Bright 1995, O'Rourke and Williamson

1999). Indeed, it is difficult to imagine a more penetrating and transforming event than the European expansion into the tropics between the sixteenth and nineteenth centuries (Crosby 1972). Nevertheless, the contemporary globalizing factors affecting agriculture, such as more unified world markets, centralized breeding, large germplasm collections, and supply of industrialized inputs, have the potential to affect local production beyond previous experience. Increasing reliance on markets for allocating land, labor, and goods amplifies the transforming power of cultural contact, migration, and crop diffusion. Only a small fraction of the world's villages do not experience regular and sustained connection with markets and other external social structures. Migration is ubiquitous and particularly important in this social intercourse that now pervades most rural areas. Moreover, infrastructure and technology that were not available in earlier periods of globalization amplify the effectiveness of these factors. It is logical to conclude that the availability of cellular telephones and Internet connections in otherwise remote areas of the world are an adumbration of transformations in every aspect of life.

The undertow of modernizing forces, especially population growth and globalization, and the syllogism that equates modernization with loss of crop diversity shape our ideas about the future of landraces, diversity, and crop evolution in Vavilov centers. These lenses admit only affirmative answers to the questions, "Is genetic erosion inevitable?", "Does improving the welfare of rural people depend on pushing their agriculture through a genetic bottleneck?", and "Does the availability of high gain inputs such as irrigation and chemical fertilizer doom local varieties?" Indeed, it is often assumed that the only possible fate of crop diversity in a globalized world with eleven billion people is to disappear.

Fault Lines in Modernization

While the syllogism connecting diversity to underdevelopment is compelling, it is also tautological. It rests on debatable and untested suppositions that diversity and tradition are inexorably coupled and that agricultural modernization everywhere will follow the path blazed by Europe and North America. The dubious logic of collapsing all societies onto a single continuum has been exposed in various branches of social science (Giddens 1990, Latour 1993, Kahn 2001), and the same challenge can be made against reducing agricultural change to a single continuum (Escobar 1995). Indeed, elements of modernization are criticized in the narratives of rural producers who confront the transformation of production guided by this concept (Scott 1985).

The Peruvian write José Maria Arguedas (1989) captures an essential message of the antipathy to the modernization theme in his poem "A Call to Certain Academics," originally in Quechua:

> They say that we do not know anything
> That we are backwardness
> That our head needs changing for a better one
> They say that some learned men are saying this about us
> These academics who reproduce themselves
> In our lives
> What is there on the banks of these rivers, Doctor?
> Take out your binoculars
> And your spectacles
> Look if you can.
> Five hundred flowers
> From five hundred different types of potato
> Grow on the terraces
> Above abysses
> That your eyes don't reach
> These five hundred flowers
> Are my brain
> My flesh.

The mistake of assuming that all agricultural systems must conform to the pattern set in Europe and North America is shown in numerous studies about the value of local practices in environments where imported technologies have failed (e.g., Richards 1985, Lansing 1991). The co-existence of landraces and new production technologies (Brush 1995) is also evidence that a unilineal technological development and replacement, as experienced in Europe and North America, is an inappropriate model for other farming systems. Moreover, on closer inspection, this model may not even apply to all farming systems in Europe and North America because it appears that crop diversity can actually be enhanced by modernization (see below).

Despite the synergistic power of population growth and globalization, the counterweights of environmental heterogeneity, local adaptation, economic strategizing, social resistance, and cultural defense have refracted the momentum toward an agriculture based on uniform seed and industrial inputs. Refraction is visible in regional disparities and local mixtures of "old" and "new" agricultural practices. Coastal areas and fertile, well-watered valley bottoms are easily distinguished from hill lands and isolated interior regions not only by income, education, and infant mortality, but also according to the dominance of markets in the organization of agriculture. Refraction is evident in inequalities in the price and availability of inputs. The facility with which world commodity prices translate into national and local prices is almost never matched by a comparable facility in providing agricultural inputs. Farmers almost everywhere sell their produce at commodity prices clustered around the world market value, but fertilizer, irrigation, improved seeds, and information are seldom available in each locality of a nation. In highland Mexico, peasant maize producers market their surplus at the world price of maize, which is often just at or below the cost of production (Perales R., Brush, and Qualset 1998). These peasant producers rely on local maize varieties because the Mexican agricultural research system has not developed improved maize for the high altitudes to compete with the landraces developed by farmers. In contrast, farmers in the low elevations of Mexico, such as in the Grijalva Depression of Chiapas (Brush, Bellon, and Schmidt 1988) or El Bajio of Guanajuato (Aguirre G., Bellon, and Smale 2000), are provided with access to commercial hybrids or improved, open-pollinated maize varieties. But, lest we think that neglect explains the patterns of maize throughout Mexico, Perales R., Brush, and Qualset (2003) point out that the region surrounding Mexico City, which is well endowed with research facilities and highly commercial maize production, remains a bastion of local varieties of the Chalqueño race.

The future of crop diversity in Vavilov centers depends on the balance between the countervailing forces of modernization and persistence of localized management, and it is not at all certain whether one of the forces will prevail. Indeed, the existing pattern of partial adoption of modern crop varieties along with retention of local types, observed as in Peru, Turkey, and Mexico, is a plausible alternative to both technological replacement and resistance. Population growth and national political agendas favor the modernization view that the combination of demand, markets, and technological inputs will penetrate traditional farming systems and replace local varieties and seed management with improved varieties derived from centralized breeding. On the other side, numerous factors favor the persistence of local varieties and local management. Included among these are difficulties in overcoming environmental, geographic, social, and cultural obstacles—marginal agronomic conditions, physical isolation, missing markets, and ethnic identification with local seeds. There is strong evidence that these obstacles are enduring, even under otherwise successful modernization (Brush 1995).

These countervailing forces have played out differently and have led to genetic erosion in some farming systems but to local persistence in others. These scenarios can be seen in the uneven diffusion of modern crop varieties during the course of the Green Revolution. Byerlee (1996) reports that 95 percent of irrigated lowland rice was planted in modern varieties by the mid-1980s. This percentage drops off dramatically as one moves to rain-fed lowlands (40 percent) and to uplands (0 percent). A comparable situation exists for the diffusion of modern wheat varieties in Turkey (Meng 1997). Zimmerer (1998) observes a similar pattern on a microscale in the Paucartambo Valley, Peru, where native potatoes thrive in higher altitudes but where they have largely disappeared in the low altitudes. This change may have historical roots in the pre-Hispanic introduction of maize to the Andes and the prominence of maize below the tuber belt.

In Mesoamerica, the situation of maize is similar but with more limited success of modern varieties. In Mexico, commercial hybrids accounted for only 19 percent of maize area in 1996, and open-pollinated improved varieties for another 1 percent (Morris and López-Pereira 1999). Modern varieties are concentrated in the well-watered and fertile lowlands, but even here the persistence of local varieties is notable. In Chiapas, for instance, Bellon and Risopoulos (2001) report an increase in the area in the landrace *Olotillo* during the decade after Bellon's original study (1990). In 1988, Olotillo maize was planted by 35.4 percent of the farmers on 10.9 percent of the maize area, and in 1997 it was planted by 54.1 percent of the farmers on 26.6 percent of the area (Bellon and Risopoulos 2001). This increase is possibly an impact of the 1992 reform of Mexican land tenure that allowed the creation of a land market. This appears to be an instance where a policy promulgated to modernize one aspect of agriculture (land tenure) actually encouraged crop diversity. As the land market developed in Vicente Guerrero after the 1992 reform, larger farms began to occupy more of the optimal valley bottom lands where the improved *Tuxpeño* varieties do well. Small-scale farmers, displaced from the valley bottom, increased their numbers and area on the hillsides where the Olotillo landrace thrives.

In Guanajuato, Aguirre G., Bellon, and Smale (2000) found that optimal environments with good infrastructure retained landraces in a majority of maize areas in spite of the availability of improved varieties. Here, the threat to maize landraces came from alternative crops. The cases of Vicente Guerrero and Guanajuato point to the co-existence of landraces and modern varieties in areas that are relatively developed in terms of markets, commercial production, and the availability of infrastructure and technological inputs. The future should be no different from the recent past in terms of balancing the countervailing forces for and against technological substitution in the cropping systems of Vavilov centers. We should expect some areas to shift away from landraces and local seed management, and others to retain them.

Comparative Advantage and the Competitiveness of Landraces

A striking but overlooked analysis of world crop production by Jennings and Cock (1977) showed that the productivity of crops is elevated by cultivating them outside of their hearths of domestication, evolution, and diversity. Overall, yields outside of the center of origin were 160 percent of the yield within. This difference cannot be simply dismissed as a reflection of developed vs. developing country agriculture because it involves comparison between all types of countries. The advantage persists in comparisons of countries at similar levels of technological development. Thus, Jennings and Cock (1977) report sunflower yields of 1.1 ton/ha in its North American center of origin and 1.5 ton/ha in Europe. High yielding (semidwarf) varieties of rice and wheat add 0.5 ton/ha in their Asian hearths but 1.5 to 2.0 ton/ha of rice in tropical America and wheat in Mexico (Jennings and Cock 1977). The reasons for the disadvantage of agriculture in centers of origin are not fully understood, but such factors as pressure of co-evolved pests and pathogens are suspected.

One lesson from the study of Jennings and Cock (1977) is that the disadvantage of low yield in Vavilov centers is exceedingly difficult to surmount and may require extraordinary resources to overcome. Another lesson is that a crop's production in its center of origin may always be "underdeveloped" in terms of its yield. These lessons suggest the plausibility of a different future in Vavilov centers from that imagined by modernization theory. This alternative future is not one of the technological and biological homogenization of agriculture in Vavilov centers but rather the persistence of local crops and seed management. The comparative advantage of non-Vavilov areas, which was signaled in Jennings and Cock's (1977) analysis, may discourage investing in modern varieties in Vavilov centers because of the availability of cheaper commodities from trade at world market prices. Other obstacles to technological replacement, such as marginal agricultural conditions, missing markets, and ethnic identification with local seeds, might actually be heightened by globalization. At the national level, participation in world grain markets is likely to become a more viable alternative for feeding increasingly urbanized populations than seeking technological modernization in the conservative farming systems of their nations. At the local level, income gained from migration, off-farm employment, and remittances, coupled with the development of food markets, is likely to reduce pressure to intensify local production and to invest in new technologies. The result is the continuing lack of technological alternatives to local crop populations that have been adapted to the marginal conditions of many Vavilov centers.

The scenario of economic integration strengthening local crops and seed management is readily observable in the maize fields of central Mexico. Mexico City, with a population of twenty million, is one of world's largest cities and the largest in the western hemisphere. The megalopolis exerts a tremendous and deeply penetrating effect on the central highlands that surround it. It is the hub of a dense network of roads, communication, and markets. Migration from surrounding provinces is omnipresent. The surrounding region has experienced increased siting of manufacturing plants *(maquiladoras)* that have increased the availability and importance of off-farm employment in villages that were purely agrarian two generations past. As the seat of the national government and Mexico's primate city, Mexico City has also served as a hub of agricultural research facilities and rural development programs. The Ministry of Agriculture has a large research facility in Texcoco on the northern rim of the megalopolis. The Colegio de Postgraduados, one of Mexico's premiere agricultural research universities, is located nearby in Montecillos, and the international maize and wheat research center, CIMMYT, is near the university. No region in Mexico has a greater concentration of maize breeders and other agricultural specialists. Some of Mexico's most ambitious agricultural and rural development projects have been undertaken in the region. An example is the Plan Puebla project that was implemented in the Puebla Valley, adjacent to the Valley of Mexico and on the other side of the volcanoes that tower over Mexico City. This project was launched by CIMMYT and was eventually taken over by the Colegio de Postgraduados (Winkelmann 1976). A main objective of Plan Puebla was to achieve the adoption of modern maize technology such as new varieties, planting densities, and fertilization. The Mexico City region, therefore, epitomizes the type of area that is privileged by the agricultural development establishment (Chambers 1983).

Despite its privileged location, however, the region surrounding Mexico City is characterized by a high degree of conservatism in terms of local maize varieties and local seed management. Perales R. (1998; Perales R., Brush, and Qualset 2003) studied maize agriculture and variety choice along a transect between Ayapango (2,400 meters above sea level) in the Amecameca Valley around the ancient Lake Chalco in the State of Mexico, and López Mateos (428 masl) in the Cuautla Valley in the State of Morelos. Ayapango is relatively close to Mexico City and is fully integrated into the regional economy of the capital. Only half of the farms surveyed depended on the farm itself for the major source of income, and most had income from salaried employment off-farm. Maize is the predominant crop in Ayapango, accounting for 93 percent of the cultivated area, and this maize is fully commercialized. In López Mateos, 70 percent of the farms surveyed depended on the farm as the major income source, and remittances were more important to farm households that included off-farm employment. Here, only 32 percent of cultivated area is in maize, and this maize is destined primarily for home use.

A striking contrast between the maize populations of Ayapango and López Mateos is the complete dominance of landraces *(Chalqueño)* in the former and the strong presence of modern varieties in the latter. At Ayapango, there was no measurable area in modern varieties within Perales R.'s farm sample. White varieties of the *Chalqueño* landrace dominated—96 percent of farmers grew them in 87 percent of their maize area. In Lopez Mateos, modern varieties accounted for 69 percent of the maize area—34 percent in commercial F_1 hybrids and 35 percent in advanced generation, open-pollinated modern varieties released by public breeding programs (Perales R. 1998; Perales R., Brush, and Qualset 1998).

The contrast between Ayapango and López Mateos dashes the predictions of modernization theory about seed source and management. Ayapango is much closer to the hub of maize research around Mexico City. Ayapango is economically more integrated into the national economy than is López Mateos, as measured by subsistence orientation of their farmers, off-farm employment, and maize commercialization. On these criteria, Ayapango is the more "modern" of the two towns. Yet, Ayapango farmers retain landraces and local seed production, while producers in López Mateos have shifted to modern varieties and commercial seed.

Several explanations for this reversal of the modernization predictions about seed management are possible (Perales R., Brush, and Qualset 2003). Maize breeders have been unable to surpass farmer selection and seed management in producing maize types that perform well in the relatively high altitudes of the Amecameca Valley, which is located close to the limits of the maize environment (Eagles and Lothrop 1994). It is reasonable to believe that Mexico will not pursue an expanded effort to breed modern maize varieties for the Amecameca Valley, so landraces should remain dominant. Mexico has increasingly turned to foreign trade to make up its deficit in national production (Brush, Bellon, and Schmidt 1988). The creation of the NAFTA free trade zone is predicted to increase dependence on U.S. maize, which is less expensive to produce than Mexican maize (De Janvry, Sadoulet, and Gordillo de Anda 1995). Other areas in Mexico, especially in the lowlands, are now the major maize producing areas, and, arguably, maize research in Mexico will emphasize these areas.

Despite the lack of modern varieties and a national market depressed by cheap imported grain, maize agriculture in the Mexican highlands will continue as long as farmers are there. The area around Lake Chalco has supplied maize to the Valley of Mexico since before the Aztecs established their capital at Tenochtitlan. Perales R. found that although the economy of maize was markedly disadvantageous to farmers, they remain strongly committed to the crop (Perales R., Brush, and Qualset 2003). This conclusion has been reached by virtually every researcher who has worked with Mexican farmers (e.g., De Janvry, Sadoulet, and Gordillo de Anda 1995). Indeed, most farmers in the cradle of maize do not envision an alternative to maize, a reflection of the historic connection between Mexican society and this crop. Clearly, maize is more than a commodity or a fungible element of land use to these farmers. Agrarian life in the Mesoamerican highlands since Neolithic times is a life lived with maize, and the tides of social and technological fashion seem to lack any erosive power to dissolve the millennial bond between the people here and maize. So, modernization is not likely to threaten the maize landraces of the Mexican highlands nor prize farmers away from the ancient attraction to the crop. Some aspects of modernization, such as increasing off-farm income, may actually strengthen the position of landraces by providing an economic subsidy to farm households that cannot compete in the world grain market but that will not abandon maize because of its role in the fabric of rural life in Mesoamerica.

Crop Diversity in Developed Societies

We rightly connect crop diversity with regions that also happen to be relatively less developed than Europe and North America in terms of the penetration of markets and the availability of technology. The idea that modernization would produce genetic erosion in agriculture came out of European and North American experience in the first half of the twentieth century, when crop breeding, mechanization, industrialized inputs, and commercialization swept away older agricultural methods and crops. Fowler and Mooney's assessment (1990) epitomizes the extrapolation of the loss of crop varieties in industrialized agriculture to developing country scenarios. However, three problems beset this extrapolation. First, as discussed above, a large problem lies in the modernization concept that the farming systems of Vavilov centers will follow a singular or European pattern in their development. Second, the evidence of genetic erosion in industrial countries is not well analyzed. There is no doubt that agricultural modernization in the United States eliminated diverse crop populations descended from pre-European times, such as maize (Mangelsdorf 1974). However, many of North America's crops, and almost all on Fowler and Mooney's (1990) list, are relatively recent European introductions. Fowler and Mooney (1990) show genetic erosion in the United

States through a comparison of vegetable varieties found in U.S. Department of Agriculture lists from 1903 and 1983, which reveal that only 3 percent have survived. No doubt data storage is incomplete and sometimes inadequate, but comparing lists drawn up at different times and for different purposes is hardly proof of extinction. The possibility exists that varieties listed in 1903 are duplicated in other accessions with different names. Moreover, the 1983 collections of the National Seed Storage Laboratory may have missed heirloom varieties that are now rare. This case reflects the unfortunate fact that industrialized countries have very poor documentation of crop diversity before the development of crop breeding and other causes of erosion.

Third, the scenario of diversity loss is unidirectional and does not hold out the possibility that this loss is, in fact, a passing episode in the development of modern agriculture. Crop evolution, like any other type of evolution, produces a continual flux in the numbers and dominance of species and varieties. As Darwin (1896) noted, infraspecific variation in domesticated plants and animals mirrors processes of variation and selection, albeit with the addition of conscious selection. Decline, collapse, and expansion of diversity characterize all types of evolution. Crop evolution, influenced by human migration into distinct environments and intensification of agriculture under population pressure, led to the accumulation of diversity following the bottleneck of crop domestication (Rindos 1984). The accumulation of diversity, however, has not been constant or without reversal. Two cases illustrate the potential folly of assigning a negative influence on crop diversity to recent agricultural modernization. The first case concerns the family of domesticated wheat (*Triticuum* spp.) and illustrates the ebb and flow of crop species and diversity long before the modern era. The second, that of peaches in California, demonstrates the ability of modern agriculture and economic organization to augment diversity.

EINKORN AND EMMER WHEAT

Cereal domestication was humankind's first experiment with plant domestication, and wheat, along with barley, ranks as the species that opened the way for agriculture during the Neolithic (Zohary and Hoph 2000). For wheat, the wild diploid *T. boeticum* (wild einkorn) and the wild tetraploid *T. dicoccoides* (wild emmer) are widely viewed as the precursors of the first domesticated wheats—einkorn *(T. monococcum)* and emmer (*T. turgidum* subsp. *dicoccum*) (Zohary and Hoph 2000). Toughening the rachis that connects the glume to the spike to suppress shattering is one of the hallmarks of domestication in wheat as in other cereals (Zohary and Hoph 2000), and nonshattering characterizes these two wheats. Hillman and Davis (1990) note that the rachis of einkorn and emmer is semitough in comparison to the fully tough rachis of durum and bread wheat. Moreover, einkorn and emmer represent early forms because they are not free-threshing and do not have the traits that later allowed the elaboration of macaroni and bread—hardness and a high gluten content. These two earliest domesticated wheats are hulled, and the kernels cannot be separated from the glumes by threshing. Rather, the grain is obtained by pounding to separate the wheat kernel from the glume, and parching to remove the hull (Nesbitt and Samuel 1996).

Einkorn and emmer wheat were domesticated roughly ten thousand years ago in Southwest Asia in the general region between Anatolia and the Fertile Crescent (Zohary and Hoph 2000). Molecular evidence suggests an Anatolian origin for both einkorn (Heun et al. 1997) and emmer (Dvorak and Luo n.d.). The domestication of the hulled wheats was followed rapidly by the diffusion of agriculture based on their cultivation throughout the Middle East and beyond to Egypt, central Asia, the Balkans, and Italy. Agriculture based on emmer was established in Egypt between 6000 and 5000 B.C.E., and emmer remained the sole wheat there for some five thousand years, until the arrival of hard wheat *(T. durum)* after the conquest of Alexander the Great in 332 B.C. (Nesbitt and Samuel 1996). Similarly, in Transcaucasia, emmer wheat dominated for thousands of years from the original diffusion of Neolithic cultures to the arrival of bread wheat *(T. aestivum)* and durum wheat. Indeed, Nesbitt and Samuel (1996) report that emmer was the most important cereal north of the Black Sea in the Ukraine and Moldovia until the Middle Ages. Emmer was likewise important in Roman Italy until the spread of free-threshing wheats in the first millennium A.D. (Spurr 1986). Hulled wheat was the grain of early Greece, although it appears to have declined in importance by classical times (Sallares 1991). While the actual wheat portrayed on Grecian pottery (see Figure 4.11) is probably free-threshing, the mythological wheat alluded to on painted vases might well have been hulled.

While there are no records of diversity of the emmer wheat populations that evolved over thousands of years and across a wide area in the Old World, they undoubtedly were characterized by great local and regional diversity. The evolution of modern durum and bread wheat, itself a story of tremendous diversification, was a catastrophe for the hulled wheats and the diversity they had accumulated since the Neolithic. By the modern era, hulled wheats had been pushed into isolated and marginal pockets of agriculture, for instance in the Küre Mountains of Kastamonu Department in northern Turkey (Karagöz 1996), and the Valnerina Valley of the Umbrian Apennines in central Italy (Perrino et al. 1996, Papa 1996).

The decline of emmer continued into modern times. In Turkey, the area planted in hulled wheat (einkorn and emmer) increased between 1948 and 1963, at which time 137,000 ha were sown in these wheats (Karagöz 1996). However, by 1993, the area had declined to 12,900 ha (Karagöz 1996). In Spain, these wheats have been present since the Neolithic but are now on the verge of extinction. Peña-Chocarro (1996) reports that land in Spain sown with hulled wheats increased during the twentieth century until 1960, when mechanization led to a dramatic decline. Today, einkorn is grown only in a single village in the mountainous province of Córdoba in Andalucía in southern Spain (Peña-Chocarro 1996). Emmer remains more viable in the northern province of Asturias, although much of it is mixed with spelt *(T. spelta),* a hexaploid, hulled wheat from central Europe that arose from crosses between emmer and the wild wheat relative *Aegilops tauschii* (Nesbitt and Samuel 1996).

In Italy, a similar situation existed until the late twentieth century—dominance in ancient times, followed by displacement by bread and macaroni wheat, and survival in isolated pockets. However, after about 1980 Italy experienced a renaissance in the cultivation of *farro,* the generic name for hulled wheat (D'Antuono and Bravi 1996). In 1997, 1,500 ha were planted with farro (Porfiri et al. 1998), and by 2000 this area reportedly had grown to 3,500 ha (Il farro 2001). Foods prepared with farro have become fashionable because the grain is organically produced and is associated with a traditional Mediterranean diet. The principal reason for the grain's return is the resurgence of local identity based partly on regional cuisine (Papa 1996). Emmer wheat is now commercialized and exported farro (Figure 11.1). It remains to be seen whether the niche market for farro will stabilize or disappear when food fashion changes. In Italy, the grain is a target of *in situ* conservation programs based on market promotion and breeding (Vazzana 1996). Distinctive regional populations of farro have been identified (Barcaccia et al. 2002), and efforts are underway to assign geographic origin appellation to farro products. The recent success of farro in Italy represents a case of market-driven reversal, albeit small, of a long-standing trend in the evolution of cultivated wheat toward elimination of hulled wheat species by free-threshing types. The recent success of marketing farro has prompted Italian wheat breeders to look for higher yielding varieties by crossing emmer (*T. turgidum* subsp. *dicoccum*) with durum wheat *(T. durum),* a step that may eventually erode emmer landraces that have survived previous modernization (O. Porfiri, personal communication). Another threat is the ambiguity of the term *farro,* which is used for both the relatively rare emmer populations and the more common spelt wheat *(T. spelta)* (Szabó and Hammer 1996). If the entire basis for cultivating farro shifted to a commercial one, this ambiguity might favor

Figure 11.1 *Farro* label, Italian export, 2002

spelt, which has been the subject of breeding programs and can be crossed with higher yielding bread wheat, to the detriment of emmer (O. Porfiri, personal communication).

The case of hulled wheat in the Mediterranean area shows that the loss of crop diversity is not only a phenomenon of modern agriculture, it also characterizes crop evolution since antiquity. More importantly, it suggests that crop populations can survive long periods as relics only to be "rediscovered" and promoted in modern agriculture. This survival is not based entirely on their value to subsistence agriculture but also on their association with particular cultural traditions and local identities. The resurgence is directly connected to meeting consumer demand in the national and international economy for distinctive, regional agricultural products. Another case of crop diversity in modern agriculture can be seen in California, but this is a case of new diversity rather than survival of ancient crops.

CALIFORNIA PEACHES

Compared to Vavilov centers, agriculture in California is in its infancy. Crop production in California is only as old as the European arrival and the creation of Spanish missions in the eighteenth century. However, the brief span of agricultural history in California is witness to rapid evolution through the adoption of new technology and integration into the national and global economy. Indeed, the evolution of California agriculture never went through a stage of peasant agriculture focused on subsistence production for the household. California producers fit Loomis's (1984) description of the American farmers whose tradition is to be constantly innovative and open to change. They have adjusted their crops and agricultural practices with alacrity to take advantage of new markets, to seek new efficiency, and to profit from the land. This includes acquiring the political means to gain subsidies from the state and federal governments for irrigation, transportation, and research.

Peaches are among the top twenty commodities in California agriculture, and California is the leading peach producing state in the United States (CDFA 2000). California produces over half the total U.S. crop (Ramming 1988). Peaches were originally domesticated in China despite their scientific name, *Prunus persica* (Persian apple). Peaches first came to California in the eighteenth century with Spanish missions (Faust and Timon 1995) and were reported as a commercial orchard crop when state record keeping began in the early twentieth century (Tadesse 2001). Peaches' recent arrival, the relatively limited number of introductions, and the establishment of commercial production have acted as bottlenecks on the genetic diversity of peaches in California. Like peaches elsewhere in the United States, the California fruit has a narrow genetic base and lineage derived from the Elberta and Hale varieties (Warburton and Bliss 1996, Werner and Okie 1998). This narrow base has been used successfully in breeding programs that have influenced peach cultivation around the world. The United States releases more peach varieties than any other nation (Strada 1996).

California peach production increased from some 3,176 ha in 1910, when record keeping began, to its peak in 1940, when 20,351 ha were planted (Tadesse 2001). Land area in the fruit declined to 14,093 ha in 1990. This production is concentrated in the San Joaquin Valley, especially in Fresno County, which is one of the most productive and technologically advanced agricultural regions in the world. Peaches are a relatively high-value crop that requires careful management, and most peach orchards in Fresno County are moderately sized compared to other U.S. and California farms. Tadesse (2001) reports that the average peach orchard in his sample of sixty-eight producers in Fresno County in 1999 was 27 ha, compared to a mean farm size in the county of 153 ha. These orchards are fully integrated into the national economy through refrigerated rail and truck transport, canning, and grower-supported research and marketing programs. Many producers own their own packing shed. Moreover, public and private breeding programs provide a steady flow of new peach cultivars.

Although measuring diversity by counting peach cultivars may conceal a narrow genetic base, the history of cultivar diversity in California contradicts the expectation that crop diversity will inevitably wane under the influence of modern agriculture and economic organization. When the California Agricultural Statistical Service began record keeping of peach cultivars in 1914, there were 6 cultivars with sufficient acreage to be reported. In 1992, the last year of reporting, there were 69 cultivars with this acreage (Tadesse 2001). In 1997, California had 204 marketable peach cultivars, 93 being freestone varieties that had at least ten thousand boxes of fresh fruit (Tadesse 2001). Sustained increase in the diversity of California peaches during the last century is observable no matter how diversity is measured. Tadesse (2001) analyzes peach diversity in several ways, including numbers of cultivars (richness), spatial distribution of cultivars (Shannon-Weaver index), and temporal diversity (turnover and life expectancy). On all measures, peach diversity has increased in California since the early twentieth century (Tadesse 2001). Figure 11.2 charts the Shannon-Weaver index of peach diversity in California orchards between 1914 and 1992. This is a standard diversity measure for the spatial distribution of different organisms.

After 1970, there was a complete change in the dominant cultivars and an increase in the addition of new cultivars to the state's peach inventory. For instance, the share of land planted in the top ten cultivars has declined from 40 percent in the 1938–62 period to 23 percent in the 1963–92 period. Higher temporal diversity over time was measured by comparing the chances of a cultivar being removed from cultivation between the early period of record keeping (1914–59) and a later one (1959–92). The chance of removal increased sixteenfold (0.25 to 4.0) between the two periods (Tadesse 2001).

Increased diversification of California peaches occurred simultaneously with the state's agricultural development. Research for breeding, orchard management, pest control, and postharvest handling has benefited peach production in the state since the nineteenth century (Tufts 1946). Commercial processing and refrigerated transportation have been available to the state's peach producers for over a century (Tufts 1946). Peach producers, like other farmers in the state, organized themselves into cooperative organizations to coordinate marketing, research, and policy beneficial to agriculture. Producers established the California Canning Peach Growers organization in 1921 to conduct collective bargaining with the canning industry (Tufts 1946). Besides full articulation with the state's commodity infrastructure, peach production also shows other signs of "modernization," including a decline in the number of farms and an increase in average farm size (Tadesse 2001).

Comparison between the diversity in California peach orchards and the "modernization" of agriculture in the state poses a counterexample to the widespread belief that development is antithetical to diversity. As with the case

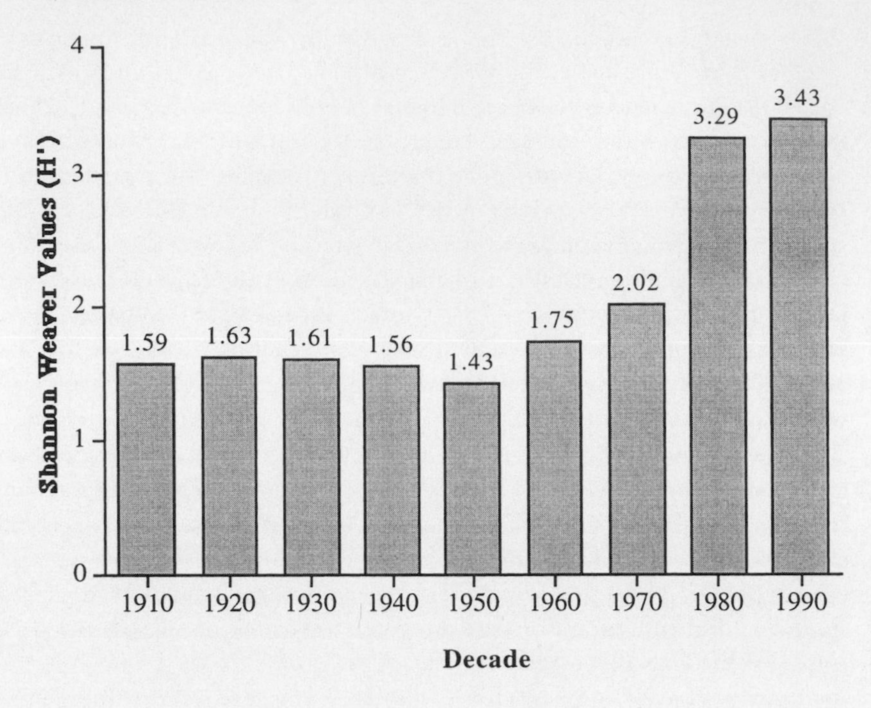

$H' = -\sum_{i=1}^{n} P_i \ln P_i$ where P_i is the proportion of the total area share planted by the ith peach cultivar.

Figure 11.2 Diversity trend (Shannon-Weaver Index) in California peaches, 1910–90 (Tadesse 2001) (Reprinted with permission from Dawit Tadesse)

of emmer wheat (farro) in Italy, the growth of California peach diversity derives from economic integration and commercialization, forces that were once believed to be antithetical to crop diversity. One of the driving forces behind diversification in California peach orchards is the desire of farmers to lengthen the harvest period. By adopting cultivars with different harvest dates, they have extended the peach harvest season from the middle of May to the end of September (Zaiger 1988). This allows growers to take advantage of dramatically higher prices at the beginning and end of the harvest season. Moreover, it helps them avoid bottlenecks in the availability of labor and equipment, thus reducing costs of production. Tadesse (2001) shows that maintaining peach diversity is a means of reducing economic risk for farmers who are fully integrated into the market.

The Challenge of Transgenic Crops

One of the newer tools in the arsenal of crop breeders is biotechnology that allows creation of genetically modified organisms (GMOs), also known as "transgenic crops." These differ from crops resulting from conventional breeding programs because they incorporate genetic material from unrelated species. Among its advantages to breeders, genetic engineering makes new and exotic traits available and allows greater specificity in adding genetic material. Diffusion and adoption of transgenic crops has been rapid in the United States, where 16.2 million ha were cultivated in transgenic crops in 2001 (James 2001). These crops have experienced steady diffusion in developing countries, whose share of the total GMO area grew to 26 percent in 2001 (James 2001). While two countries, Argentina and China, account for most of the hectares in transgenic crops in developing areas, they are now established in many countries. Two traits, herbicide tolerance and insect resistance (from the *Bacillus thuringiensis* bacterium, Bt), dominate transgenic crops, although numerous other traits are being developed with this technology, including disease resistance, salt tolerance, improved protein quality, vitamin enhancement, and pharmaceutical properties (Dunwell 1999).

Just as the rise of crop breeding after the discovery of the Mendelian laws of inheritance was a logical extension of the age-old search for more productive crops, so, too, is the shift to transgenic methods a normal step in crop breeding. Vavilov recognized that the new applied botany would benefit from collecting and using genetic material from an ever widening geographic and biological scale, a recognition that spurred more systematic plant collection, evaluation, and conservation in many countries. Progressing from selection within local populations to crosses among populations, crosses of crop lines from distant regions, and wide crosses across species within a crop's lineage, crop geneticists have continually extended the range of diversity and complexity of the pedigrees of the crops (Poehlman 1995). One clear difference between conventional crop breeding and genetic engineering is the opposition that transgenic crops have engendered (Hobbelink 1991, Lappé and Bailey 1998).

Three threats to the genetic resources of Vavilov centers and the stewards of those resources are attributed to transgenic crops: (1) genetic erosion or contamination, (2) environmental degradation, and (3) social inequity. Displacement of local crop populations by GMOs, or introduction of transgenic germplasm and unwanted traits into landraces in Vavilov centers, are variations on the threat of genetic erosion from agricultural modernization. Transgenic methods have spread rapidly to crop breeding programs in industrial countries, and within a few years genetically modified varieties of soybean, maize, oilseed rape, and cotton became the dominant types in the United States (Dunwell 2000). The pace of this change suggests that transgenic crops may succeed where more traditional crop breeding methods have not. Mexico is a case in point. Mexican maize has proven to be remarkably resistant to genetic replacement by improved types, especially in highland areas (Aquino, Carrión, and Calvo 1999; Perales R., Brush, and Qualset 2003). Despite government restrictions, however, traces of transgenic maize have been reported to be present in maize landraces of Oaxaca, a state regarded as agriculturally conservative (Quist and Chapela 2001). Notwithstanding the novelty of transgenic maize and the remoteness of Oaxaca, this technology has penetrated crop populations that otherwise have been relatively impervious to genetic replacement.

The diffusion of GMOs into Vavilov centers has been limited by governmental uncertainty about biological safety and the lack of breeding programs using the new methodology, limitations that may be short-lived if transgenic crops are perceived as beneficial by governments, consumers, and farmers (Herrera-Estrella 1999). The ability of transgenic crops to accelerate genetic erosion is conditioned by the same factors that have affected the diffusion of modern crop varieties in Vavilov centers. Heterogeneous or marginal environmental conditions and agricultural economies with small, subsistence-oriented farms and missing markets are likely to be as unfavorable to transgenic crops as they are to modern varieties from conventional breeding programs. These areas are unlikely to be the focus of breeding programs or investment in transgenic technology. Self-fertilizing cereal crops are subject to more rapid displacement than outcrossing crops but are less likely to acquire traits through infraspecific hybridization. If expressed, unwelcome traits acquired by infraspecific hybridization with landraces will be selected against. It is uncertain whether the introduction of transgenic traits that are not observed by farmers will remain in their crop populations, especially if there is a constant flow-through of new seed stock that does not have these traits. Traits such as tolerance to salt and cold, or disease resistance, may give transgenic varieties extraordinary fitness and thus make them especially competitive against traditional varieties, but the human ecology of agricultural biodiversity suggests that single traits rarely lead to crop population replacement in heterogeneous regions of crop diversity (Bellon 1996). Diffusion of genetically engineered crops is more likely in areas where modern varieties have already replaced traditional varieties than in areas where traditional crops have remained competitive. In sum, it is too early to assess the genetic erosion threat of transgenic crops, but their threat is likely to be similar to that from conventional breeding.

Among the major threats to the agricultural environment are transfer of GMO traits to wild and weedy populations, the rapid development of insect resistance, and negative impacts on nontarget organisms (Krimsky and Wurbel 1996; Rissler and Mellon 1996; Altieri 2000; Dale, Clarke, and Fontes 2002; NRC 2002). The presence of wild and weedy crop relatives is a defining characteristic of Vavilov centers, and gene flow between cultivated and noncultivated species in these habitats appears to be common (Ellstrand, Prentice, and Hancock 1999). The potential of transgenes to cross into wild and weedy relatives is not disputed (NRC 2002), but whether this poses a novel or unusual risk is uncertain. The National Research Council report (NRC 2002) on environmental effects of transgenic crops notes that gene transfer is also associated with conventional breeding and nature in general. Indeed, the human genome contains "a surprisingly large number of insertions of bacterial DNA" (NRC 2002, 48). The NRC report observes that "presently there are no data to suggest that the extremely low rate of natural horizontal transfer should change with transgenic organisms" (NRC, 67). The transfer of genes from genetically engineered plants to wild and weedy populations does not necessarily lead to increased weediness or "super weeds." Traits developed by transgenic methodology, such as herbicide resistance, are also known to have arisen by natural evolution (Dale, Clarke, and Fontes 2002). Moreover, single traits rarely confer unusual competitive advantage.

The rapid development of insect resistance to the Bt trait is cited as the second environmental danger of transgenic crops. The evolution of pest resistance to Bt is accepted as a likelihood of the deployment of transgenic crops with this trait (NRC 2002), and, in the United States, strategies to avert or delay this have been implemented. One strategy is to plant buffer areas around transgenic crops that will serve as refuges, but some question the sufficiency of these buffers and the extent of their use (Altieri 2000; Dale, Clarke, and Fontes 2002). The evolution of Bt resistance also may result in the use of less environmentally benign pesticides (Dale, Clarke, and Fontes 2002). Bt is currently allowed for pest control by certified organic farmers, so this group would be negatively impacted by the evolution of Bt-resistant pests. Such evolution, of course, is possible without transgenic crops because of the natural presence of Bt.

Impact on nontarget species, especially from the transfer of insecticidal traits, is an additional undesirable environmental consequence. Five different categories of nontarget species are recognized as potential victims: (1) beneficial species such as pollinators and natural enemies of pests, (2) nontarget pests, (3) soil organisms, (4) species of conservation value such as butterflies, and (5) biodiversity (NRC 2002). Effects on the fourth category have been the most widely discussed. Losey, Rayor, and Carter (1999) reported on their laboratory experiment showing that large amounts of pollen from transgenic maize was toxic to monarch butterfly larvae. This finding was reconfirmed by Jesse and Obrycki (2000) for naturally occurring levels of pollen from transgenic maize with the Bt trait. While these effects have not been observed in field conditions where transgenic crops are grown, risk to nontarget species is widely accepted as a potential hazard (NRC 2002).

The third general threat from transgenic crops is social inequity. While the ecological effects have been widely discussed and researched, the negative social effects of transgenic crops have received little attention. One problem is that the use of these crops is largely confined to a few industrial countries and to commercial cropping systems in even fewer developing countries (James 2001). Because social impact cannot be modeled in the laboratory, conjecture about potential impact is most appropriately evaluated by reference to other agricultural technologies, such as the semidwarf varieties of the Green Revolution, which have been exhaustively studied (Brush 2001). The analogy between the Green Revolution and transgenic crops is warranted because both are products of crop science laboratories rather than farmers' experimentation. The primary difference is that the Green Revolution crops were created as public goods while transgenic ones are more often proprietary. Direct and indirect negative consequences of the Green Revolution have been studied. Direct consequences include losses suffered by particular social groups, such as small-scale farmers and farm laborers. Indirect consequences emphasize the loss of position by small-scale vis-à-vis larger-scale farms and have received more attention. A special category of indirect effects results from the prevalence of private research and intellectual property.

The case that the Green Revolution was harmful to small-scale farmers has been made based on country case studies (e.g., Frankel 1971), comparative assessments (e.g., Griffin 1974), and detailed ethnographic accounts of single communities or regions (e.g., Scott 1985). The crux of the critical social assessment is that the new crop varieties were biased in favor of landlords and against peasants (Griffin 1974). This bias results in loss of wages, displacement from land, unemployment, and higher rents, and the bias in turn leads to rural violence (Frankel 1971, Scott 1985).

The reality of scale bias and its negative consequences have been contested by an extensive number of studies (e.g., Hayami and Kikuchi 2000, Hazell and Ramasay 1991, Herdt 1987). Long-term data show that small farms adopted the new technology at comparable levels to large farms but that they lagged in the normal diffusion curve (Evenson and Gollin 2003). In India, the long-term data show that small farms tended to be favored by the new technology (Hazell and Ramasay 1991). The determinants of who benefits were shown to be multiple, complex, and dynamic rather than the simple dichotomy of peasants versus landlords who are fixed in permanent socioeconomic positions. Among the regular beneficiaries of the Green Revolution across different countries were rural laborers and consumers. Farms of all sizes failed to reap the full benefit of the Green Revolution because of the declining position of farmers relative to urban workers in developing economies, which is predicted by Engle's Law. The perception of bias is widely observed (e.g., Scott 1985), but numerous complicating factors stand in the way of confirming that the Green Revolution benefited landlords at the expense of the peasantry. The Green Revolution experience, therefore, does not categorically portend negative social impacts from transgenic crops.

The proprietary nature of transgenic crops makes them dissimilar to the Green Revolution and, therefore, possibly a cause of social inequity. Unlike the earlier technology, which existed as a public good, biotechnology is marked by a prevalence of private firms, patents, and licensing (Kenney 1986, Pray and Umali-Deininger 1998). Thompson (1997) outlines two social inequities that might result from the proprietary nature of transgenic crops: (1) farmers will be deprived of rightful compensation for property they already own, and (2) farmers will be deprived of important future economic opportunities.

The possibility that the creation of transgenic crops will deprive farmers of compensation reflects the idea of "biopiracy," meaning that patents involving genetic resources acquired as common heritage goods represent a taking of farmers' property. As argued above, the idea that farmers have some form of natural property right, or one that is culturally unique, is not supported by the utilitarian basis of the property system or the ethnographic record. Historical trends since establishing property over plants have been to expand coverage geographically and to new categories of plants, and to increase the rights of breeders. This tendency runs counter to the idea that plants should be exempt from intellectual property because of their nature and/or importance to human welfare. Moreover, the expansion of plant breeders' rights has not in theory eroded the distinction between products *of* nature and products *from* nature (Bozicevic 1987), and the open nature of landrace populations weighs heavily toward viewing them as products *of* nature. The argument of lost compensation because of patenting is, therefore, difficult to sustain.

The possibility that farmers will be deprived of future economic opportunities might be realized in different ways. The Enola Bean patent is based on conventional crop breeding methods and typifies the negative impact on farmers from patents issued for crops that are undifferentiated from farmer varieties. This precedent violates the rules of the game of the patent system, but it seems unlikely that transgenic crops will increase the likelihood of similar inequities. However, another consequence of the shift to private research and transgenic crops is likely. This consequence is that the rise of private research will cause or accelerate a decline in public funding and research for agricultural development directed to the poor. The decline in publicly supported maize research following the rise of the hybrid maize industry (Kloppenburg 1988b) shows the effect of a shift in agricultural technology on the research system. Until recently, the balance of research for developing countries was public, but this appears to be shifting (Pray and Umali-Deininger 1998). The possibility of bias that hangs over public research (Chambers 1983) is yet more probable for private research. The high investment required for developing transgenic crops means that companies working in developing countries are likely to target commercial crops and farm sectors. The negative aspect of this is that private research expenditures will offset public ones, leaving the small-farm, subsistence sector without a research system working on its behalf.

In sum, the diffusion of transgenic crops in Vavilov centers is too limited to provide a sound basis for judging whether they will have adverse consequences on the genetic resource base in traditional crops, on the agricultural environment, or on small-scale farmers. The diffusion of transgenic crops is likely to be similar to that of modern varieties from conventional breeding programs, and transgenic crops are likely to be most important in areas already heavily affected by modern varieties. The most problematic effects appear to be on nontarget species and in diverting the agricultural research system away from serving the interests of small farmers. In Vavilov centers, this diversion might, for instance, remove funds from *in situ* conservation programs such as collaborative plant breeding.

Summary

My inquiry into the nature of crop diversity began above the *ceja de montaña*—the "eyebrow of the jungle"—in northern Peru. There, I encountered the extraordinary diversity of potatoes that persisted despite the forces that were arrayed against traditional agriculture—new technology, poverty, population pressure, economic integration. Crop diversity promised to be a useful key to interpreting the relationship between ancient farming cultures and their natural environments. Because potato diversity is both a biological and a social phenomenon, it seemed to be just the right tool for investigating the human ecology of the Andes, a factor that could be quantified across time and space to elucidate the interaction between people and their environment in a changing world. Of course, the complexity of both crop diversity and human ecology frustrated these grandiose but naïve plans, and I was content to understand something of the nature of potatoes and their social context in Peru. Indeed, it now seems like hubris to believe that a single scientific project can merge two strong traditions of inquiry, one conducted by crop scientists and the other by social scientists. In the Andes, these scientific traditions each have nearly a century of experience and hundreds of practitioners. Nevertheless, merger of biological and social scientists into a single research framework is ultimately necessary for full understanding of crop evolution.

Although the human ecology of the Andes can be partially understood only through studying potato diversity, my experience in Peru laid the foundation for research on other crops and cultures and alerted me to aspects of crop biology and agrarian systems elsewhere. One lesson from my initial work in Peru was that the generalized processes of social change and crop evolution are refracted by local context into very different patterns. I was first prompted to study the diffusion of modern crop varieties into ancient agricultural systems because the potato farmers of Uchucmarca persisted in planting local types even while they were adopting improved potato varieties. Anthropological training had instructed me about syncretism that fused native American and European symbols and beliefs into a common religion, and my generation of anthropology graduate students was drawn toward the idea of cultural resistance. However, neither syncretism nor resistance adequately describes the modern evolution of Andean potato agriculture. This evolution is characterized by adoption of technology from different sources to meet local agronomic and social needs. To borrow Lévi-Strauss's (1966) famous image, peasants, like other premodern people, are "bricoleurs" (handymen) who assemble their world from experience and the bits and pieces of life that are available to them. Unlike crop breeders, peasants don't have a fixed idea of a crop archetype or a theoretical framework to assess yield. Instead, they are experimenters, bricoleurs who cobble effective agricultural technology together from different sources. The peasant farmers I have worked with are eager to try any new seed and to depose any that doesn't meet their needs. This fluidity applies to crops that are both local and exotic, and traditional and improved.

Of course, the landscapes and resources of peasant farming are rarely uniform or subject to a single technology. "Bricolage" in these agricultural systems often results in a mélange of crops and crop types as well as other technologies for producing and storing food. It is inappropriate, however, to imagine these mixtures as congeries of undifferentiated elements in a heap. More appropriately, the mélange is an assembly of discrete elements chosen and kept for specific roles in household production. Owing to the natural heterogeneity of the Andes, these criteria are particularly evident in limiting the domination of any technology or seed over wide areas. As a result, the pattern of partial adoption of new seeds was observable. Maintenance of local types did not derive from some premodern mentality or insistence on tradition, but rather from a rational choice that local, diverse potatoes were more suited to the Andean environment.

The idea of partial adoption of modern technology led to another lesson—that the very complexity of peasant agriculture was an asset in the survival of crops that might otherwise be thought of as obsolete relics. Hence, the complexity of peasant agriculture creates numerous opportunities and purchase for conserving diversity. Once *in situ* conservation became an acceptable idea, numerous strategies soon became available to enhance the supply and demand of landraces. It is inappropriate to think of any contemporary farmers as "traditional" in the sense of blindly following ancestral ways, because crop selection, like all agricultural production, involves active decision making in the present. However, the cases of the Valley of Mexico, Turkey, Italy, and California are reminders that farmers who are unambiguous participants in the modern world make decisions to maintain ancient crops and diversity.

The survival of crop diversity because of partial adoption and consumer demand for specialized crops provides the lesson that conservation can succeed by adding incremental value to landraces in selected environments and farming systems rather than posing landraces as an alternative to modern crops for all farmers. In other words, conservation of crops should follow the pattern of wildlife conservation to target key areas and populations. *In situ* conservation will not succeed as a policy for an entire agricultural sector because there are too many interests at stake in agricultural intensification—consumers, governments, agricultural scientists, and farmers. Moreover, there is sufficient uncertainty about the ability of landraces to produce sufficient food to feed the present and future population.

Research by me and by others has shown a surprising degree of persistence of crop diversity, and this fact has given support to the idea that landraces and crop diversity can survive in a world with eleven billion people. Scientific and political communities have accepted the mandate that conservation of crop genetic resources should extend beyond gene banks and botanical gardens to include farmers who have inherited and maintained the legacy of crop evolution. Virtually all of the changes that affect agriculture in Vavilov centers today are variations on themes that are very old—population increase, incorporation into larger cultural, political, and economic systems, and the appearance of new technologies, including new crops and varieties. These changes can have profound impacts on the biological and cultural foundations of agriculture, but they can also be met with local resilience and ingenuity that permit the survival of biological and cultural resources. I have tried to show here that resilience and survival of these resources are vigorous parts of contemporary agriculture in Vavilov centers, where domestication began. No crystal ball reveals how long the pattern of resilience and persistence of landraces and ancient evolutionary processes will continue.

Although rapid and radical climate change might eliminate the effective environments of these ancient crop populations, farmers in Vavilov centers should continue to cope, as they and their ancestors did, with the many forces that buffet them. Peasant agriculture has survived in the past by accommodation to new technology and economic alternatives while maintaining, to some degree, local crops. Whether or not their descendants and ours will enjoy the pleasures of diverse crops with strong local accents must remain an unanswered question. However, we are beginning to understand that preserving crop diversity in the hearths of crop domestication and evolution is an emblem of our common history and an obligation to future generations.

Bibliography

Adams, I. 1948. Rice cultivation in Asia. *American Anthropologist* 50: 256–82.

Agrawal, A. 1995. Dismantling the divide between indigenous and scientific knowledge. *Development and Change* 26: 413–39.

Aguirre Beltran, G. 1979. *Regions of Refuge*. Washington, DC: Society for Applied Anthropology.

Aguirre G., J. A., M. R. Bellon, and M. Smale. 2000. A regional analysis of maize biological diversity in southeastern Guanajuato, Mexico. *Economic Botany* 54: 60–72.

Alcorn, J. B. 1984. *Huastec Mayan Ethnobotany*. Austin: University of Texas Press.

Alemayehu, F., and J. E. Parlevliet. 1997. Variation between and within Ethiopian barley landraces. *Euphytica* 94: 183–89.

Alford, W. P. 1995. *To Steal a Book Is an Elegant Offense: Intellectual Property Law in Chinese Civilization*. Stanford: Stanford University Press.

Ali, A. M. S. 1987. Intensive paddy agriculture in Shyampur, Bangladesh. In *Comparative Farming Systems*, edited by B. L Turner II and S. B. Brush, 276–305. New York: Guilford Press.

Allard, R. W. 1970. Population structure and sampling methods. In *Genetic Resources in Plants–Their Exploration and Conservation*, edited by O. H. Frankel and E. Bennett, 97–113. Oxford: Blackwell Scientific Publications, International Biological Programme Handbook No. 11.

——. 1988. Genetic change associated with the evolution of adaptedness in cultivated plants and their wild progenitors. *Journal of Heredity* 79: 225–38.

Allen, C. J. 1988. *The Hold Life Has: Coca and Cultural Identity in an Andean Community*. Washington, DC: Smithsonian Institution Press.

Allen, R. C. 1983. Collective invention. *Journal of Economic Behavior and Organization* 4: 1–24.

Almekinders, C., and W. de Boef, eds. 2000. *Encouraging Diversity: The Conservation and Development of Plant Genetic Resources*. London: Intermediate Technology Publications.

Almekinders, C. J. M., and A. Elings. 2001. Collaboration of farmers and breeders: Participatory crop improvement in perspective. *Euphytica* 122: 425–38.

Altieri, M. A. 2000. The ecological impacts of transgenic crops on agroecosystem health. *Ecosystem Health* 6: 13–23.

Altieri, M. A., and L. C. Merrick. 1987. *In situ* conservation of crop genetic resources through maintenance of traditional farming systems. *Economic Botany* 41: 86–96.

Ammerman, A. J., and L. L. Cavalli-Sforza. 1984. *The Neolithic Transition and the Genetics of Populations in Europe*. Princeton: Princeton University Press.

Anderson, E. 1946. Maize in Mexico: A preliminary survey. *Annals of the Missouri Botanical Garden* 33: 147–247.

———. 1947. Field studies of Guatemalan maize. *Annals of the Missouri Botanical Garden* 34: 433–51.

———. 1952. *Plants, Man, and Life*. Boston: Little Brown and Company.

Anderson, E., and H. C. Cutler. 1942. Races of *Zea mays*: I. Their recognition and classification. *Annals of the Missouri Botanical Garden* 29: 69–89.

Anderson, J. R., and P. R. B. Hazell. 1989. *Variability in Grain Yields*. Baltimore: Johns Hopkins University Press.

Aquino, P., F. Carrión, and R. Calvo. 1999. Selected wheat statistics. In *CIMMYT 1998–1999 World Wheat Facts and Trends, Global Research in a Changing World: Challenges and Achievements*, edited by P. L. Pingali, 33–43. Mexico: CIMMYT.

Arguedas, J. M. 1989. A call to certain academics. Poem translated from Quechua to English by W. Rowe. Personal communication, W. Rowe, University of London. Also, New Internationalist Issue 197–July 1989, available at http://www.oneworld.org/ni/issue197/keynote.htm.

Arndt, H. W. 1987. *Economic Development: The History of an Idea*. Chicago: University of Chicago Press.

Asfaw, Z. 2000. The barleys of Ethiopia. In *Genes in the Field: On-Farm Conservation of Crop Diversity*, edited by S. B. Brush, 77–107. Boca Raton, FL: Lewis Publishers.

Ashby, J. A., and L. Sperling. 1995. Institutionalizing participatory, client-driven research and technology development in agriculture. *Development and Change* 26: 753–70.

Atran, S. 1987. Origin of the species and genus concepts: An anthropological perspective. *Journal of the History of Biology* 20: 195–279.

Atran, S., D. Medin, N. Ross, E. Lynch, J. Coley, E. U. Ek, and V. Vapnarsky. 1999. Folk ecology and commons management in the Maya lowlands. *Proceedings of the National Academy of Sciences USA* 96: 7598–603.

Autrique, E., M. M. Nachit, P. Monneveux, S. D. Tanksley, et al. 1996. Genetic diversity in durum wheat based on RFLPS, morphophysiological traits, and coefficient of parentage. *Crop Science* 36: 735–42.

Baenziger, P. S., R. A. Kleese, and R. F. Barnes, eds. 1993. *Intellectual Property Rights: Protection of Plant Materials*. Madison, WI: Crop Science Society of America.

Baker, D. 1991. Reorientation, not reversal: African farmer-based experimentation. *Journal for Farming Systems Research and Extension* 2: 125–47.

Baker, H. G. 1965. Characteristics and modes of origins of weeds. In *Genetics of Colonizing Species*, edited by G. L. Stebbins, 17–168. New York: Academic Press.

——. 1970. Taxonomy and the biological species concept in cultivated plants. In *Genetic Resources in Plants—Their Exploration and Conservation*, edited by O. H. Frankel and E. Bennett, 49–68. Oxford: Blackwell Scientific Publications, International Biological Programme Handbook No. 11.

Baker, P. T., and M. A. Little, eds. 1976. *Man in the Andes: A Multidisciplinary Study of High-Altitude Quechua*. Stroudsburg, PA: Dowden, Hutchinson & Ross.

Barcaccia, G., L. Molinari, O. Porfiri, and F. Veronesi. 2002. Molecular characterization of emmer (*Triticum dicoccon* Schrank) Italian landraces MD4.MD112 *Genetic Resources & Crop Evolution* 49: 415–26.

Barker, R., E. C. Gabler, and D. Winkelmann. 1981. Long term consequences of technological change on crop yield stability: The case for cereal grain. In *Food Security for Developing Countries*, edited by A. Valdés, 53–78. Boulder, CO: Westview Press.

Baron, R. C., ed. 1987. *The Garden and Farm Books of Thomas Jefferson*. Golden, CO: Fulcrum.

Barrett, C. B., and T. J. Lybbert. 2000. Is bioprospecting a viable strategy for conserving tropical ecosystems? *Ecological Economics* 34: 293–300.

Barth, F. 1969. *Ethnic Groups and Boundaries: The Social Organization of Culture Difference*. Boston: Little, Brown.

Barton, J. H., and W. E. Siebeck. 1994. *Material Transfer Agreements in Genetic Resources Exchange—The Case of the International Agricultural Research Centers*. Rome: International Plant Genetic Resources Institute.

Bassett, M. J., R. Lee, C. Otto, and P. E. McClean. 2002. Classical and molecular genetic studies of the strong greenish yellow seedcoat color in 'wagenaar' and 'enola' common bean. *Journal of the American Society for Horticultural Science* 127: 50–55.

Bekele, E. 1983. Some measures of gene diversity analysis on land race populations of Ethiopian barley. *Hereditas* 98: 127–43.

Belay, G., E. Bechere, D. Mitiku, A. Merker, and S. Tsegyae. 1997. Patterns of morphological diversity in tetraploid wheat (*Triticum turgidum* L.) landraces from Ethiopia. *Acta Agricultura Scandinavica* (Section B., Soil and Plant Science) 47: 221–28.

Bellon, M., J.-L. Pham, and M. T. Jackson. 1997. Genetic conservation: A role for rice farmers. In *Plant Genetic Conservation: The* In Situ *Approach*, edited by N. Maxted, B. V. Ford-Lloyd, and J. G. Hawkes, 263–89. London: Chapman & Hall.

Bellon, M. R. 1990. The ethnoecology of maize production under technological change. Ph.D. diss., University of California, Davis (University Microfilms, Ann Arbor, MI).

——. 1991. The ethnoecology of maize variety management: A case study from Mexico. *Human Ecology* 19: 389–418.

——. 1996. The dynamics of crop infraspecific diversity: A conceptual framework at the farmer level. *Economic Botany* 50: 26–39.

——. 2001. Demand and supply of crop infraspecific diversity on farms: Towards a policy framework for on-farm conservation. *Economics Working Paper 01–01*. Mexico: CIMMYT.

Bellon, M. R., and S. B. Brush. 1994. Keepers of maize in Chiapas, Mexico. *Economic Botany* 48: 196–209.

Bellon, M. R., and J. Risopoulos. 2001. Small-scale farmers expand the benefits of improved maize germplasm: A case study from Chiapas, Mexico. *World Development* 29: 799–811.

Bellon, M. R., M. Smale, A. Aguirre, S. Taba, F. Aragón, J. Díaz, and H. Castro. 2000. Identifying appropriate germplasm for participatory breeding: An example from the central valleys of Oaxaca, Mexico. *CIMMYT Economics Working Paper 00–03*. Mexico: CIMMYT.

Bellon, M. R., and J. E. Taylor. 1993. Farmer soil taxonomy and technology adoption. *Economic Development and Cultural Change* 41: 764–86.

Bennett, E., ed. 1967. *Record of the FAO Technical Conference on the Exploration, Utilization and Conservation of Plant Genetic Resources*. Rome: Food and Agricultural Organization of the U.N.

———. 1970. Adaptation in wild and cultivated plant populations. In *Genetic Resources in Plants—Their Exploration and Conservation,* edited by O. H. Frankel and E. Bennett, 115–29. Oxford: Blackwell Scientific Publications, International Biological Programme Handbook No. 11.

Benz, B. F. 2001. Archaeological evidence of teosinte domestication from Guilá Naquitz, Oaxaca. *Proceedings of the National Academy of Sciences of the United States of America* 98: 2104–6.

Bérard, L., and P. Marchenay. 1996. Tradition, regulation, and intellectual property: Local agricultural products and foodstuffs in France. In *Valuing Local Knowledge: Indigenous People and Intellectual Property Rights*, edited by S. Brush and D. Stabinsky, 230–43. Washington, DC: Island Press.

Berlin, B. 1976. The concept of rank in ethnobiological classification: Some evidence from Aguaruna folk botany. *American Ethnologist* 3: 381–99.

———. 1992. *Ethnobiological Classification: Principles of Categorization of Plants and Animals in Traditional Societies*. Princeton: Princeton University Press.

Berlin, B., D. E. Breedlove, and P. H. Raven. 1974. *Principles of Tzeltal Plant Classification*. New York: Academic Press.

Bertonio, P. L. 1612. *Vocabulario de la Lengua Aymara*. La Paz, Bolivia: Centro de Estudios de la Realidad Económica y Social (reprinted 1984).

Boserup, E. 1965. *The Conditions of Agricultural Growth*. Chicago: Aldine.

Boster, J. S. 1985. Selection for perceptual distinctiveness: Evidence from Aguaruna cultivars. *Economic Botany* 39: 310–25.

Boyd, R., and P. J. Richerson. 1985. *Culture and the Evolutionary Process*. Chicago: University of Chicago Press.

———. 1995. Why does culture increase human adaptability. *Ethology and Sociobiology* 16: 125–43.

Bozicevic, Karl. 1987. Distinguishing "products of nature" from products derived from nature. *Journal of the Patent and Trademark Office Society* 69: 415–26.

Bray, F. 1984. Agriculture. In *Science and Civilization in China*, edited by J. Needham, vol. VI, part 2. Cambridge: Cambridge University Press.

——. 1986. *The Rice Economies: Technology and Development in Asian Societies*. Berkeley: University of California Press.

Brennan, J. P., and D. Byerlee. 1991. The rate of crop varietal replacement on farms: Measures and empirical results for wheat. *Plant Varieties and Seeds* 4: 99–106.

Bretting, P. K., and D. N. Duvick. 1997. Dynamic conservation of plant genetic resources. *Advances in Agronomy* 61: 1–51.

Brockway, L. H. 1979. *Science and Colonial Expansion: The Role of the British Royal Botanic Gardens*. New York: Academic Press.

Brokensha, D., M. Warren, and O. Werner. 1980. *Indigenous Knowledge Systems and Development*. Lanham, MD: University Press of America.

Brookfield, H. C. 2001. *Exploring Agrodiversity*. New York: Columbia University Press.

Brown, A. H. D. 1999. The genetic structure of crop landraces and the challenge to conserve them *in situ* on farms. In *Genes in the Field: On-Farm Conservation of Crop Diversity*, edited by S. B. Brush, 29–48. Boca Raton, FL: Lewis.

Brown, A. H. D., M. T. Clegg, A. L. Kahler, and B. S. Weir, eds. 1990. *Plant Population Genetics, Breeding and Genetic Resources*. Sunderland, MA: Sinauer Associates.

Brown, G. M. 1990. Valuation of genetic resources. In *The Preservation and Valuation of Biological Resources*, edited by G. H. Orians et al., 203–28. Seattle: University of Washington Press.

Brown, M. F. 1998. Can culture be copyrighted? *Current Anthropology* 39: 193–221.

Brown, W. L., and H. F. Robinson. 1992. The status, evolutionary significance and history of eastern Cherokee maize. *Maydica* 37: 29–39.

Browning, J. A. 1981. The agro-ecosystem — natural ecosystem dichotomy and its impact on phyotpathological concepts. In *Pests, Pathogens, and Vegetation*, edited by J. M. Thresh, 159–72. Boston: Pitman Advanced Publishing Program.

Brunel, G. R. 1975. Variation in Quechua folk biology. Ph.D. diss., University of California, Berkeley.

Brush, S. B. 1977. *Mountain, Field, and Family: The Economy and Human Ecology of an Andean Valley*. Philadelphia: University of Pennsylvania Press.

——. 1989. Rethinking crop genetic resource conservation. *Conservation Biology* 4: 19–29.

——. 1992. Ethnoecology, biodiversity, and modernization in Andean potato agriculture. *Journal of Ethnobiology* 12: 161–85.

——. 1993. Indigenous knowledge of biological resources and intellectual property rights: The role of anthropology. *American Anthropologist* 95: 653–86.

——. 1995. *In situ* conservation of landraces in centers of crop diversity. *Crop Science* 35: 346–54.

——. 1996. Is common heritage outmoded? In *Valuing Local Knowledge: Indigenous People and Intellectual Property Rights*, edited by S. Brush and D. Stabinsky, 143–64. Washington, DC: Island Press.

——, ed. 1999. *Genes in the Field: On-Farm Conservation of Crop Diversity.* Boca Raton, FL: Lewis.

——. 2001. Genetically modified organisms in peasant farming: Social impact and equity. *Indiana Journal of Global Legal Studies* 9: 135–62.

Brush, S. B., M. Bellon, and E. Schmidt. 1988. Agricultural development and maize diversity in Mexico. *Human Ecology* 16: 307–28.

Brush, S. B., H. J. Carney, and Z. Huamán. 1981. Dynamics of Andean potato agriculture. *Economic Botany* 35: 70–85.

Brush, S. B., R. Kesseli, R. Ortega, P. Cisneros, K. S. Zimmerer, and C. Quiros. 1995. Potato diversity in the Andean center of crop domestication. *Conservation Biology* 9: 1189–98.

Brush, S. B., and E. Meng. 1998. Farmers' valuation and conservation of crop genetic resources. *Genetic Resources and Crop Evolution* 45: 139–50.

Brush, S. B., and D. Stabinsky, eds. 1996. *Valuing Local Knowledge: Indigenous People and Intellectual Property Rights*. Washington, DC: Island Press.

Brush S. B., J. E. Taylor, and M. Bellon. 1992. Biological diversity and technology adoption in Andean potato agriculture. *Journal of Development Economics* 39: 365–87.

Brush, S. B., and B. L. Turner II. 1987. The nature of farming systems and views of their change. In *Comparative Farming Systems*, edited by B. Turner and S. Brush, 11–48. New York: Guilford Press.

Bulmer, R. 1970. Which came first, the chicken or the egg-head? In *Échanges et communications*, edited by J. Pouillon and P. Miranda, 1069–91. The Hague: Mouton.

——. 1974. Folk biology in the New Guinea highlands. *Social Science Information* 13: 9–28.

Burdon, J. J., and A. M. Jarosz. 1990. Disease in mixed cultivars, composites, and natural plant populations: Some epidemiological and evolutionary consequences. In *Plant Population Genetics, Breeding and Genetic Resources,* edited by A. H. D. Brown, M. T. Clegg, A. L. Kahler, and B. S. Weir, 215–28. Sunderland, MA: Sinauer Associates.

Burtt, B. L. 1970. Infraspecific categories in flowering plants. *Biological Journal of the Linnean Society* 2: 233–38.

Byerlee, D. 1996. Modern varieties, productivity, and sustainability: Recent experience and emerging challenges. *World Development*. 24: 697–718.

Byerlee, D., and T. Husain. 1993. Agricultural research strategies for favoured and marginal areas — the experience of FSR in Pakistan. *Experimental Agriculture* 29: 155–71.

California Department of Food and Agriculture Resource Directory. 2000. Agricultural Statistical Review 1999, http://www.cdfa.ca.gov/docs/CAStats.pdf.

California Tree Fruit Agreement (CTFA). 1996. *Annual Report — 1996*. Reedley, CA: California Tree Fruit Agreement.

Cancian, F. 1972. *Change and Uncertainty in a Peasant Economy: The Maya Corn Farmers of Zinacantan*. Stanford: Stanford University Press.

Carter, W. E., and M. Mamani P. 1982. *Irpa Chico: Individuo y comunidad en la cultura Andina*. La Paz: Libreria-Editorial Juventud.

Ceccarelli, S., E. Acevadeo, and S. Grando. 1991. Breeding for yield stability in unpredictable environments: Single traits, interaction between traits, and architecture of genotypes. *Euphytica* 56: 169–85.

Ceccarelli, S., and S. Grando. 1999. Barley landraces from the Fertile Crescent: A lesson for plant breeders. In *Genes in the Field: On-Farm Conservation of Crop Diversity*, edited by S. B. Brush, 51–76. Boca Raton, FL: Lewis.

Chambers, R. 1983. *Rural Development: Putting the Last First*. London: Longman.

Chambers, P., and J. Jiggins. 1987. Agricultural research for resource-poor farmers, part I: Transfer of technology and farming systems research. *Agricultural Administration and Extension* 27: 35–52.

Chang, T. T. 1985. Crop history and genetic conservation: Rice—a case study. *Iowa State Journal of Research* 59: 425–55.

Chapman, C. G. D. 1985. *Genetic Resources of Wheat: A Survey and Strategy for Collecting*. Rome: International Board for Plant Genetic Resources Secretariat.

Charles, D. 2001. Seeds of discontent. *Science* 294: 772–75.

Chayanov, A. V. 1966. *The Theory of Peasant Economy,* edited by Daniel Thorner et al. Homewood, IL: American Economic Association, Irwin.

CIMMYT. 1993. *1992/93 CIMMYT World Wheat Facts and Trends. The Wheat Breeding Industry in Developing Countries: An Analysis of Investments and Impacts*. Singapore: CIMMYT.

Clawson D. 1985. Harvest security and interspecific diversity in traditional tropical agriculture. *Economic Botany* 39: 56–67.

Clayton, W. D. 1983. Tropical grasses. In *Genetic Resources of Forage Plants*, edited by J. G. McIvor and R. A. Bray, 39–46. Melbourne: CSIRO.

Cleveland, D. A., and S. C. Murray. 1997. The world's crop genetic resources and the rights of indigenous farmers. *Current Anthropology* 38: 477–515.

Cleveland, D. A., D. Soleri, and S. E. Smith. 1994. Do folk crop varieties have a role in sustainable agriculture? *BioScience* 44: 740–51.

Coase, R. H. 1960. The problem of social cost. *Journal of Law and Economics* 3: 1–44.

Cochrane, W. W. 1993. *The Development of American Agriculture: A Historical Analysis*, 2d ed. Minneapolis: University of Minnesota Press.

Cohen, J. E. 1995. *How Many People Can the Earth Support?* New York: Norton.

Cole, J. W., and E. R. Wolf. 1974. *The Hidden Frontier: Ecology and Ethnicity in an Alpine Valley*. New York: Academic Press.

Comaroff, J., and J. Comaroff. 1993. *Modernity and Its Malcontents: Ritual and Power in Postcolonial Africa*. Chicago: University of Chicago Press.

Commission on Intellectual Property Rights (CIPR). 2002. *Integration of Intellectual Property Rights and Development Policy*. London: Report of the Commission on Intellectual Property Rights.

Conklin, H. C. 1957. *Hanunóo Agriculture: A Report on an Integral System of Shifting Cultivation in the Philippines*. Rome: Food and Agricultural Organization of the United Nations.

——. 1972. *Folk Classification: A Topically Arranged Bibliography of Contemporary and Background References Through 1971*. New Haven: Yale University, Department of Anthropology.

Convention on Biological Diversity. 1994. *Convention on Biological Diversity: Texts and Annexes*. Geneva: Interim Secretariat for the Convention on Biological Diversity, Geneva Executive Center.

Coombe, R. J. 1996. Embodied trademarks: Mimesis and alterity on American commercial frontiers. *Cultural Anthropology* 11: 202–24.

Correa, C. M. 2000. Options for the implementation of farmers' rights at the national level. Working Papers No 8, Trade Related Agenda, Development and Equity. Geneva: South Centre.

Cox, T. S. 1991. The contribution of introduced germplasm to the development of U.S. wheat cultivars. In *Use of Plant Introductions in Cultivar Development*, part 1, edited by H. L. Shands and L. E. Wiesner, 25–47. Madison, WI: Crop Science Society of America, CSSA Special Publication No. 17.

——. 1998. Deepening the wheat gene pool. *Journal of Crop Production* 1: 1–25.

Cox, T. S., J. P. Murphy, and D. M. Ridgers. 1986. Changes in genetic diversity in red winter wheat regions of the United States. *Proceedings of the National Academy of Sciences USA* 83: 5583–86.

Cox, T. S., and D. Wood. 1999. The nature and role of crop biodiversity. In *Agrobiodiversity: Characterization, Utilization and Management*, edited by D. Wood and J. M. Lenné, 35–58. Wallingford, UK: CABI.

Crosby, A. W. 1972. *The Columbian Exchange: Biological and Cultural Consequences of 1492*. Westport, CT: Greenwood.

Cunningham, F. X. 1981. The common heritage: an overview of the international laws that call for sharing global and celestial wealth. *Foreign Service Journal* (July/August): 13–15.

Cunningham, I. S. 1984. *Frank N. Meyer, Plant Hunter in Asia*. Ames: Iowa State University Press.

Dale, P. J., B. Clarke, and E. M. G. Fontes. 2002. Potential for the environmental impact of transgenic crops. *Nature Biotechnology* 20: 567–74.

Dalrymple, D. 1986. *Development and Spread of High-Yielding Rice Varieties in Developing Countries*. Washington, DC: USAID.

Dalrymple, D. G. 1988. Changes in wheat varieties and yields in the United States, 1919–1984. *Agricultural History* 62: 20–36.

D'Antuono, L. F., and R. Bravi. 1996. The hulled wheats industry: Present developments and impact on genetic resources conservation. In *Hulled Wheats. Proceedings of the First International Workshop on Hulled Wheats*, 21–22 July 1995, Castelvecchio Pascoli, Tuscany, Italy, edited by S. Padulosi, K. Hammer, and J. Heller, 221–33. Rome: International Plant Genetic Resources Institute.

Darwin, C. 1859. *On the Origin of Species by Means of Natural Selection*. London: John Murray.

——. 1896. *The Variation of Animals and Plants Under Domestication*, 2 vols. New York: D. Appleton and Company. (Orig. pub. 1868.)

David, J. L., M. Zivy, M. L. Cardin, and P. Brabant. 1997. Protein evolution in dynamically managed populations of wheat: Adaptive responses to macro-environmental conditions. *Theoretical and Applied Genetics* 95: 932–41.

Dean, W. 1987. *Brazil and the Struggle for Rubber: A Study in Environmental History*. Cambridge: Cambridge University Press.

De Candolle, A. 1914. *Origin of Cultivated Plants*. Translation of 1882 French revision of 1855 original. New York: D. Appleton and Company.

de Janvry, A., M. Fafchamps, and E. Sadoulet. 1991. Peasant household behaviour with missing markets: Some paradoxes explained. *The Economic Journal* 101: 1400–17.

de Janvry, A., E. Sadoulet, and G. Gordillo de Anda. 1995. NAFTA and Mexico's maize producers. *World Development* 23: 1349–62.

de Janvry, A., E. Sadoulet, and L. W. Young. 1989. Land and labour in Latin American agriculture from the 1950s to the 1980s. *Journal of Peasant Studies* 16: 398–424.

Demissie, A., and Å. Bjørnstad. 1996. Phenotypic diversity of Ethiopian barleys in relation to geographical regions, altitude range, and agro-ecological zones: As an aid to germplasm collection and conservation. *Hereditas* 124: 17–29.

——. 1997. Geographical, altitude and agro-ecological differentiation on isozyme and hordein genotypes of landrace barleys from Ethiopia: Implications to germplasm conservation. *Genetic Resources and Crop Evolution* 44: 43–55.

Dennis, J. V. 1987. Farmer management of rice variety diversity in northern Thailand. Ph.D. diss., Cornell University (University Microfilms, Ann Arbor, MI).

Dennis, P. A. 1987. *Inter-Village Conflict in Oaxaca.* New Brunswick, NJ: Rutgers University Press.

Dewalt, B. R. 1985. Farming systems research. *Human Organization* 44: 106–14.

Diaz, G., and A. Rodgers. 1993. *The Codex Borgia: A Full Color Restoration of the Ancient Mexican Manuscript.* New York: Dover.

Dodds, K. S. 1962. Classification of cultivated potatoes. In *The Potato and Its Wild Relatives*, edited by D. S. Correl, 517–39. Renner, TX: Texas Research Foundation.

Doebley, J. F. 1990. Molecular evidence and the evolution of maize. *Economic Botany.* 44 (S): 6–27.

Doebley, J. F., M. M. Goodman, and C. W. Stuber. 1985. Isozyme variation in the races of maize from Mexico. *American Journal of Botany* 72: 629–39.

Donald, C. M., and J. Hamblin. 1984. Thc convergent evolution of annual seed crops in agriculture. *Advances in Agronomy* 36: 97–143.

Donald, D. C. 1968. The breeding of crop ideotypes. *Euphytica* 17: 385–403.

Dougherty, J. W. D. 1978. Salience and relativity in classification. *American Ethnologist* 5: 66–80.

Dove, M. R. 1996. Center, periphery, and biodiversity: A paradox of governance and a development challenge. In *Valuing Local Knowledge: Indigenous People and Intellectual Property Rights*, edited by S. B. Brush and D. Stabinsky, 41–67. Washington, DC: Island Press.

Dunwell, J. M. 1999. Transgenic crops: The next generation, or an example of 20/20 vision. *Annals of Botany* 84: 269–77.

——. 2000. Transgenic approaches to crop improvement. *Journal of Experimental Botany* 51: 487–96.

Duvick, D. N. 1984. Genetic diversity in major farm crops on the farm and in reserve. *Economic Botany* 38: 161–78.

Dvorak, J., and M.-C. Luo. n.d. Identification of the site of emmer wheat domestication points to southeastern Turkey as the cradle of agriculture in the Fertile Crescent. Manuscript, Department of Agronomy and Range Science, University of California, Davis.

Eagles, H. A., and J. E. Lothrop. 1994. Highland maize from Central Mexico — its origin, characteristics, and use in breeding programs. *Crop Science* 34: 11–19.

Elias, M., L. Rival, and D. McKey. 2000. Perception and management of cassava (*Manihot esculenta* Cranz) diversity among Makushi Amerindians of Guayana (South America). *Ethnobiology* 20: 239–65.

Ellen, R., and H. Harris. 2000. Introduction. In *Indigenous Environmental Knowledge and Its Transformations*, edited by R. Ellen, P. Parkes, and A. Bicker, 1–34. Amsterdam: Harwood Academic Publishers.

Ellen, R., P. Parkes, and A. Bicker, eds. 2000. *Indigenous Environmental Knowledge and Its Transformations*. Amsterdam: Harwood Academic Publishers.

Ellstrand, N. C., H. C. Prentice, and J. F. Hancock. 1999. Gene flow and introgression from domesticated plants into their wild relatives. *Annual Review of Ecology and Systematics* 30: 539–63.

Elton, C. S. 1927. *Animal Ecology*. London: Sidgwick & Jackson.

Engel, E. 1970. Exploration of the Chilca Canyon, Peru. *Current Anthropology* 11: 55–58.

Epstein, T. S. 1962. *Economic Development and Social Change in South India*. Manchester, UK: Manchester University Press.

Escobar, A. 1994. *Encountering Development: The Making and Unmaking of the Third World*. Princeton: Princeton University Press.

Esquinas-Alcazar, J. 1998. Farmers' rights. In *Agricultural Values of Plant Genetic Resources*, edited by R. E. Evenson, D. Gollin, and V. Santaniello, 207–18. Wallingford, UK: CABI.

Etkin, N. L., ed. 1994. *Eating on the Wild Side: The Pharmacologic, Ecologic, and Social Implications of Using Noncultigens*. Tucson: University of Arizona Press.

Eubanks, M. W. 2001. The mysterious origin of maize. *Economic Botany* 55: 492–514.

Evans, L. T. 1993. *Crop Evolution, Adaptation and Yield*. Cambridge: Cambridge University Press.

Evenson, R. E., and D. Gollin. 1997. Genetic resources, international organizations, and improvement in rice varieties. *Economic Development and Cultural Change* 45: 471–500.

———. 2003. Assessing the impact of the green revolution, 1960–2000. *Science* 300: 758–61.

Eyzaguirre, P., and M. Iwanaga, eds. 1996. *Participatory Plant Breeding. Proceedings of a Workshop on Participatory Plant Breeding*, 26–29 July 1995, Wageningen, The Netherlands. Rome: International Plant Genetic Resources Institute.

Fairchild, D. G. 1938. *The World Was My Garden: Travels of a Plant Explorer*. New York: Scribner's Sons.

Farmer, B. H., ed. 1977. *Green Revolution?: Technology and Change in Rice-Growing Areas of Tamil Nadu and Sri Lanka*. Boulder, CO: Westview Press.

Farrington, J. 1988. Whither farming systems research? *Development Policy Review* 6: 323–32.

Faust, M., and B. Timon. 1995. Origin and dissemination of peach. *Horticultural Reviews* 17: 331–79.

Feder, G., R. E. Just, and D. Zilberman. 1985. Adoption of agricultural innovations in developing countries: A survey. *Economic Development and Cultural Change* 33: 255–98.

Feldman, M., F. G. H. Lupton, and T. E. Miller. 1995. Wheats. In *Evolution of Crop Plants*, 2d ed., edited by J. Smartt and N. W. Simmons, 184–92. London: Longman.

Ferguson, J. 1990. *The Anti-Politics Machine: "Development," Depoliticization, and Bureaucratic Power in Lesotho*. Cambridge: Cambridge University Press.

Ferroni, M. A. 1979. The urban bias of Peruvian food policy: Consequences and alternatives. Ph.D. diss., Cornell University, Ithaca, NY.

Findlen, P. 1994. *Possessing Nature: Museums, Collecting, and Scientific Culture in Early Modern Italy*. Berkeley: University of California Press.

Fitzgerald, D. 1990. *The Business of Breeding: Hybrid Corn in Illinois*. Ithaca, NY: Cornell University Press.

Food and Agricultural Organization of the United Nations (FAO). 1987. Extract of the twenty-second session of the FAO Conference, Rome, 5–23 November 1983. Resolution 8/83—International Undertaking on Plant Genetic Resources and Annex. CPGR/87/Inf. 3. FAO, Rome.

———. 1998. *The State of the World's Plant Genetic Resources for Food and Agriculture*. Rome: FAO.

———. 2001. *International Treaty on Plant Genetic Resources of Food and Agriculture*. Commission on Genetic Resources for Food and Agriculture. Rome: FAO, http://www.fao.org/ag/cgrfa/itpgr.htm.

———. 2003. *International Treaty on Plant Genetic Resources for Food and Agriculture*. Rome: FAO Legal Office, http://www.fao.org/Legal/TREATIES/033s-e.htm.

Fowler, C. 1994. *Unnatural Selection: Technology, Politics, and Plant Evolution*. Langhorne, PA: Gordon and Breach.

Fowler, C., and P. Mooney. 1990. *Shattering: Food, Politics and the Loss of Genetic Diversity*. Tucson: University Arizona Press.

Fowler, C., M. Smale, and S. Gaiji. 2001. Unequal exchange? Recent transfers of agricultural resources and their implications for developing countries. *Development Policy Review* 19: 181–204.

Frankel, F. 1971. *India's Green Revolution: Economic Gains and Political Costs*. Princeton: Princeton University Press.

Frankel, O. H. 1970. Genetic conservation in perspective. In *Genetic Resources in Plants—Their Exploration and Conservation*, edited by O. H. Frankel and E. Bennett, 469–89. Oxford: Blackwell Scientific Publications, International Biological Programme Handbook No. 11.

———, ed. 1973. *Survey of Crop Genetic Resources in Their Centres of Diversity*, First Report. Rome: Food and Agricultural Organization of the U.N.

Frankel, O. H., and E. Bennett, eds. 1970. *Genetic Resources in Plants—Their Exploration and Conservation*. Oxford: Blackwell Scientific Publications, International Biological Programme Handbook No. 11.

Frankel O. H., A. H. D. Brown, and J. J. Burdon. 1995. *The Conservation of Plant Biodiversity*. Cambridge: Cambridge University Press.

Frankel, O. H., and J. G. Hawkes, eds. 1975. *Crop Genetic Resources for Today and Tomorrow*. International Biological Programme 2. Cambridge: Cambridge University Press.

Frankel, O. H., and M. E. Soulé. 1981. *Conservation and Evolution*. Cambridge: Cambridge University Press.

Franquemont, C., T. Plowman, E. Franquemont, S. R. King, C. Niezgoda, W. Davis, and C. R. Sperling. 1990. The Ethnobotany of Chincero, an Andean Community in Southern Peru. In *Fieldiana, Botany,* new series No. 24. Chicago: Field Museum of Natural History.

Freeman, D. 1955. *Iban Agriculture: A Report on the Shifting Cultivation of Hill Rice by the Iban of Sarawak*. Colonial Research Studies No. 18. London: H. M. Stationary Office.

———. 1970. *Report on the Iban*. London: Athlone Press.

Gade, D. W. 1975. *Plants, Man and the Land in the Vilcanota Valley of Peru*. Vol. 6, *Biogeographica*. The Hague: Dr. W. Junk B.V. Publishers.

———. 1999. *Nature and Culture in the Andes*. Madison, WI: University of Wisconsin Press.

Geertz, C. 1963. *Agricultural Involution: The Process of Ecological Change in Indonesia*. Berkeley: University of California Press.

Gepts, P. 1998. Origin and evolution of common bean: Past events and recent trends. *Hortscience* 33: 1124–30.

Gepts, P., and F. A. Bliss. 1986. Phaseolin variability among wild and common beans from Colombia. *Economic Botany* 40: 469–78.

Geyer, M., and C. Bright. 1995. World history in a global age. *American Historical Review* 100: 1034–60.

Giddens, A. 1990. *The Consequences of Modernity*. Stanford: Stanford University Press.

Glacken, C. J. 1967. *Traces on the Rhodian Shore: Nature and Culture in Western Thought from Ancient Times to the End of the Eighteenth Century*. Berkeley: University of California Press.

Goland, C. 1993. Field scattering as agricultural risk management—a case study from Cuyo Cuyo, Department of Puno, Peru. *Mountain Research and Development* 13: 317–38.

Gollin, D. 1998. Valuing farmers' rights. In *Agricultural Values of Plant Genetic Resources*, edited by R. E. Evenson, D. Gollin, and V. Santaniello, 233–48. Wallingford, UK: CABI.

Gollin, M. A., and S. A. Laird. 1996. Global Politics, Local Actions: The Role of National Legislation in Sustainable Biodiversity Prospecting. *Boston University Journal of Science and Technology Law* Lexus: 2 *B. U. J. Sci. and Tech. L.* 16.

Golte, J. 1980. *La racionalidad de la organización Andina*. Lima: Instituto de Estudios Peruanos.

Gonzales, T. 1999. The cultures of seed in the Peruvian Andes. In *Genes in the Field: On-Farm Conservation of Crop Diversity,* edited by S. B. Brush, 217–38. Boca Raton, FL: Lewis.

Goodman, D. 1975. The theory of diversity-stability relationships in ecology. *The Quarterly Review of Biology* 50: 237–66.

Goodman, M. M. 1985. Exotic maize germplasm: Status, prospects, and remedies. *Iowa State Journal of Research* 59: 497–527.

———. 1988. The history and evolution of maize. *CRC Critical Reviews in Plant Sciences*. 7: 197–220.

Goodman, M. M., and R. McK. Bird. 1977. The races of maize IV: Tentative grouping of 219 Latin American races. *Economic Botany* 31: 204–21.

Gotelli, N. J. 1998. *A Primer of Ecology*, 2d ed. Sunderland, MA: Sinaur Associates, Inc.

Gourou, P. 1955. *The Peasants of the Tonkin Delta: A Study in Human Geography*. New Haven: Human Relations Area Files. (original: 1936. *Les Paysans du Delta Tonkin*. Paris: Editions d'Art et d'Histoire.)

Grace, J. B. 1990. On the relationship between plant traits and competitive ability. In *Perspectives on Plant Competition*, edited by J. B. Grace and D. Tilman, 51–65. San Diego: Academic Press.

Greaves, T. A. 1996. Tribal rights. In *Valuing Local Knowledge: Indigenous People and Intellectual Property Rights*, edited by S. B. Brush and D. Stabinsky, 25–40. Washington, DC: Island Press.

Greaves, T., ed. 1994. *A Sourcebook on Intellectual Property and Indigenous Knowledge*. Washington, DC: Society for Applied Anthropology.

Griffin, K. 1974. *The Political Economy of Agrarian Change: An Essay on the Green Revolution*. Cambridge: Harvard University Press.

Griliches, Z. 1957. Hybrid corn: an exploration in the economics of technological change. *Econometrica* 25: 501–22.

Grist, D. H. 1936. *An Outline of Malayan Agriculture*. Malayan Planting Manual No. 2. Kuala Lumpur: Department of Agriculture, Straits Settlements and Federated Malay States.

———. 1986. *Rice*. 6th edition. London: Longman. (Orig. pub. 1953.)

Grobman, A., W. Salhuana, and R. Sevilla (in collaboration with P. C. Mangelsdorf). 1961. *Races of Maize in Peru, Their Origins, Evolution and Classification*. Washington, DC: National Academy of Sciences, National Research Council.

Guaman Poma de Ayala, F. 1936. *Nueva Cronónica y Buen Gobierno*. Paris: Institut d'Ethnologie.

Guarino, L. 1995. Assessing the threat of genetic erosion. In *Collecting Plant Genetic Diversity: Technical Guidelines*, edited by L. Guarino, V. R. Rao, and R. Reid, 67–74. Wallingford, UK: CABI.

Hammer, K., H. Knüpffer, L. Xhuveli, and P. Perrino. 1996. Estimating genetic erosion in landraces—two case studies. *Genetic Resources and Crop Evolution* 43: 329–36.

Hanks, L. M. 1972. *Rice and Man: Agricultural Ecology in Southeast Asia*. Chicago: Aldine Atherton.

Hanski, I. A. 1998. Metapopulation dynamics. *Nature* 396: 41–49.

Hanski, I. A., and M. E. Gilpin, eds. 1997. *Metapopulation Biology: Ecology, Genetics, and Evolution*. San Diego: Academic Press.

Hanski, I., and D. Sinberloff. 1997. The metapopulation approach, its history, conceptual domain, and application to conservation. In *Metapopulation Biology: Ecology, Genetics and Evolution*, 5–26. San Diego: Academic Press.

Hardin, G. 1960. The competitive exclusion principle. *Science* 131: 1292–97.

———. 1968. The tragedy of the commons. *Science* 162: 1243–48.

Harlan, H. V., and M. L. Martini. 1936. Problems and results of barley plant breeding. *USDA Yearbook of Agriculture*, 303–46. Washington, DC: U.S. Government Printing Office.

Harlan, J. R. 1950. Collection of crops plants in Turkey. *Agronomy Journal* 42: 258–59.

———. 1970. Evolution of cultivated plants. In *Genetic Resources in Plants—Their Ex-*

ploration and Conservation, edited by O. H. Frankel and E. Bennett, 19–32. Oxford: Blackwell Scientific Publications, International Biological Programme Handbook No. 11.

——. 1975. Our vanishing genetic resources. *Science* 188: 618–21.

——. 1992. *Crops and Man*. 2d. ed. Madison, WI: American Society of Agronomy, Crop Science Society of America.

——. 1995. *The Living Fields: Our Agricultural Heritage*. Cambridge: Cambridge University Press.

Harris, D. R. 1989. The evolutionary continuum of people–plant evolution. In *Foraging and Farming: The Evolution of Plant Exploitation*, edited by D. R. Harris and G. C. Hillman, 11–16. London: Unwin Hyman.

Harris, D. R., and G. C. Hillman, eds. 1989. *Foraging and Farming: The Evolution of Plant Exploitation*. London: Unwin Hyman.

Harris, J. M. 1996. World agricultural futures: Regional sustainability and ecological limits. *Ecological Economics* 17: 95–115.

Hawkes, J. G. 1947. On the origin and meaning of South American Indian potato names. *Journal of the Linnean Society (Botany)* 53: 205–50.

——. 1978. Biosystematics of the potato. In *The Potato Crop: The Scientific Basis for Improvement*, edited by P. M. Harris, 15–69. London: Chapman & Hall.

——. 1983. *The Diversity of Crop Plants*. Cambridge: Harvard University Press.

——. 1990. *The Potato: Evolution, Biodiversity, and Genetic Resources*. London: Belhaven.

Hawkes, J. G., and J. P. Hjerting. 1989. *The Potatoes of Bolivia: Their Breeding Value and Evolutionary Relationships*. Oxford: Clarendon Press.

Hawley, A. H. 1950. *Human Ecology: A Theory of Community Structure*. New York: Ronald Press.

Hayami, Y., and M. Kikuchi. 2000. *A Rice Village Saga*. Lanham, MD: Barnes & Noble.

Hazell, P. B. R. 1982. Instability in Indian Foodgrain Production. *IFPRI Research Report No. 30*. Washington, DC: International Food Policy Research Institute.

——. 1989. Changing patterns of variability in world cereal production. In *Variability in Grain Yields: Implication for Agricultural and Research and Policy in Developing Countries*, edited by J. R. Anderson and P. B. R. Hazell, 13–34. Baltimore: Johns Hopkins University Press.

Hazell, P. B. R., and C. Ramasay, eds. 1991. *The Green Revolution Reconsidered: The Impact of High-Yielding Rice Varieties in South India*. Baltimore: Johns Hopkins University Press.

Heiser, C. B. 1973. *Seed to Civilization: The Story of Man's Food*. San Francisco: W. H. Freeman.

Heller, M. A. 1998. The tragedy of the anticommons: Property in the transition from Marx to markets. *Harvard Law Review* 111: 622–88.

Heller, M. A., and R. S. Eisenberg. 1998. Can patents deter innovation? The anticommons in biomedical research. *Science* 280: 698–701.

Herdt, R. 1987. A retrospective view of technological and other farming changes in Philippine rice farming, 1965–1982. *Economic Development and Cultural Change* 35: 329–49.

Hernández, X. E. 1972. Exploración etnobotánica en maíz. *Fitotecnica Latinoamericana* (Caracas, Venezuela) 8: 46–51.

——. 1973a. Consumo humano de maíz y el aprovechamiento de tipos con alto valor nutritivo. *Memoria del simposio sobre desarrollo y utilización de maíces de alto valor nutritivo,* June 1972. Colegio de Postgraduados, Escuela Nacional de Agricultural, Chapingo, México. Secrataria de Agricultura y Ganaderia, 1973, 149–56.

——. 1973b. Genetic resources of primitive varieties of Mesoamerica: *Zea* spp., *Phaseolus* spp., *Capsicum* spp., and *Cucurbita* spp. In *Survey of Crop Genetic Resources in Their Centres of Diversity, First Report*, edited by O. H. Frankel, 76–115. Rome: Food and Agricultural Organization of the United Nations.

——. 1985. Maize and man in the greater southwest. *Economic Botany* 39: 416–30.

Herrera-Estrella, L. 1999. Transgenic plants for tropical regions: Some considerations about their development and their transfer to the small farmer. *Proceedings of the National Academy of Sciences of the United States of America* 96: 5978–81.

Heun, M., R. Schäfer-Pregl, D. Klawan, R. Castagna, M. Accerbi, B. Borghi, and F. Salamini. 1997. Site of einkorn wheat domestication identified by DNA fingerprinting. *Science* 278: 1312–14.

Hillman, G. C., and M. S. Davis. 1990. Measured domestication rates in wild wheats and barley under primitive cultivation, and their archaeological implications. *Journal of World Prehistory* 4: 157–222.

Hobbelink, H. 1991. *Biotechnology and the Future of World Agriculture*. London: Zed Books.

Hoffner, H. A. 1974. *Alienta Herhaeorum: Food Production in Hittite Asia Minor*. New Haven: American Oriental Society.

Hohfeld, W. N. 1913. Some fundamental legal concepts as applied to judicial reasoning. *The Yale Law Journal* 23: 16–59.

Hoppin, J. C. 1919. *A Handbook of Attic Red-Figured Vases, Signed by or Attributed to the Various Masters of the Sixth and Fifth Centuries* B.C., vol. II. Cambridge, MA: Harvard University Press.

Horkheimer, H. 1973. *Alimentación, y obtención de alimentos en el Peru prehispanico*. Lima: Universidad Nacional Mayor de San Marcos.

Horton, D. 1984. *Social Scientists in Agricultural Research: Lessons from the Mantaro Valley Project, Peru*. Ottawa: IDRC.

Hosaka, K. 1995. Successive domestication and evolution of the Andean potatoes as revealed by chloroplast DNA restriction endonuclease analysis. *Theoretical and Applied Genetics* 90: 356–63.

Huamán, Z. 1986. Conservation of Potato Genetic Resources at CIP. *CIP Circular* 14: 1–7.

Huamán, Z., and D. M. Spooner. 2002. Reclassification of landrace populations of cultivated potatoes (*Solanum* sect. *Petota*). *American Journal of Botany* 89: 947–65.

Huamán, Z., J. T. Williams, W. Salhuana, and L. Vincent. 1977. *Descriptors of the Cultivated Potato*. Rome: International Board for Plant Genetic Resources.

Humboldt, A. V. 1807. *Essai sur la Geographie de Plantes*. Repr., London: Society for the Bibliography of Natural History, 1959.

——. 1817. *De Distributione geographica plantarum secundum coeli temperiem et altitudinem montium*. Paris: Prolegomena.

Hutchinson, G. E. 1957. Concluding remarks. *Cold Spring Harbor Symposium on Quantitative Biology* 22: 415–27.

Huxley, J. 1942. *Evolution, The Modern Synthesis*. New York: Harper & Brothers.

Hyland, H. L. 1977. History of U.S. plant introduction. *Environmental Review* 4: 26–33.

Il farro. 2001. Products of Italy: About our original farro. http://www.ilfarro.com/pages/about.htm.

Iltis, H. 1983. From teosinte to maize: The catastrophic sexual transmutation. *Science* 222: 886–93.

India. 1999. The protection of plant varieties and farmers' rights bill, 2000. Act No. 123, 1999. http://www.grain.org/brl/india-pvp-2000-act-en.cfm.

International Board for Plant Genetic Resources (IBPGR). 1985. *Ecogeographical Surveying and* In Situ *Conservation of Crop Relatives*. Rome: International Board for Plant Genetic Resources Secretariat.

International Institute for Sustainable Development (IISD). 2001. Summary of the 6th Extraordinary Session of the Commission on Genetic resources for Food and Agriculture: 24 June–1 July 2001. *Earth Negotiations Bulletin* 9 (197), http://www.iisd.ca/biodiv/ExCGRFA-6/.

Isbell, B. J. 1978. *To Defend Ourselves: Ecology and Ritual in an Andean Village*. Austin: University of Texas Press.

Iskandar, J., and R. Ellen. 1999. *In situ* conservation of rice landraces among the Baduy of West Java. *Journal of Ethnobiology* 19: 97–125.

Izikowitz, K. G. 1951. *Lamet: Hill Peasants in French Indochina*. Etnologiska Studier No. 17. Göteborg, Sweden: Etnografisk Museet.

Jackson, L. E., and G. W. Koch. 1997. The ecophysiology of crops and their wild relatives. In *Ecology in Agriculture*, edited by L. E. Jackson, 3–38. San Diego: Academic Press.

James, C. 2001. Global Status of Commercialized Transgenic Crops: 2001. *ISAAA Briefs No. 24: Preview*. Ithaca, NY: ISAAA.

Jarvis, D., and T. Hodgkin. 1999. Farmer decision making and genetic diversity: Linking multidisciplinary research to implementation on-farm. In *Genes in the Field: On-Farm Conservation of Crop Diversity*, edited by S. B. Brush, 261–78. Boca Raton, FL: Lewis.

Jennings, P. R., and J. H. Cock. 1977. Centres of origin of crops and productivity. *Economic Botany* 31: 51–54.

Jesse, L. C. H., and J. J. Obrycki. 2000. Field deposition of Bt transgenic corn pollen: Lethal effects on the monarch butterfly. *Oecologia* 125: 241–48.

Johns, T. 1990. *With Bitter Herbs They Shall Eat It: Chemical Ecology and the Origins of Human Diet and Medicine*. Tucson: University of Arizona Press.

Johns, T., and S. L. Keen. 1986. Ongoing evolution of the potato on the altiplano of western Bolivia. *Economic Botany* 40: 409–24.

Johnson, A. 1974. Ethnoecology and planting practices in a swidden agricultural system. *American Ethnologist* 1: 87–101.

Johnsson, M. 1986. *Food and Culture Among the Bolivian Aymara: Symbolic Expressions of Social Relations*. Uppsala Studies in Cultural Anthropology 7. Stockholm: Almqvist and Wiksell International.

Jondle, R. J. 1989. Overview and status of plant proprietary rights. In *Intellectual Prop-*

erty Rights Associated with Plants, edited by B. E. Caldwell and J. A. Schillinger, 5–15. Madison, WI: Crop Science Society of America, ASA Special Publication No. 52.

Jordan, A. 1994. Paying the incremental costs of global environmental protection–the evolving role of the GEF. *Environment* 36: 12ff.

Joshi, A., and J. R. Whitcombe. 1996. Farmer participatory crop improvement. II. Participatory varietal selection, a case study in India. *Experimental Agriculture* 32: 461–77.

Joyner, C. C. 1986. Legal implications of the concept of the common heritage of mankind. *International and Comparative Law Quarterly* 35: 190–99.

Kahn, J. S. 2001. Anthropology and modernity. *Current Anthropology* 42: 651–80.

Kanbur, S. M. R., and T. Sandler with K. M. Morrison. 1999. The future of development assistance: Common pools and international public goods. *Policy Essay* No. 25. Washington, DC: Overseas Development Council.

Karagöz, A. 1996. Agronomic practices and socioeconomic aspects of emmer and einkorn cultivation in Turkey. In *Hulled Wheats. Proceedings of the First International Workshop on Hulled Wheats*, 21–22 July 1995, Castelvecchio Pascoli, Tuscany, Italy, edited by S. Padulosi, K. Hammer, and J. Heller, 172–77. Rome: International Plant Genetic Resources Institute.

Kaul, I., I. Grunberg, and M. A. Stern, eds. 1999. *Global Public Goods: International Cooperation in the 21st Century*. New York: Oxford University Press.

Kenney, M. 1986. *Biotechnology: The University-Industrial Complex*. New Haven: Yale University Press.

Kent, A., and H. Lancour. 1972. *Copyright: Current Viewpoints on History, Laws, Legislation*. New York: Bowker.

Kimpel, J. A. 1999. Freedom to operate: Intellectual property protection in plant biology and its implications for the conduct of research. *Annual Review of Phytopathology* 37: 29–51.

King, S. R., T. J. Carlson, and K. Moran. 1996. Biological diversity, indigenous knowledge, drug discovery, and intellectual property rights. In *Valuing Local Knowledge: Indigenous People and Intellectual Property Rights*, edited by S. B. Brush and D. Stabinsk, 167–85. Washington, DC: Island Press.

Kjellqvist, E. Near East. 1973. In *Survey of Crop Genetic Resources in Their Centers of Diversity, First Report*, edited by O. H. Frankel, 9–21, 25–44. Rome: Food and Agricultural Organization of the United Nations.

Kloppenburg, J., ed. 1988a. *Seeds and Sovereignty: Use and Control of Plant Genetic Resources*. Durham, NC: Duke University Press.

Kloppenburg, J. R. Jr. 1988b. *First the Seed: The Political Economy of Plant Biotechnology 1492–2000*. Cambridge: Cambridge University Press.

Kloppenburg, J. Jr., and D. L. Kleinman. 1987. The plant germplasm controversy. *BioScience* 37: 190–98.

Klose, N. 1950. *America's Crop Heritage: The History of Foreign Plant Introduction by the Federal Government*. Ames: Iowa State College Press.

Kramer, M. H. 1997. *John Locke and the Origins of Private Property: Philosophical Explorations of Individualism, Community, and Equality*. Cambridge: Cambridge University Press.

Kresovich, S., and J. R. McFerson. 1992. Assessment and management of plant genetic diversity: Considerations on intra- and interspecific variation. *Field Crops Research* 29: 185–204.

Krimsky, S., and R. P. Wurbel. 1996. *Agricultural Biotechnology and the Environment: Science, Policy, and Social Issues*. Urbana: University of Illinois Press.

Kroeber, A. L. 1939. *Cultural and Natural Areas of Native North America*. Berkeley: University of California Press.

Krutilla, J. V. 1967. Conservation reconsidered. *American Economic Review* 57: 777–86.

Kunstadter, P., E. C. Chapman, and S. Sabharsi, eds. 1978. *Farmers in the Forest: Economic Development and Marginal Agriculture in Northern Thailand*. Honolulu: University Press of Hawaii.

La Barre, W. 1947. Potato taxonomy among the Aymara Indians of Bolivia. *Acta Americana* 5: 83–103.

Lakhanpal, T. N. 1989. Crop diseases and catastrophes. In *Plants and Society*, edited by M. S. Swaminathan and S. L. Kochhar, 420–51. London: Macmillan.

Lambert, D. H. 1985. *Swamp Rice Farming: The Indigenous Pahang Malay Agricultural System*. Boulder, CO: Westview Press.

Lansing, J. S. 1991. *Priests and Programmers: Technologies of Power in the Engineered Landscape of Bali*. Princeton: Princeton University Press.

Lappé, F. M., and B. Bailey. 1998. *Against the Grain: Biotechnology and the Corporate Takeover of Food*. Monroe, ME: Common Courage Press.

Latour, B. 1993. *We Have Never Been Modern*. Translated by C. Porter. Cambridge: Harvard University Press.

Laughlin, R. M. 1975. *The Great Tzotzil Dictionary of San Lorenzo Zinacantan*. Smithsonian Contributions to Anthropology No. 19. Washington, DC: Smithsonian Institution Press.

León-Portilla, M. 1988. El Maiz: nuestro sustento, su realidad divina y humana en Mesoamerica. *América Indígena* 48: 477–502.

Lesser, W. H. 1994a. An approach to securing rights to indigenous knowledge. Appendix D. In *Workshop Report: Intellectual Property Rights and Indigenous Peoples*, 19–25. Geneva: International Academy of the Environment.

———. 1994b Valuation of plant variety protection certificates. *Review of Agricultural Economics* 16: 231–38.

———. 1998. *Sustainable Use of Genetic Resources Under the Convention on Biological Diversity: Exploring Access and Benefit Sharing Issues*. Wallingford, UK: CABI.

Levin, R. C., A. K. Klevorkick, R. R. Nelson, and S. G. Winter. 1987. Appropriating the returns from industrial research and development. *Brookings Papers on Economic Activity* 3: 783–820.

Levins, R. 1969. Some demographic and genetic consequences of environmental heterogeneity for biological control. *Bulletin of the Entomological Society of America* 15: 237–40.

Lévi-Strauss, C. 1966. *The Savage Mind*. Chicago: University of Chicago Press.

Lipton, M. 1989. *New Seeds and Poor People*. London: Unwin Hyman.

Lipton, M., and R. Longhurst. 1985. *Modern Varieties, International Agricultural Research, and the Poor*. CGIAR Study Paper 2. Washington, DC: World Bank.

Litman, J. 1990. The public domain. *Emory Law Journal* 39: 965–1023.

Long, N., and B. Roberts. 1984. *Miners, Peasants and Entrepreneurs: Regional Development in the Central Highlands of Peru*. Cambridge: Cambridge University Press.

Loomis, R. S. 1984. Traditional agriculture in America. *Annual Review of Ecology and Systematics* 15: 449–78.

Losey, J. E., L. S. Rayor, and M. E. Carter. 1999. Transgenic pollen harms monarch larvae. *Nature* 399: 214.

Loskutov, I. G. 1999. *Vavilov and His Institute: A History of the World Collection of Plant Genetic Resources in Russia*. Rome: International Plant Genetic Resources Institute.

Louette, D. 1999. Traditional management of seed and genetic diversity: What is a landrace? In *Genes in the Field: On-Farm Conservation of Crop Diversity*, edited by S. Brush, 109–42. Boca Raton, FL: Lewis.

Louette D., A. Charrier, and J. Berthaud. 1997. *In situ* conservation of maize in Mexico, genetic diversity and maize seed management in a traditional community. *Economic Botany* 51: 20–39.

Louette, D., and M. Smale. 2000. Farmers' seed selection practices and traditional maize varieties in Cuzalapa, Mexico. *Euphytica* 113: 25–41.

Lovejoy, A. O. 1936. *The Great Chain of Being: A Study of the History of an Idea*. Cambridge: Harvard University Press.

Machlup, F., and E. Penrose. 1950. The patent controversy in the nineteenth century. *Journal of Economic History* X: 1–29.

Maffi, L., ed. 2001. *On Biocultural Diversity: Linking Language, Knowledge, and the Environment*. Washington, DC: Smithsonian Institution Press.

Mallon, F. E. 1983. *The Defense of Community in Peru's Central Highlands: Peasant Struggle and Capitalist Transition, 1860–1940*. Princeton: Princeton University Press.

Mangelsdorf, P. C. 1974. *Corn: Its Origin, Evolution, and Improvement*. Cambridge: Harvard University Press.

Markman, R. H., and P. T. Markman. 1992. *The Flayed God: The Mesoamerican Mythological Tradition: Sacred Texts and Images from Pre-Columbian Mexico and Central America*. San Francisco: Harper.

Marshall, D. R. 1977. The advantages and hazards of genetic homogeneity. In *The Genetic Basis of Epidemics in Agriculture*, edited by P. R. Day. *Annals of the New York Academy of Sciences* 287: 1–20.

Marshall, D. R., and A. H. D. Brown. 1975. Optimum sampling strategies in genetic conservation. In *Crop Genetic Resources for Today and Tomorrow*, edited by O. H. Frankel and J. G. Hawkes, 53–80. Cambridge: Cambridge University Press, International Biological Programme 2.

Marten, G. O., ed. 1986. *Traditional Agriculture in Southeast Asia*. Boulder, CO: Westview Press.

Mastenbroek, C. 1988. Plant breeders' rights, an equitable legal system for new plant cultivars. *Experimental Agriculture* 24: 15–30.

Masuda, S., I. Shimada, and C. Morris, eds. 1985. *Andean Ecology and Civilization: An Interdisciplinary Perspective on Andean Ecological Complementarity*. Tokyo: University of Tokyo Press.

Maxted, N. B., V. Ford-Lloyd, and J. G. Hawkes, eds. 1997. *Plant Genetic Conservation: The* In Situ *Approach*. London: Chapman & Hall.

Maxted, N., L. Guarino, L. Myer, and E. A. Chiwona. 2002. Towards a methodology for on-farm conservation of plant genetic resources. *Genetic Resources and Crop Evolution* 49: 31–46.

May, R. M. 1973. *Stability and Complexity in Model Ecosystems*. Princeton: Princeton University Press.

Mayer, E. 1979. *Land Use in the Andes*. Lima: Centro Internacional de la Papa.

———. 2002. *The Articulated Peasant: Household Economies in the Andes*. Boulder, CO: Westview Press.

Mayer, E., and C. Fonseca. 1979. *Sistemas agrários en la cuenca del río cañete (departamento de Lima)*. Lima: Oficina Nacional de Evaluación de Recursos Naturales.

Mayer, E. (with M. Glave). 2002. Alguito para ganar ('a little something to earn'): Profits and losses in peasant economies. In *The Articulated Peasant: Household Economies in the Andes*, 205–38. Boulder, CO: Westview Press.

Mays, T. D., K. Mazan, E. J. Asebey, M. R. Boyd, and G. M. Cragg. 1996. Quid pro quo: Alternatives for equity and conservation. In *Valuing Local Knowledge: Indigenous People and Intellectual Property Rights*, edited by S. Brush and D. Stabinsky, 259–80. Washington, DC: Island Press.

McArthur, R. H., and E. O. Wilson. 1967. *The Theory of Island Biogeography*. Princeton: Princeton University Press.

McCann, K. S. 2000. The diversity-stability debate. *Nature* 405: 228–33.

McCay, B. J., and J. M. Acheson, eds. 1987. *The Question of the Commons: The Culture and Ecology of Communal Resources*. Tucson: University of Arizona Press.

McNeely, J. A. 1988. *Economics and Biological Diversity: Developing and Using Economic Incentives to Conserve Biological Resources*. Gland, Switzerland: International Union for Conservation of Nature and Natural Resources.

Mehra, S. 1981. Instability in Indian agriculture in the context of the new technology. *IFPRI Research Report No. 25*. Washington, DC: International Food Policy Research Institute.

Meilleur, B. A. 1996. Selling Hawaiian crop cultivars. In *Valuing Local Knowledge: Indigenous People and Intellectual Property Rights*, edited by S. Brush and D. Stabinsky, 244–58. Washington, DC: Island Press.

Mendelsohn, R. 2000. The market value of farmers' rights. In *Agriculture and Intellectual Property Rights: Economic, Institutional and Implementation Issues in Biotechnology*, edited by V. Santaniello, R. E. Evenson, D. Zilberman, and G. A. Carlson, 121–26. Wallingford, UK: CABI.

Meng, E. 1997. Land allocation decisions and *in situ* conservation of crop genetic resources: The case of wheat landraces in Turkey. Ph.D. diss., University of California, Davis (University Microfilms, Ann Arbor, MI).

Meng, E. C. H., J. E. Taylor, and S. B. Brush. 1998. Implications for the conservation of wheat landraces in Turkey form a household model of varietal choice. In *Farmers,*

Gene Banks, and Crop Breeding: Economic Analyses of Diversity in Wheat, Maize, and Rice, edited by M. Smale, 127–42. Boston: Kluwer.

Merrill-Sands, D. M., S. D. Biggs, R. J. Bingen, P. T. Ewell, J. L. McAllister, and S. V. Poats. 1991. Institutional considerations in strengthening on-farm client-oriented research in national agricultural research systems: Lessons from a nine-country study. *Experimental Agriculture* 27: 343–73.

Merrill-Sands, D. M., P. Ewell, S. D. Biggs, and J. McAllister. 1989. Issues in institutionalizing on-farm client-oriented research: A review of experiences from nine national agricultural research systems. *Quarterly Journal of International Agriculture* 28: 279–300.

Merges, R. P., and R. R. Nelson. 1994. On limiting or encouraging rivalry in technical progress: The effect of patent scope decisions. *Journal of Economic Behavior and Organization*. 25: 1–24.

Merton, R. K. 1973. *The Sociology of Science: Theoretical and Empirical Investigations*. Chicago: University of Chicago Press.

Mitchell, S. 1993. *Anatolia: Land, Men, and Gods in Asia Minor*. Oxford: Clarendon Press.

Moerman, M. 1968. *Agricultural Change and Peasant Choice in a Thai Village*. Berkeley: University of California Press.

Mooney, P. R. 1979. *Seeds of the Earth: A Private or Public Resource?* Ottawa: Inter Pares.

———. 1983. The law of the seed: Another development and plant genetic resources. *Development Dialogue* 1983: 1–2. Uppsala: Dag Hammarskjöld Foundation.

Mooney, P. 1996. Viewpoint of non-governmental organizations. In *Agrobiodiversity and Farmers' Rights*, edited by M. S. Swaminathan, 40–43. Delhi: Konark Publishers.

Moran, E. F. 1982. *Human Adaptability: An Introduction to Ecological Anthropology*. Boulder, CO: Westview Press.

Moran, K., S. R. King, and T. J. Carlson. 2001. Biodiversity prospecting: Lessons and prospects. *Annual Review of Anthropology* 30: 505–26.

Morley, S. G. 1946. *The Ancient Maya*. Stanford: Stanford University Press.

Morris, M. L., and M. A. López-Pereira. 1999. Impacts of maize breeding research in Latin America, 1966–1997. Mexico: CIMMYT.

Murra, J. V. 1960. Rite and crop in the Inca state. In *Culture in History: Essays in Honor of Paul Radin*, edited by S. Diamond, 393–407. New York: Columbia University Press.

———. 1972. El 'control vertical' de un maximo de pisos ecológicos en la economia de las sociedades Andinas. In *Visita de la provincia de Leon de Huanuco en 1562, Ortiz de Zuniga, I., visitador*, 429–76. Huanuco, Peru: Universidad Nacional Hermilio Valdizan.

Muséo Nacional de Culturas Populares. 1982. *El Maíz, fundamento de la cultural popular Mexicana*. Mexico: Muséo Nacional de Culturas Populares.

Musgrave, T., and W. Musgrave. 2000. *An Empire of Plants: People and Plans That Changed the World*. London: Cassell.

Nabhan, G. P. 1985. Native crop diversity in aridoamerica: Conservation of regional gene pools. *Economic Botany* 39: 387–99.

———. 1989. *Enduring Seeds: Native American Agriculture and Wild Plant Conservation*. San Francisco: North Point Press.

National Research Council (NRC). 1972. *Genetic Vulnerability of Major Crops*. Washington, DC: National Academy of Sciences.

———. 1989. *Lost Crops of the Incas: Little Known Plants of the Andes with Promise of Worldwide Cultivation*. Washington, DC: National Academy Press.

———. 1993. *Managing Global Genetic Resources: Agricultural Crop Issues and Policies*. Washington, DC: National Academy Press.

———. 2002. *Environmental Effects of Transgenic Plants*. Washington, DC: National Academy Press.

Nazarrea-Sandoval, V. D. 1995. *Local Knowledge and Agricultural Decision Making in the Philippines*. Ithaca, NY: Cornell University Press.

Nee, S. 1994. How populations persist. *Nature* 367: 123–24.

Nee, S., R. M. May, and M. P. Hassel. 1997. Two-species metapopulation models. In *Metapopulation Biology: Ecology, Genetics, and Evolution*, edited by I. Hanski and D. Simberloff, 123–47. San Diego: Academic Press.

Nesbitt, M. 1995. Plants and people in ancient Anatolia. *Biblical Archaeologist* 58: 68–81.

Nesbitt, M., and D. Samuel. 1996. From staple crop to extinction? The archaeology and history of the hulled wheats. In *Hulled Wheats. Proceedings of the First International Workshop on Hulled Wheats*, 21–22 July 1995, Castelvecchio Pascoli, Tuscany, Italy, edited by S. Padulosi, K. Hammer, and J. Heller, 41–100. Rome: International Plant Genetic Resources Institute.

Netting, R. M. 1968. *Hill Farmers of Nigeria: Cultural Ecology of the Kofyar of the Jos Plateau*. Seattle: University of Washington Press.

———. 1974. Agrarian ecology. *Annual Review of Anthropology* 3: 21–56.

———. 1993. *Smallholders, Householders: Farm Families and the Ecology of Intensive, Sustainable Agriculture*. Stanford: Stanford University Press.

Nigh, R. 2002. Maya medicine in the biological gaze: Bioprospecting research as herbal fetichism. *Current Anthropology* 43:451–477.

OAU (Organization of African Unity). 2000. African model legislation for the protection of the rights of local communities, farmers and breeders, and for the regulation of access to biological resources. OAU Model Law, Algeria 2000, http://www.grain.org/brl/oau-model-law-en.cfm.

Ochoa, C. 1975. Potato collecting expeditions in Chile, Bolivia and Peru, and the genetic erosion of indigenous cultivars. In *Crop Genetic Resources for Today and Tomorrow*, edited by O. H. Frankel and J. G. Hawkes, 167–73. Cambridge: Cambridge University Press, International Biological Programme 2.

Ochoa, C. M. 1990. *The Potatoes of South America., Bolivia*. Translated by D. Ugent. Cambridge: Cambridge University Press.

Odek, J. O. 1994. Bio-piracy: Creating proprietary rights in plant genetic resources. *Journal of Intellectual Property Law* 2: 141–81.

Odum, E. P. 1953. *Fundamentals of Ecology*. Philadelphia: Saunders.

Ohnuki-Tierney, E. 1993. *Rice as Self: Japanese Identities Through Time*. Princeton: Princeton University Press.

Orlove, B. S. 1978. *Alpacas, Sheep and Men: The Wool Export Economy and Regional Society of Southern Peru*. New York: Academic Press.

Orlove, B. S., and R. Godoy. 1986. Sectoral fallow systems in the central Andes. *Journal of Ethnobiology* 6: 169–204.

O'Rourke, K. H., and J. G. Williamson. 1999. *Globalization and History: The Evolution of a Nineteenth-Century Atlantic Economy*. Cambridge: MIT Press.

Ostrom, E., R. Gardner, and J. Walker. 1994. *Rules, Games, and Common-Pool Resources*. Ann Arbor: University of Michigan Press.

Padoch, C., J. M. Ayres, M. Pinedo-Vasquez, and A. Henderson, eds. 1999. *Varzea: Diversity, Development, and Conservation of Amazonia's Whitewater Floodplains*. Bronx, NY: New York Botanical Garden Press.

Pannell, J. R., and B. Charlesworth. 2000. Effects of metapopulation processes on measures of genetic diversity. *Philosophical Transactions of the Royal Society of London B* 355: 1851–64.

Papa, C. 1996. The 'farre de Montelione': Landrace and representation. In *Hulled Wheats. Proceedings of the First International Workshop on Hulled Wheats*, 21–22 July 1995, Castelvecchio Pascoli, Tuscany, Italy, edited by S. Padulosi, K. Hammer, and J. Heller, 154–71. Rome: International Plant Genetic Resources Institute.

Paterniani, E. 1969. Selection for reproductive isolation between two populations of maize, *Zea mays* L. *Evolution* 23: 534–47.

Peattie, K. 1995. *Environmental Marketing Management: Meeting the Green Challenge*. London: Pitman.

Peet, R. K. 1974. The measurement of species diversity. *Annual Review of Ecology and Systematics* 5: 285–307.

Peña-Chocarro, L. 1996. *In situ* conservation of hulled wheat species: The case of Spain. In *Hulled Wheats. Proceedings of the First International Workshop on Hulled Wheats*, 21–22 July 1995, Castelvecchio Pascoli, Tuscany, Italy, edited by S. Padulosi, K. Hammer, and J. Heller, 129–46. Rome: International Plant Genetic Resources Institute.

Perales R., H. R. 1998. Conservation and evolution of maize in Amecameca and Cuautla valleys of Mexico. Ph.D. diss., University of California, Davis (University Microfilms International, Ann Arbor, MI).

Perales R., H. R., S. B. Brush, and C. O. Qualset. 1998. Agronomic and economic competitiveness of maize landraces and *in situ* conservation in Mexico. In *Farmers, Gene Banks, and Crop Breeding: Economic Analyses of Diversity in Wheat, Maize, and Rice*, edited by M. Smale, 109–26. Boston: Kluwer.

——. 2003. Maize landraces of central Mexico: An altitudinal transect. *Economic Botany* 57: 7–20.

Perrino, P., G. Laghetti, L. F. D'Antuono, M. Al Ajlouni, M. Kantertay, A. T. Szabó, and K. Hammer. 1996. Ecogeographic distribution of hulled wheat species. In *Hulled Wheats. Proceedings of the First International Workshop on Hulled Wheats*, 21–22 July 1995, Castelvecchio Pascoli, Tuscany, Italy, edited by S. Padulosi, K. Hammer, and J. Heller, 101–19. Rome: International Plant Genetic Resources Institute.

Peeters, J. P., and J. T. Williams. 1984. Towards better use of genebanks with special reference to information. *Plant Genetic Resources Newsletter* No. 60: 22–32.

Phan Trieu, G. 2001. A study of rice landraces of the Mnong people in Daklak Province, Central Highlands, Vietnam. Master's thesis, University of California, Davis.

Pickersgill, B., and C. B. Heiser. 1978. Origins and distribution of plants domesticated in the New World tropics. In *Origins of Agriculture*, edited by C. A. Reed, 208–36. The Hague: Mouton.

Pimm, S. L., 1986. Community stability and structure. In *Conservation Biology*, edited by M. E. Soulé, 309–29. Sunderland, MA: Sinauer Associates.

Piperno, D. R., and K. V. Flannery. 2001. The earliest archaeological maize (*Zea mays* L.) from highland Mexico: New accelerator mass spectrometry dates and their implications. *Proceedings of the National Academy of Sciences of the United States* 98: 2101–3.

Pistorius, R. 1997. *Scientists, Plants, and Politics: A History of the Plant Genetic Resources Movement*. Rome: International Plant Genetic Resources Institute.

Plucknett, D. L., N. J. H. Smith, J. T. Williams, and N. M. Anishetty. 1987. *Gene Banks and the World's Food*. Princeton: Princeton University Press.

Poehlman, J. M. 1995. How crop improvement developed. In *The Literature of Crop Science*, edited by W. C. Olsen, 1–18. Ithaca, NY: Cornell University Press.

Popovskii, M. A. 1984. *The Vavilov Affair*. Hamden, CT: Archon Books.

Porfiri, O., L. F. D'Antuono, P. Codianni, L. Mazza, and R. Castagna. 1998. Genetic variability of a hulled wheats collection for agronomic and quality characteristics. In *Proceedings of the 3rd International Triticeae Symposium* (4–8 May 1997, Aleppo, Syria), 387–92. Enfield, NH: Science Publishers.

Porfiri, O., R. Torricelli, D. D. Silveri, R. Papa, G. Barcaccia, and V. Negri. n.d. The *Triticeae* genetic resources of central Italy: Collection, evaluation and conservation. Unpublished manuscript.

Posey, D. A. 1985. Indigenous management of tropical forest ecosystems: The case of the Kayapó Indians of the Brazilian Amazon. *Agroforestry Systems* 3: 139–58.

———. 1990. Intellectual property rights and just compensation for indigenous knowledge. *Anthropology Today* 6: 13–16.

Pray, C. E., and D. Umali-Deininger. 1998. The private sector in agricultural research: Will it fill the gap? *World Development* 16: 1127–48.

Prescott-Allen, R., and C. Prescott-Allen. 1990. How many plants feed the world? *Conservation Biology* 4: 365–74.

Qualset, C. O., A. B. Damania, A. C. A. Zanatta, and S. B. Brush. 1997. Locally based crop plant conservation. In *Plant Genetic Conservation: The* In Situ *Approach*, edited by N. Maxted, B. V. Ford-Lloyd and J. G. Hawkes, 160–75. London: Chapman & Hall.

Quiros, C., S. B. Brush, D. S. Douches, K. S. Zimmerer, and G. Huestis. 1990. Biochemical and folk assessment of variability of Andean cultivated potatoes. *Economic Botany*: 44: 254–66.

Quist, D., and I. H. Chapela. 2001. Transgenic DNA introgressed into traditional maize landraces in Oaxaca, Mexico. *Nature* 414: 541–43.

Rabinowitz, D., C. R. Linder, R. Ortega, D. Begazo, M. Murguia, D. S. Douches, and C. F. Quiros. 1990. High levels of interspecific hybridization between *Solanum sparsipilum* and *S. stenotomum* in experimental plots in the Andes. *American Potato Journal* 67: 73–81.

Ragan, W. H. 1905. *Nomenclature of the Apple: Catalog of Known Varieties Referred to in American Publications from 1804–1904*. Washington, DC: USDA Bureau of Plant Industry Bulletin 56.

Raman, K. V., E. Zimnoch-Guzowska, and N. Zoteyeva. 2000. The Vavilov Research

Institute's potato collection survived the siege of Leningrad–will it survive 21st century economic restructuring? *Diversity* 16: 12–15.

Ramming, D. W. 1988. California peach cultivars, breeding objectives, needs. In *The Peach,* edited by N. F. Childers and W. B. Sherman, 89–96. Gainesville, FL: Horticultural Publications.

Rappaport, R. A. 1967. *Pigs for the Ancestors: Ritual in the Ecology of a New Guinea People.* New Haven: Yale University Press.

Recinos, A. 1950. *Popol Vuh: The Sacred Book of the Ancient Quiché Maya,* translated by Delia Goetz and Sylvanus G. Morley. Norman, OK: University of Oklahoma Press.

Redfield, R. 1941. *The Folk Culture of Yucatan.* Chicago: University of Chicago Press.

Redford, K. H., and C. Padoch, eds. 1992. *Conservation of Neotropical Forests: Working from Traditional Resource Use.* New York: Columbia University Press.

Reid, W. V., S. Laird, C. Meyer, R. Gámez, A. Sittenfeld, et al. 1993a. *Biodiversity Prospecting: Using Resources for Sustainable Development.* Washington, DC: World Resources Institute.

——. 1993b. A new lease on life. In *Biodiversity Prospecting: Using Resources for Sustainable Development*, edited by W. V. Reid et al., 1–52. Washington, DC: World Resources Institute.

Rice, E., M. Smale, and J.-L. Blanco. 1998. Farmer's use of improved seed selection practices in Mexican maize: Evidence and issues from the Sierra de Santa Marta. *World Development* 26: 1625–40.

Richards, P. 1985. *Indigenous Agricultural Revolution: Ecology and Food Production in West Africa.* Boulder, CO: Westview Press.

——. 1986. *Coping with Hunger: Hazard and Experiment in an African Rice Farming System.* London: Alwin and Unwin.

Rindos, D. 1984. *The Origins of Agriculture: An Evolutionary Appraisal.* New York: Academic Press.

——. 1989. Darwinism and its role in the explanation of domestication. In *Foraging and Farming: The Evolution of Plant Exploitation*, edited by D. R. Harris and G. C. Hillman, 27–41. London: Unwin Hyman.

Rissler, J., and M. Mellon. 1996. *The Ecological Risks of Engineered Crops.* Cambridge: MIT Press.

Robinson, J. 1956. *The Accumulation of Capital.* London: Macmillan.

Rogers, E. M. 1969. *Modernization Among Peasants: The Impact of Communication.* New York: Holt, Rinehart and Winston.

Root, W., and R. de Rochemont. 1976. *Eating in America: A History.* New York: William Morrow.

Rose, M. 1993. *Authors and Owners: The Invention of Copyright.* Cambridge: Harvard University Press.

Rosengarten, F. 1991. *Wilson Popenoe: Agricultural Explorer, Educator, and Friend of Latin America.* Lawai, Kauai, HI: National Tropical Botanical Garden.

Rosenzweig, M. L. 1995. *Species Diversity in Time and Space.* Cambridge: Cambridge University Press.

Ryerson, K. A. 1933. History and significance of the foreign plant introduction work of the United States Department of Agriculture. *Agricultural History* 7: 110–28.

Sahagún, Fr. B. de. 1979 (1577). *Florentine Codex–General History of the Things of*

New Spain, translated by A. O. Anderson and C. E. Dibble. Santa Fe, NM: The School of American Research.

Sahai, S. 2001. India: Plant variety protection, farmers' rights bill adopted. *TWN Online*, Third World Network, http://www.twnside.org.sg/title/variety.htm.

Salaman, R. N. 1926. *Potato Varieties*. Cambridge: Cambridge University Press.

———. 1985. *The History and Social Influence of the Potato*. New introd. and corrections by J. G. Hawkes. Cambridge: Cambridge University Press. (Orig. pub. 1949.)

Salick, J., N. Cellinese, and S. Knapp. 1997. Indigenous diversity of cassava: Generation, maintenance, use and loss among the Amuesha, Peruvian upper Amazon. *Economic Botany* 51: 6–19.

Sallares, R. 1991. *The Ecology of the Ancient Greek World*. Ithaca, NY: Cornell University Press.

Sambatti, J. B. M., P. Martins, and A. Ando. 2001. Folk taxonomy and evolutionary dynamics of cassava: A case study in Ubatuba, Brazil. *Economic Botany* 55: 93–105.

Sanchez G., J. 1994. Modern variability and patterns of maize movement in Mesoamerica. In *Corn and Culture in the Prehistoric New World*, edited by S. Johannessen and C. A. Hastorf, 135–56. Boulder, CO: Westview Press.

Sanchez G., J., and M. M. Goodman. 1992. Relationships among Mexican races of maize. *Economic Botany* 46: 72–85.

Sanchez G., J., M. M. Goodman, and C. W. Stuber. 2000. Isozymatic and morphological diversity in the races of maize of Mexico. *Economic Botany* 54: 43–59.

Sandstrom, A. R. 1991. *Corn Is Our Blood: Culture and Ethnic Identity in a Contemporary Aztec Indian Village*. Norman: University of Oklahoma Press.

Sauer, J. D. 1994. *Historical Geography of Crop Plants: A Select Roster*. Boca Raton, FL: CRC Press.

Schele, L. 1974. Observations on the cross motif at Palenque. In *Primera Mesa Redonda de Palenque Part 1*, edited by M. G. Robertson, 41–70. Pebble Beach, CA: Robert Louis Stephenson School.

Schiemann, E. 1948. *Weizen, Roggen, Gerste. Systematik, Geschhte und Verwendung*. Jena, Germany: Verlang von Gustav Fischer.

Scott, J. C. 1976. *The Moral Economy of the Peasant: Rebellion and Subsistence in Southeast Asia*. New Haven: Yale University Press.

———. 1985. *Weapons of the Weak: Everyday Forms of Peasant Resistance*. New Haven: Yale University Press.

Sedjo, R. A. 1988. Property rights and the protection of plant genetic resources. In *Seeds and Sovereignty: The Use and Control of Plant Genetic Resources*, edited by J. R. Kloppenburg Jr., 293–314. Durham, NC: Duke University Press.

Sevilla, R. 1994. Variation in modern Andean maize and its implications for prehistoric patterns. In *Corn and Culture in the Prehistoric New World*, edited by S. Johannessen and C. A. Hastorf, 219–44. Boulder, CO: Westview Press.

Shands, H. L. 1991. Complementary on *in-situ* and *ex-situ* germplasm conservation from the standpoint of the future user. *Israel Journal of Botany* 40: 521–28.

Shigeta, M. 1990. Folk *in-situ* conservation of ensete (*Ensete ventricosum* (Welw.) E. E. Cheesman): Towards the interpretation of indigenous agricultural science of the Ari, southwestern Ethiopia. *African Study Monograph* 10: 93–107.

Shiva, V. 1991. *Violence of the Green Revolution: Third World Agriculture, Ecology and Politics*. London: Zed Books.

———. 1997. *Biopiracy: The Plunder of Nature and Knowledge*. Boston: South End Press.

Simmonds, N. W. 1962. Variability in crop plants, its use and conservation. *Biological Reviews* 37: 442–65.

———. 1985. Farming systems research: a review. Technical Paper No. 43, World Bank, Washington, DC.

Singh, A. J., and D. Byerlee. 1990. Relative variability in wheat yields across countries and over time. *Journal of Agricultural Economics* 41: 21–32.

Singh, M., S. Ceccarelli, and S. Grando. 1999. Genotype *x* environment interaction of crossover type: Detecting its presence and estimating the crossover point. *Theoretical and Applied Genetics* 99: 988–95.

Smale, M. 1997. The green revolution and wheat genetic diversity: Some unfounded assumptions. *World Development* 25: 1257–69.

Smale, M., J. Hartell, P. W. Heisey, and B. Senauer. 1998. The contributions of genetic resources and diversity to wheat production in the Punjab of Pakistan. *American Journal of Agricultural Economics* 80: 482–93.

Smale, M., M. P. Reynolds, M. Wharburton, B. Skovmand, R. Trethowan, R. P. Singh, I. Ortiz-Monasterio, and J. Crossa. 2002. Dimensions of diversity in modern spring bread wheat in developing countries from 1965. *Crop Science* 42: 1766–79.

Smale, M., D. Soleri, D. A. Cleveland, D. Louette, E. Rice, J.-L. Blanco, and A. Aguirre. 1998. Collaborative plant breeding as an incentive for on-farm conservation of genetic resources: Economic issues from studies in Mexico. In *Farmers, Gene Banks, and Crop Breeding: Economic Analyses of Diversity in Wheat, Maize, and Rice*, edited by M. Smale, 239–57. Boston: Kluwer.

Smith, J. S. C. 1986. Genetic diversity within the Corn Belt dent racial complex of maize (*Zea mays* L.). *Maydica* 31: 349–67.

Smith, M. E., F. Castillo G., and F. Gómez. 2001. Participatory plant breeding with maize in Mexico and Honduras. *Euphytica* 122: 551–65.

Soleri, D., and D. A. Cleveland. 2001. Farmers' genetic perceptions regarding their crop populations: An example with maize in the central valleys of Oaxaca, Mexico. *Economic Botany* 55: 106–28.

Soleri, D., and S. E. Smith. 1995. Morphological and phenological comparisons of two Hopi maize varieties conserved *in situ* and *ex situ*. *Economic Botany* 49: 56–77.

Soleri, D., S. E. Smith, and D. A. Cleveland. 2000. Evaluating the potential for farmer and plant breeder collaboration: A case study of farmer maize selection in Oaxaca, Mexico. *Euphytica* 116: 41–57.

Soltis, J., Boyd, R., and P. J. Richerson. 1995. Can group-functional behaviors evolve by cultural group selection: An empirical test. *Current Anthropology* 36: 473–94.

Soulé, M. E., ed. 1986. *Conservation Biology: The Science of Scarcity and Diversity*. Sunderland, MA: Sinauer Associates.

Spagnoletti Zeuli, P. L., and C. O. Qualset. 1987. Geographical diversity for quantitative spike characters in a world collection of durum wheat. *Crop Science* 27: 235–41.

Spencer, J. E. 1954. *Land and People in the Philippines*. Berkeley: University of California Press.

——. 1966. *Shifting Cultivation in Southeastern Asia*. University of California Publications in Geography, vol. 19. Berkeley: University of California Press.

Sperling, L., M. E. Loevinsohn, and B. Ntabomvura. 1993. Rethinking the farmer's role in plant breeding: Local bean experts and on-station selection in Rwanda. *Experimental Agriculture* 29: 509–19.

Spurr, M. S. 1986. Arable cultivation in Roman Italy, c. 200 B.C.–c. A.D. 100. *Journal of Roman Studies Monographs*, no. 3. London: Society for the Promotion of Roman Studies.

Stahpit, B. R., K. D. Joshi, and J. R. Whitcombe. 1996. Farmer participatory crop improvement. III. Participatory plant breeding, a case study for rice in Nepal. *Experimental Agriculture* 32: 479–96.

Standhill, G. 1976. Trends and deviations in the yield of the English wheat crop during the last 750 years. *Agro-Ecosystems* 3: 1–10.

Steward, J. H. 1955. *Theory of Culture Change*. Urbana: University of Illinois Press.

Strada, D. G. 1996. Peach and nectarine cultivars in the world from 1980 to 1992. Proceedings of the 3rd International Peach Symposium, Beijing, 3–10 September 1993, edited by C. Fideghelli and F. Grassi (*Acta Horticulturae* 374: 43–51). Leuven, Belgium: International Society for Horticultural Science.

Sutlive, V. H. 1978. *The Iban of Sarawak*. Arlington Heights, IL: AHM.

Szabó, A. T., and K. Hammer. 1996. Notes on the taxonomy of farro: *Triticum monococcum, T. dicoccon,* and *T. spelta*. In *Hulled Wheats. Proceedings of the First International Workshop on Hulled Wheats*, 21–22 July 1995, Castelvecchio Pascoli, Tuscany, Italy, edited by S. Padulosi, K. Hammer, and J. Heller, 2–40. Rome: International Plant Genetic Resources Institute.

Tadesse, D. 2001. The dynamics of diversity in modern agriculture: The case of peaches in California. Ph.D. diss., University of California, Davis (University Microfilms Inc., Ann Arbor, MI).

Tapia N., M., and J. A. Flores Ochoa. 1984. *Pastoreo y pastizales de los Andes del sur del Perú*. Lima: Instituto Nacional de Investigación y Promoción Agropecuaría.

ten Kate, K., and S. A. Laird. 1999. *The Commercial Use of Biodiversity: Access to Genetic Resources and Benefit-Sharing*. London: Earthscan.

Teni, R. 1989. Cosmological importance of corn among Mayan peoples. *Northeast Indian Quarterly* 6: 14–19.

Tesfaye, T., B. Getachew, and M. Woerde. 1991. Morphological diversity in tetraploid wheat landrace populations from the central highlands of Ethiopia. *Hereditas* 114: 171–76.

Teshome, A., L. Fahrig, J. K. Torrance, J. D. Lambert, T. J. Arnason, and B. R. Baum. 1999. Maintenance of sorghum (*Sorghum bicolor,* Poaceae) landrace diversity by farmers' selection in Ethiopia. *Economic Botany* 53: 79–88.

Theophrastus. 1916. *Enquiry into Plants and Minor Works on Odours and Weather Signs,* translated by A. Hort. London: W. Heinemann.

Thomas, R. B. 1973. *Human Adaptation to a High Andean Energy Flow System*. Occasional Papers in Anthropology No. 7. University Park: Pennsylvania State University, Department of Anthropology.

Thompson, P. B. 1997. *Food Biotechnology in Ethical Perspective*. London: Blackie Academic & Professional.

Tilman, D. 1982. *Resource Competition and Community Structure*. Princeton: Princeton University Press.

——. 1994. Competition and biodiversity in spatially structured habitats. *Ecology* 75: 2–16.

——. 1996. Biodiversity: Population versus ecosystem stability. *Ecology* 77: 350–63.

Tilman, D., R. M. May, C. L. Lehman, and M. A. Nowak. 1994. Habitat destruction and the extinction debt. *Nature* 371: 65–66.

Tilman, D., and S. Pacala. 1993. The maintenance of species richness in plant communities. In *Species Diversity in Ecological Communities: Historical and Geographical Perspectives*, 13–25. Chicago: University of Chicago Press.

Tin, H. Q., T. Berg, and Å. Børnstad. 2001. Diversity and adaptation in rice varieties under static (*ex situ*) and dynamic (*in situ*) management. *Euphytica* 122: 491–502.

Trimble, S. W., and P. Crosson. 2000. Land use: U.S. soil erosion rates—myth and reality. *Science* 289: 248–50.

Tripp, R. 1991. The farming systems research movement and on-farm research. In *Planned Change in Farming Systems: Progress in On-Farm Research*, edited by R. Tripp, 3–16. Chichester, UK: John Wiley & Sons.

Tripp, R., P. Anandajayasekeram, D. Byerlee, and L. Harrington. 1990. Farming systems research revisited. In *Agricultural Development in the Third World*, edited by C. K. Eicher and J. M. Staatz, 384–99. Baltimore: Johns Hopkins University Press.

Troll, C., ed. 1968. *Geo-ecology of the Mountainous Regions of the Tropical Americas*. Bonn: Dummler in Kommission.

Tufts, W. P. 1946. *The Rich Pattern of California Crops*. Reprinted from *California Agriculture*. Berkeley: University of California Press.

Ucko, P. J., and G. W. Dimbleby, eds. 1969. *The Domestication and Exploitation of Plants and Animals*. London: Duckworth.

Ugent, D. 1970. The potato. *Science* 170: 1161–66.

Universidad Nacional San Cristobal de Huamanga. 1983. *Diagnóstico técnico agropecuario de las comunidades campesinas de Arizona, Qasanqay Ayacucho 1983*. Lima: Instituto Interamericano de Cooperación para la Agricultura (IICA).

Urban, F., and M. Trueblood. 1990. *World Population by Country and Region, 1950–2050*. Economic Research Service, Staff Report No. AGES 9024., U. S. Department of Agriculture, Washington, DC.

Valdivia, R., V. Choquehuanca, J. Reinoso, and M. Holle. 1998. Identification of the dynamics of biodiversity microcenters of Andean tubers in the circumlacustrine highlands, Puno, Peru. In *Strengthening the Scientific Basis of* In Situ *Conservation of Agricultural Diversity On-Farm,* edited by D. Jarvis and T. Hodgkin, 19–20. Rome: International Plant Genetic Resources Institute.

Vavilov, N. 1926. *Studies on the Origin of Cultivated Plants.* Leningrad: Institute of Applied Botany and Plant Improvement.

——. 1951. The Origin, Variation, Immunity, and Breeding of Cultivated Plants. *Chronica Botanica* 13: 1–366 (translated from Russian by K. S. Chester).

——. 1992. Origin and Geography of Cultivated Plants. Translated by Doris Löve. Cambridge: Cambridge University Press.

Vayda, A. P., and R. A. Rappaport. 1968. Ecology, cultural and noncultural. In *Introduction to Cultural Anthropology*, edited by J. A. Clifton, 447–97. Boston: Houghton Mifflin.

Vázquez, Francisco, 1937. *Crónica de la provincia del Santísimo nombre de Jesús de Guatemala de la orden de N. Seráfico padre San Francisco en el reino de la Nueva España*, vol. XIV, 2d ed. Lázaro Lamadrid, Guatemala: Biblioteca "Guatemala" de la Sociedad de Geografía e Historia de Guatemala.

Vazzana, C. 1996. The role of farmers' associations in safeguarding populations of farro in Italy. In *Hulled Wheats. Proceedings of the First International Workshop on Hulled Wheats*, 21–22 July 1995, Castelvecchio Pascoli, Tuscany, Italy, edited by S. Padulosi, K. Hammer, and J. Heller, 147–52. Rome: International Plant Genetic Resources Institute.

Visser, B., D. Eaton, N. Louwaars, and J. Engles. 2000. Transaction costs of germplasm exchange under bilateral agreements. Document No. GFAR/00/17-04-04, *Global Forum on Agricultural Research*, FAO, Rome, www.egfar.org/documents/conference/GFAR_2000/gfar1702.PDF.

Vogel, J. H. 1994. *Genes for Sale: Privatization as a Conservation Policy*. New York: Oxford University Press.

Vogt, E. Z. 1976. *Tortillas for the Gods: A Symbolic Analysis of Zinacanteco Rituals*. Cambridge: Harvard University Press.

Wallace, H. A., and W. L. Brown. 1988. *Corn and Its Early Fathers*, rev. ed. Ames: Iowa State University Press.

Warburton, M., and F. Bliss. 1996. Genetic diversity in peach (*Prunus persica* L. Batch) revealed by random amplyfied polymorphic DNA (RAPD) markers and compared to inbreeding coefficients. *Journal of American Horticultural Science* 121: 1012–19.

Warman, A. 1989. Maize as organizing principle: How corn shaped space, time and relationships in the New World. *Northeast Indian Quarterly* 6: 20–27.

Wasik, J. F. 1996. *Green Marketing and Management: A Global Perspective*. Cambridge, MA: Blackwell Business.

Watson, A. M. 1983. *Agricultural Innovation in the Early Islamic World. The Diffusion of Crops and Farming Techniques*. Cambridge: Cambridge University Press.

Weberbauer, A. 1945. *El Mundo vegetal de los Andes Peruanos, estudio fitogeografico*. Lima: Estacíon Experimental Agricola de la Molina, Direccíon de Agricultura, Ministerio de Agricultura.

Weismantel, M. J. 1988. *Food, Gender and Poverty in the Ecuadorean Andes*. Philadelphia: University of Pennsylvania Press.

Wellhausen, E. J., A. Fuentes, and A. Hernandez Corzo. 1957. *Races of Maize in Central America,* Publication 511. Washington, DC: National Academy of Sciences, National Research Council.

Wellhausen, E., J. Roberts, L. M. Roberts, and E. Hernández X. 1952. *Races of Maize in Mexico, Their Origin, Characteristics, and Distribution*. Cambridge, MA: The Bussey Institution, Harvard University.

Werner, D. J., and W. R. Okie. 1998. A history and description of the *Prunus persica* plant introduction collection. *Horticultural Science*. 33: 787–93.

Whitcombe, J., and A. Joshi. 1996. Farmer participatory approaches for varietal breeding and selection and linkages with the formal sector. In *Participatory Plant Breeding*, edited by P. Eyzaguirre and M. Iwanaga, 57–65. Rome: International Plant Genetic Resources Institute.

Whitcombe, J. R., A. Joshi, K. D. Joshi, and B. R. Sthapit. 1996. Farmer participatory crop improvement. I. Varietal selection and breeding methods and their impact on biodiversity. *Experimental Agriculture* 32: 445–60.

Whittaker, R. H., and S. Levin, eds. 1975. *Niche Theory and Application*. Benchmark Papers in Ecology 3. Stroudsburg, PA: Dowden, Hutchinson & Ross.

Wilkes, H. G. 1985. Teosinte: The closest relative of maize revisited. *Maydica* 30: 209–23.

———. 1989. Maize: Domestication, racial evolution, and spread. In *Foraging and Farming*, edited by D. R. Harris and G. C. Hillman, 440–55. London: Unwin-Hyman.

———. 1991. *In situ* conservation of agricultural systems. In *Biodiversity: Culture, Conservation and Ecodevelopment*, edited by M. L. Oldfield and J. B. Alcorn, 86–101. Boulder, CO: Westview Press.

———. 1993. Germplasm collections: Their use, potential, social responsibility, and genetic vulnerability. In *International Crop Science I*, edited by D. R. Buxton, R. Shibles, R. A. Forsberg, B. L. Blad, K. H. Asay, G. M. Paulsen, and R. F. Wison, 445–50. Madison, WI: Crop Science Society of America.

Wilson, W. M., and D. L. Dufour. 2002. Why "bitter" cassava? Productivity of "bitter" and "sweet" cassava in a Tukanoan Indian settlement in the northwestern Amazon. *Economic Botany* 56: 41–48.

Winkelmann, D. 1976. *The Adoption of New Maize Technology in Plan Puebla, Mexico*. Mexico: CIMMYT.

Winterhalder, B. P., and R. B. Thomas. 1978. *Geoecology of Southern Highland Peru: A Human Adaptation Perspective*. Boulder: University of Colorado, Institute of Arctic and Alpine Research.

Wolfe, M. S. 1985. The current status and prospects of multiline cultivars and variety mixtures for disease resistance. *Annual Review of Phytopathology* 23: 251–73.

Wood, D., and J. M. Lenné, eds. 1999. *Agrobiodiversity: Characterization, Utilization and Management*. Wallingford, UK: CABI.

Woolley, C. L. 1928. *The Sumerians*. Oxford: Clarendon Press.

World Resources Institute, World Conservation Union, and United Nations Environment Programme. 1992. *Global Biodiversity Strategy*. Washington, DC: World Resources Institute.

Zaiger, F. 1988. Objectives, experiences of a California private peach breeder. In *The Peach*, edited by N. F. Childers and W. B. Sherman, 100–5. Gainesville, FL: Horticultural Publications.

Zeven, A. C. 1996. Results of activities to maintain landraces and other material in some European countries *in situ* before 1945 and what we may learn from them. *Genetic Resources and Crop Evolution* 43: 337–41.

———. 1998. Landraces: A review of definitions and classifications. *Euphytica* 104: 127–39.

———. 1999. The traditional inexplicable replacement of seed and seed ware of landraces and cultivars: A review. *Euphytica* 110: 181–91.

Zhu, Y. Y., H. Chen, J. Fan, Y. Y. Wang, Y. Li, J. Chen, J. X. Fan, S. Yang, L. Hu, H. Leung, T. W. Mew, P. S. Teng, Z. Wang, and C. C. Mundt. 2000. Genetic diversity and disease control in rice. *Nature* 406: 718–22.

Zimmerer, K. S. 1991. The regional biogeography of native potato cultivars in highland Peru. *Journal of Biogeography* 18: 165–78.

———. 1994. Local soil knowledge—answering basic questions in highland Bolivia. *Journal of Soil And Water Conservation* 49: 29–34.

———. 1996. *Changing Fortunes: Biodiversity and Peasant Livelihood in the Peruvian Andes*. Berkeley: University of California.

———. 1998. The ecogeography of Andean potatoes. *BioScience* 48: 445–54.

Zimmerer, K. S., and K. R. Young, eds. 1998. *Nature's Geography: New Lessons for Conservation in Developing Countries*. Madison: University of Wisconsin Press.

Zohary, D., and M. Hopf. 2000. *Domestication of Plants in the Old World: The Origin and Spread of Cultivated Plants in West Asia, Europe, and the Nile Valley*, 3d ed. New York: Oxford University Press.

Index